The Outer Limits of Reason

The Outer Limits of Reason

What Science, Mathematics, and Logic Cannot Tell Us

Noson S. Yanofsky

The MIT Press
Cambridge, Massachusetts
London, England

MIT Press books may be purchased at special quantity discounts for business or sales promotional use. For information, please email special_sales@mitpress.mit.edu or write to Special Sales Department, The MIT Press, 55 Hayward Street, Cambridge, MA 02142.

This book was set in Stone Sans and Stone Serif by Toppan Best-set Premedia Limited, Hong Kong. Printed and bound in the United States of America.

Library of Congress Cataloging-in-Publication Data

Yanofsky, Noson S., 1967–
The outer limits of reason : what science, mathematics, and logic cannot tell us / Noson S. Yanofsky.
pages cm
Includes bibliographical references and index.
ISBN 978-0-262-01935-4 (hardcover : alk. paper)
1. Knowledge, Theory of. 2. Science—Philosophy. 3. Mathematics—Philosophy. I. Title.
Q175.32.K45Y36 2013
001.01—dc23
2012050531

10 9 8 7 6 5 4 3 2 1

Please email all comments and criticisms to noson@sci.brooklyn.cuny.edu.

For
Shayna Leah,
Hadassah, and Rivka

Table of Contents

Preface ix

Acknowledgments xi

1 Introduction 1

2 Language Paradoxes 15

2.1 Liar! Liar! 15

2.2 Self-Referential Paradoxes 19

2.3 Naming Numbers 26

3 Philosophical Conundrums 31

3.1 Ships, People, and Other Objects 31

3.2 Hangin' with Zeno and Gödel 41

3.3 Bald Men, Heaps, and Vagueness 50

3.4 Knowing about Knowing 57

4 Infinity Puzzles 65

4.1 Sets and Sizes 66

4.2 Infinite Sets 69

4.3 Anything Larger? 76

4.4 Knowable and Unknowable 85

5 Computing Complexities 97

5.1 Some Easy Problems 98

5.2 Some Hard Problems 109

5.3 They're All Connected 121

5.4 Almost Solving Hard Problems 129

5.5 Even Harder Problems 131

6 Computing Impossibilities 135

6.1 Algorithms, Computers, Machines, and Programs 136

6.2 To Halt or Not to Halt? 139

6.3 They're All Connected 146
6.4 A Hierarchy of the Unknown 152
6.5 Minds, Brains, and Computers 157

7 Scientific Limitations 161
7.1 Chaos and Cosmos 161
7.2 Quantum Mechanics 175
7.3 Relativity Theory 214

8 Metascientific Perplexities 235
8.1 Philosophical Limitations of Science 235
8.2 Science and Mathematics 252
8.3 The Origin of Reason 272

9 Mathematical Obstructions 297
9.1 Classical Limits 298
9.2 Galois Theory 304
9.3 Harder Than Halting 309
9.4 Logic 320
9.5 Axioms and Independence 331

10 Beyond Reason 339
10.1 Summing Up 339
10.2 Defining Reason 345
10.3 Peering Beyond 349

Notes 355
Bibliography 379
Index 393

Preface

The more we learn about the world, and the deeper our learning, the more conscious, specific, and articulate will be our knowledge of what we do not know, our knowledge of our ignorance.[1]
—Karl Popper

A man's got to know his limitations.
—Harry Callahan, *Magnum Force* (1973)

Everything should be made as simple as possible, but not simpler.
—Attributed to Albert Einstein

With understanding comes ambivalence. Once we know something, we often find it boring and trite. On the other hand, the mysterious and unknown fascinates us and holds our attention. That which we do not know or understand is what interests us, and what we *cannot* know intrigues us even more. This book explores topics that reason tells us we cannot know because they are beyond reason.

Many books convey the amazing facts that science, mathematics, and reason have revealed to us. There are also books that cover topics that science, mathematics, and reason have not yet fully explained. This book is a little different. Here we study what science, mathematics, and reason tell us *cannot* be revealed. What cannot be predicted or known? What will never be understood? What are the limitations of computers, physics, logic, and our thought processes? What is beyond the bounds of reason? This book aims to answer some of these questions and is full of ideas that challenge our deep-seated beliefs about the universe, our rationality, and ourselves.

Along the way we will study simple computer problems that would take trillions of centuries to solve; consider perfectly formed English sentences that have no meaning; learn about different levels of infinity; leap into the bizarre and wonderful world of the quantum; discuss specific problems that computers can never solve; befriend butterflies that bring about blizzards; ponder particles that simultaneously dance at different parties; hear about paradoxes and self-referential paradoxes; see what relativity theory tells us about our naive notions of space, time, and causality; understand Gödel's famous theorems about the limitations of logic; discover certain problems in mathematics and physics that are impossible to solve; explore the very nature of science, mathematics, and reason; wonder why the universe seems perfect for human beings; and examine the complex relationship between our mind, reason, and the physical universe. We will also attempt to peek beyond the borders of reason and see what, if anything, is out there. These and many other fascinating topics will be presented in a way that is clear and comprehensible.

While exploring these various limitations in diverse areas, we will see that many of the limitations have a similar pattern. These patterns will be investigated in order to better understand the structure of reason and its limits.

This book is not a compendium of all the diverse examples in which limitations of reason are found. Rather, our goal is to understand why these boundaries arise and why reason cannot extend beyond them. Several representative limitations in each area are selected and discussed in depth.

Rather than just listing the limitations, I aim to explain them or at least provide the intuition of why a particular area is beyond reason. It is important to realize that this book is not meant to be speculative or to have a New Age orientation. Nor is it a history book in which I gloss over the meaning of ideas in order to focus on their chronological development. This is a popular science book that will gradually and clearly explain the ideas presented.

Since I accept Stephen Hawking's dictum that every equation halves the number of readers, very few equations are found in this book. However, I do believe in the power of diagrams, charts, and graphs to simplify complex ideas. My goal is clarity.

Each chapter deals with a different area: science, mathematics, language, philosophy, and so on. These chapters are arranged from concrete to abstract. I start with simple problems of everyday language and move on to straightforward philosophical questions, ending with the abstract world of mathematics. For the most part, the chapters are independent of each other and can be read in any order. Readers are encouraged to begin with topics that most interest them. (The unifying theme of self-referential paradoxes is found in chapters 2, 4, 6, and 9.)

Acknowledgments

In a sense, this book is a collaborative effort with my friends and colleagues in the Computer and Information Science Department at Brooklyn College. They have read the chapters, corrected my mistakes, chided me when I was being silly, and encouraged me when I was stuck. They have given me the warm intellectual environment that made this book possible. I thank them!

Many Brooklyn College faculty members were kind enough to read and comment on several of these chapters: Jonathan Adler, David Arnow, George Brinton, Samir Chopra, Jill Cirasella, Dayton Clark, Eva Cogan, Jim Cox, Scott Dexter, Keith Harrow, Danny Kopec, Yedidyah Langsam, Matthew Moore, Rohit Parikh, Simon Parsons, Michael Sobel, Aaron Tenenbaum, and Paula Whitlock. Their comments have made this a far better book.

Some chapters were used in a course given at Brooklyn College. Great benefit was derived from student participation and the give-and-take of a classroom dialog. I thank the students for listening, arguing, helping, and correcting. Many students at Brooklyn College and the Graduate Center of the City University of New York read and commented on earlier drafts: Firat Atagün, Can Başkent, Hubert Bennett, Greg Benson, Be Birchall, Raila C. Brejt, Fatemeh Choopani, Simon Dexter, Aline Elmann, Madelene Feingold, Terri Grosso, Miriam Gutherc, Jay Jankelewicz, Matthew P. Johnson, Joel Kammet, Tatiana Kedel, Wai Khoo, Karen Kletter, Erdal Köse, Michael Lampis, Shalva Landy, Holly Lo Voi, Jon Lo Voi, Matthew Meyer, Valia Mitsou, Jordi Navarrette, Shoshana Neuburger, Hadassah Norowitz, Nicole Reilly, Artur Sahakyan, Connor Savage, Angela Shatashwili, Alex Sverdlov, Stanislav Turzhavskiy, Fredda Weinberg, and Karol Wysocki. I am indebted to them all.

Numerous other people looked over parts of the book and have been helpful: Ros Abramsky, Samson Abramsky, Marcia Barr, Michael Barr,

Rebecca Barr, Adam Brandenburger, Richard Churchill, Melvin Fitting, Leopold Flatto, Robert J. Fogelin, Chaim Goodman-Strauss, Ariel Halpert, Eliyahu Hershfeld, Ellen Hershfeld, Faigy Hershfeld, Pinchas Hershfeld, Shai Hershfeld, Yitzchok Hershfeld, Michael Hicks, Joshua Honigwachs, Roman Kossak, Klaas Landsman, André Lebel, Raphael Magarik, Camille Martin, Jolly Mathen, Rochel Moskowitz, Larry Moss, Naftoli Neuburger, Janos Pach, Carol Parikh, Suri Raber, N. Raja, Barbara Rifkind, Andrei Rodin, Ariel Ropek, Evan J. Siegel, Michael Vitz, Sharon Yanofsky, and Mark Zelcer. Their criticism, comments, and helpful ideas are deeply appreciated.

I thank Robbert Dijkgraaf for permission to use his artwork in figure 8.6. I also thank C. Goodman-Strauss for permission to use many of his diagrams in section 9.3. My beautiful and talented daughter Hadassah was very helpful with some of the diagrams.

James DeWolf, Marc Lowenthal, Marcy Ross, and the whole MIT Press team have been very helpful in getting this book into shape. Thank you.

Karen Kletter remains the world's greatest editor and proofreader. Thank you, Karen!

However, at the end of the day, I am the sole cause of any errors that may remain.

Several other, more general debts should be acknowledged. I am grateful to my friend and research partner Ralph Wojtowicz of Baker Mountain Research Corporation for supporting my other research while I was working on this book.

In the spring of 1987 I had a chance encounter with Dr. Avi Rabinowitz on a street corner in Jerusalem. Avi is a brilliant physicist bristling with creativity and enthusiasm. We eventually became traveling companions and good friends. There are few topics that Avi cannot discuss in depth. Conversations with him usually proceed at the speed of light. We've had many intense conversations while climbing mountains in Greece and watching ridiculous science fiction movies. He is a true mentor and friend. Every page in this book contains ideas that I have discussed with him over the years. (He would probably disagree with most of what I wrote.) His influence is immense and I am forever appreciative.

Over the past few years, three people who enriched my life have passed away. They added much to my education and hence much to this book.

During my senior year at Brooklyn College, Professor Chaya Gurwitz supervised me in a research project, thereby introducing me to the rigors of higher mathematics and computer science. She taught me how to read an academic paper, to put my ideas into action, and to analyze the results. This experience piqued my interest in attending graduate school. She graciously invited me to her home for many meals, where I also became friends with her husband and eight wonderful children. She continued to guide me as a graduate student, as a teacher, as a colleague, and as a person until her untimely passing in 2008. I am truly indebted to her.

My dissertation advisor, *mon maître*, Alex Heller, was a distinguished professor in the Department of Mathematics at the Graduate Center of the City University of New York. He was a kind, gentle man. Although I graduated in 1996, we continued to meet once or twice a week until a few days before his sad passing in 2008. (In a sense, he gave me twelve years of postdoctoral research.) Our conversations meandered from mathematics, to politics, to morality, to philosophy, to history, and so on. He was an amazing genius and his range of knowledge was astonishing; in mathematics, however, it was particularly striking. While speaking with him, one got the impression that he had a magnificently clear vision of the entire structure of mathematics before him. It was a privilege to study under him and to be befriended by him.

A word must be said about Professor Heller's unique method of mentoring. After learning much from him during two years of coursework, I was disinvited from attending any more of his classes. He said that I had gained enough from him. From then until his passing, even though we talked about mathematics constantly, he never *taught* me a scintilla of mathematics. My job was to present my work or what I was studying. His job was to find flaws in my presentation or my understanding. He would harangue me about developing a correct definition, indicate where my proofs had failed, and point out when I was not exact. Although a gentleman—always with a kind word—his method of mentoring was intimidating and disheartening, to say the least. He never articulated this sink-or-swim philosophy. Nevertheless, he had a valid point: it was my job to learn mathematics and I had to struggle with it on my own. I am forever grateful for his confidence in me and for the independence he insisted I develop. He was the greatest of teachers.

I had the benefit of having a world-class mathematician as a neighbor. Professor Leon Ehrenpreis lived a few blocks from me and I took advantage of it by visiting him on a regular basis. My Friday-night visits were always greeted with a warm, welcoming smile. Besides being a first-rate mathematician, he was also a rabbinic scholar, marathon runner, handball player, classical pianist, and father of eight. This Renaissance man's breadth of knowledge was truly astounding. I have many fond memories of sitting at his kitchen table chatting about the subtleties of Hebrew grammar, the edge-of-the-wedge theorem, raising children, the consequences of the Kochen-Specker theorem, the role of cows in the Book of Genesis, hypergeometric functions, and many other topics. Professor Ehrenpreis was blessed with the most pleasant disposition and always had a kind, encouraging word. I learned much from him. He passed away in August of 2010.

All three are painfully missed.

This book is dedicated to my wife, Shayna Leah, whose warmth and loving support made this work possible, and to my daughters, Hadassah and Rivka, who fill our home and hearts with laughter and joy. My love and gratitude toward them are limitless.

1 Introduction

Human reason, in one sphere of its cognition, is called upon to consider questions, which it cannot decline, as they are presented by its own nature, but which it cannot answer, as they transcend every faculty of the mind.[1]

—Immanuel Kant (1724–1804)

As the circle of light increases, so does the circumference of darkness.[2]

—Attributed to Albert Einstein

Zorba: Why do the young die? Why does anybody die?

Basil: I don't know.

Zorba: What's the use of all your damn books if they can't answer that?

Basil: They tell me about the agony of men who can't answer questions like yours.

Zorba: I spit on this agony!

—*Zorba the Greek* (1964)

A civilization can be measured by how much progress its science and technology have made. The more advanced their science and technology are, the more advanced their civilization is. Our civilization is deemed more advanced than what we call primitive societies because of all the technological progress we have made. In contrast, if an alien civilization visited Earth, we would be considered primitive, almost by definition, since they have mastered interstellar space travel while we have not. The reason for using science and technology as a measuring stick is that these activities are the only aspect of culture that builds on itself. What was done by one generation is used by the next generation. This was expressed nicely by one of the greatest scientists of all time, Isaac Newton (1643–1727), who is quoted as saying, "If I have seen further it is only by standing on the shoulders of giants." This constant accumulated progress makes science a good measuring stick to compare civilizations. In contrast to science and technology, other areas of culture, such as the arts, human relations, literature, politics, morality, and so on, do not build on themselves.[3]

Another way to measure a civilization is by the extent to which it has banished unscientific and irrational ideas. We are more advanced today because we have cast alchemy into the wastebasket of silly dreams and study only chemistry. Centuries of treatises on astrology have been deemed nonsense while we retain our study of astronomy. As a civilization progresses, it subjects its beliefs and mythologies to logical analysis and disregards what is not within the bounds of reason.

The tool a civilization uses to make this progress is reason. Rationality and reason are the methodologies used by a society to advance. When a culture acts reasonably it will progress. When it deviates from reason, or steps beyond the limits of reason, it stagnates or regresses.

Reason comes in many forms. In broad (and perhaps inexact) terms, science is the language that we use to describe and predict the physical and measurable universe. The more abstract mathematics can be split into two areas: applied mathematics is the language of science, and pure mathematics is the language of reason. Logic is also a language of reason. Since science, technology, reason, rationality, logic, and mathematics are all intimately connected to each other, much of what I say about one will usually be true about all. At times I will just use the word *reason* to describe them all.

Philosophers have reflected and argued for centuries about what humans can and cannot know. The branch of philosophy that deals with human knowledge and its limitations is called *epistemology*. While the ideas of such philosophers are fascinating, their work will not be our central focus. Instead, we will be interested in what scientists, mathematicians, and current researchers have to tell us about the limits of human knowledge and reason.

One of the most amazing aspects of modern science, mathematics, and rationality is that they have matured to the level where they are able to see their own limits. As of late, scientists and mathematicians have joined philosophers in discussing the limitations of man's ability to know the world. These scientific limitations of reason are the central subject of this book.

The following is a cute little puzzle that gives a taste of what it means for reason to describe a limitation.[4] The puzzle is loads of fun, is worth pondering, and is also strongly recommended as a challenge at any cocktail party. Take a normal 8-by-8 chessboard and some dominoes that are of size 2-by-1. Try to cover the chessboard with the dominoes. There are sixty-four

Figure 1.1
Covering a chessboard with dominoes

squares on the chessboard and each domino covers two squares, so thirty-two dominoes will be needed. There are millions of ways to perform this task. Figure 1.1 shows how we might start the process.

That was pretty easy. Now let's try something a little more challenging. Put two queens on the opposite corners of the chessboard. Try to cover all the squares except the ones with queens, as in figure 1.2. There are sixty-two squares that need to be covered, which means thirty-one dominoes will be required. Try it!

After trying this problem for a while and not being able to cover every square, you might consider showing it to others—in particular, puzzle fans. They will have a similar experience. You might want to get a computer to work on the problem since a machine can quickly try many possibilities. There are millions, if not billions, of possible ways to try to start placing the dominoes on the board. Nevertheless, there is no way anyone or any computer will ever finish this task.

The reason why this simple problem of placing thirty-one dominos on a chessboard seems so hard is because *it cannot be done*. It is not a hard problem; it is an *impossible* problem. It is actually easy to explain why. Every domino is 2-by-1 and hence must cover a black-and-white square on the chessboard. The original board in figure 1.1 had thirty-two black squares and thirty-two white squares that needed to be covered. There was total symmetry on the board. By contrast, the board in figure 1.2 has thirty

Figure 1.2
Covering a chessboard minus two opposing corners

black squares and thirty-two white squares that need to be covered. The symmetry has been broken. There is no way anyone is going to be able to cover these sixty-two squares with dominoes where each covers one black square and one white square. Move the queens so that one is on a black square and one is on a white square. Now try it.

This small puzzle has a lot of nice features. It is easy to explain, easy to attempt a solution, and a computer can be used to try to solve the problem. Nevertheless it cannot be solved. It is not that we are not smart enough to solve the problem or that the problem is beyond the capabilities of present-day technology, but that it cannot be solved at all. It is not someone's opinion that this problem cannot be solved. Rather, it is a fact about the world. Reason dictates that there is a limitation in our ability to solve this problem. The best part of this problem is that the explanation of why it is unsolvable is easy to explain. Once it is stated, you are totally convinced and find it trivial.

This book will demonstrate a myriad of such unsolvable problems and limitations.

Rather than giving an orderly synopsis of each chapter, I'll provide a classification of the types of limitations the book covers. For every type of limitation, I will look at examples found in the different chapters. This will give the book a more unified structure.

Examples of limitations abound. Computer scientists have shown that there are many tasks that computers cannot perform in a reasonable amount of time (chapter 5). They have also shown that there are certain tasks that computers cannot perform in any amount of time (chapter 6). Physicists discuss how complex the world is and how some phenomena are just too complicated for science and mathematics to predict (section 7.1). Mathematicians have identified certain types of equations that cannot be solved by normal means (section 9.2). Logicians have proved that there are limitations to the power of proofs. They have described logical statements that are true but cannot be proved (section 9.4). Philosophers of language have shown that our ability to describe the world we live in is limited (chapter 2).

There are other types of limitations that, in a sense, are deeper. These are limitations that show that our naive intuition about the world we live in and our relationship to that world is faulty. The way we think about the universe and its properties has to be updated. Our very assumption that there is an objective definition of a particular physical object needs to be reevaluated (section 3.1). The classical philosopher Zeno has shown that our usual notions of space, time, and motion demand a deeper analysis (section 3.2). Quantum mechanics has taught us that the relationship between the knower and the known is not simple. This branch of physics has also shown us that the world is more interconnected than previously thought (section 7.2). Researchers have shown that our simple intuitions about infinity are faulty and need to be adjusted (chapter 4.) Relativity theory has demonstrated that our notions of space, time, and causality are wrong and need to be corrected. Physicists have shown that there is no objective measure of length or duration (section 7.3). The very relationship between us, our world, and the science and mathematics that we use to describe the world is not simple (chapter 8). All these topics, and many more, are explored in depth in the pages that follow.

The limitations just mentioned are demonstrated in myriad ways. One of the more interesting ways is with paradoxes. The word comes from the Greek *para-* "contrary to" and *doxa* "opinion." The *Oxford English Dictionary* gives many overlapping definitions, including

- A statement or tenet contrary to received opinion or belief. (For example, "Second-hand smoke is not so bad for you." "Democracy is not always the best form of government.")
- An apparently absurd or self-contradictory statement or proposition, or a strongly counter-intuitive one, which investigation, analysis or explanation may nevertheless prove to be well-founded or true. (For example, "In

the long-run, the stock market is a bad place to invest." "Standing is more strenuous than walking.")

To us, the most important definition will be

- An argument, based on (apparently) acceptable premises and using (apparently) valid reasoning, which leads to a conclusion that is against sense, logically unacceptable, or self-contradictory.

Such paradoxes will be our main concern. Here one has a premise or makes an assumption and using valid logic derives a falsehood. We might envision this paradox or derivation as

assumption ⇒ falsehood.

Since falsehoods cannot occur and since our derivation followed valid logic, the only conclusion is that our assumption was not true. In a way, the paradox is a test to see if an assumption is a legitimate addition to reason. If one can use valid reason and the assumption to derive a falsehood, then the assumption is wrong. The paradox shows that we have stepped beyond the boundaries of reason. A paradox in this sense is a pointer to an incorrect view. It points to the fact that the assumption is wrong. Since the assumption is wrong, it cannot be added to reason. This is a limitation of reason.

The type of falsehood that we will mostly encounter is a contradiction. By a contradiction I mean a fact that is shown to be both true and false. This is written as

assumption ⇒ contradiction.

Since the universe does not have contradictions, there must be something wrong with the assumption. For example, in chapter 6, we will see that if we assume that a computer can perform a certain task, then we can derive a contradiction about certain computers. Since there are no contradictions about physical objects like computers, there must be something wrong with our assumption.

Such paradoxes work the same way as a commonly found mathematical proof. A "proof by contradiction" or in Latin, *reductio ad absurdum* ("reduction to the absurd"), is as follows. If you want to show that some statement is true, simply assume that the statement is false and derive a contradiction:

statement is false ⇒ contradiction.

Since contradictions are not permitted in the exact world of mathematical reasoning, it must be that the assumption was incorrect, and the statement is, in fact, true. A simple example is the mathematical proof that the square root of 2 is not a rational number (section 9.1). If we assume that the square root of 2 *is* a rational number, then we derive a contradiction. From this we conclude that the square root of 2 is not a rational number. In section 4.3 I show that if we assume two particular sets are the same size, we can derive a contradiction. From this we conclude that one of the sets is larger than the other. Proofs by contradiction are ubiquitous.

One need not derive a full-fledged contradiction for a paradox. All that is needed is to derive a fact that is different from observation or simply false:

assumption $\Rightarrow$ false fact.

Once again, because we derived something false, our assumption must be in error. Zeno's paradoxes are examples of this type (section 3.2). Zeno assumes something and then proceeds to show that movement is impossible. Anyone who has ever walked down the street knows that movement occurs all the time and hence the assumption is false. The difficulty with Zeno's paradoxes is to identify the bad assumptions.

Many times paradoxes arise and highlight previously hidden assumptions. It could be that these assumptions are so deep within us that we do not even consider them (for example, that space is continuous and not discrete, or that physical objects have exact definitions). Such paradoxes will be a challenge to our intuitions about the universe we live in. By showing that our intuitions are false, we can disregard them and be propelled forward. The American philosopher Willard Van Orman Quine (1908–2000) eloquently wrote:

> The argument that sustains a paradox may expose the absurdity of a buried premise or of some preconception previously reckoned as central to physical theory, to mathematics or to the thinking process. Catastrophe may lurk, therefore, in the most innocent-seeming paradox. More than once in history the discovery of paradox has been the occasion for major reconstruction at the foundation of thought.[5]

This method of exploring paradoxes and looking for their assumptions will be one of our focuses throughout the book.

Particular types of paradoxes play a major role in the tale we tell. Self-referential paradoxes are paradoxical situations that come from a system where the objects of the system can deal with / handle / manipulate themselves. The classic example of a self-referential paradox is the so-called *liar paradox*. Consider the English sentence:

"This sentence is false."

If the sentence is true, then the sentence is, in fact, false because it says so. If the sentence is false, then since the sentence expresses its own falsehood, the sentence is true. This is a genuine contradiction. The problem arises from the fact that English sentences have the ability to describe true and false statements about themselves. For example, "This sentence has five words" is a legitimate English sentence that expresses something true about itself. In contrast, "This sentence has six words" is a false statement about itself. We will see that whenever a system can discuss properties about itself, a paradoxical situation can occur. We will find that language, thought, sets, logic, math, and computers are all systems with the ability to deal with themselves. Within each of these areas, the potential for self-reference will lead to paradoxes and hence some type of limitation. The amazing fact is that although these areas are very different, the form of the paradoxes are the same.

Another method of describing a limitation is by piggybacking on an already established limitation. Before I explain what this is all about, let's discuss some mountain climbing. Mount Everest is 29,000 feet high and Mount McKinley is "only" 20,000 feet high. The following fact seems obvious: if you can climb Mount Everest, then you can most definitely (*a fortiori*) climb Mount McKinley. We write this as

climbing Everest ⇒ climbing McKinley.

If you are able to climb Mount McKinley, you would feel great pride. We write this as

climbing McKinley ⇒ pride.

Putting the two implications together, we get

climbing Everest ⇒ climbing McKinley ⇒ pride,

which leads to the obvious conclusion that if you are able to climb Mount Everest, you would feel great pride. Now let us look at the dark side of mountain climbing. Suppose your doctor told you that bad things might happen to you if you try to climb Mount McKinley. We write this as

climbing McKinley ⇒ bad.

This is expressing a limitation of your abilities: you should not climb Mount McKinley. Combining this implication with the first one, gives us

climbing Everest ⇒ climbing McKinley ⇒ bad.

This states the obvious fact that if you should refrain from climbing Mount McKinley, then you most definitely should refrain from climbing Mount Everest. In other words, the obvious implication that

climbing Everest $\Rightarrow$ climbing McKinley

can be used to transfer or piggyback a known limitation about climbing Mount McKinley into a limitation about climbing Mount Everest. I use these simple ideas in the following pages.

Now let us use this intuition about mountain climbing to understand the general concept of one limitation piggybacking on another limitation. Imagine that a limitation was established by a contradiction as follows:

assumption A $\Rightarrow$ contradiction.

That is, assumption A cannot be correct because we can derive a contradiction from it. Now consider assumption B. If we can show that from assumption B, we can derive assumption A, that is,

assumption B $\Rightarrow$ assumption A,

then we have

assumption B $\Rightarrow$ assumption A $\Rightarrow$ contradiction.

To elaborate, if assumption B is correct, then assumption A is correct and since we already established that assumption A is not correct, we conclude that assumption B also cannot be correct. This is called a *reduction*: one assumption was reduced to another. With a reduction there is a transfer of already-known limitations to other areas.

Examples of reductions are found throughout the book:

- I show that if a certain problem takes a long time for a computer to solve, then other harder problems will take a longer time for a computer to solve (section 5.3).
- I show that if a certain problem cannot be solved by a computer, than harder problems also cannot be solved by a computer (section 6.3).
- I use similar methods to show that certain simply stated math problems are unsolvable (section 9.3).
- Other comparable reductions are found in our discussion of logic (section 9.5).

A few words about contradictions. The physical universe does not permit any contradictions:

- A certain molecule cannot both be hydrochloric acid and not hydrochloric acid.
- It cannot both be Monday and not Monday simultaneously in the same place.
- The diagonal of a square cannot be the same length as its side.

Similarly science, which is a description of the physical universe, also cannot express contradictions:

- It cannot be that the formulas $E = mc^2$ and $E \neq mc^2$ are both true.
- A calculation about a chemical process cannot be both true and not true.
- A prediction cannot predict two incompatible events.

If there were a contradiction in science, it would not be an exact description of the contradiction-free universe. Similarly for mathematics and logic: to the extent that they are used in describing the universe and science, they cannot contain any contradictions.

There is, however, a place where contradictions do occur: inside the human mind. We are all fraught with contradictions; we desire contradictory things; we believe contradictory ideas; and we predict contradictory events. Anyone who has ever been in a relationship knows the feeling of simultaneously being in love with a person and hating them. We desire to have our cake and to be thin. As the Queen says to Alice in *Through the Looking Glass*, "Why, sometimes I've believed as many as six impossible things before breakfast." The human mind is not a perfect machine. We are conflicted and confused. Similarly, human language, which expresses states of mind, must also have contradictions. There is nothing strange when we state "I love her and I hate her." It is not unusual for a person to express a desire to be thin while having another piece of cake.[6]

When we meet a paradox in the physical world and derive a contradiction, we know that there must be something wrong with the assumption of the paradox. However, when we meet a contradiction in the realm of human thought or in human language, then we need not abandon the assumption. More subtlety is possible. Why not permit the contradiction? Consider the liar paradox discussed earlier. Why not simply say that the sentence

This sentence is false.

is both true and false or perhaps meaningless? It is only an English sentence and many English sentences express contradictions. Similarly, the belief

This belief is false.

is both true and false. Why not permit such contradictory beliefs in our already-confused minds?

The relationship between the contradiction-free universe and our feeble human minds and languages raises many more interesting questions. How is it that the human mind can understand any part of the universe? How can a language formulated by human beings describe the universe? Why does science work? Why is mathematics so good at describing science and the universe? Do the laws of science have an external existence or are they only in our mind? Can there be a final description of the universe—that is, will science ever complete its mission and end? Are the truths of science and mathematics time dependent or culturally dependent? How can human beings tell when a scientific theory is true? As Albert Einstein wrote, "The eternal mystery of the world is its comprehensibility.[7] These and a host of other questions from the philosophy of science and mathematics are addressed in chapter 8.

Between the contradiction-free universe and the contradiction-laden human mind, a landscape full of vagueness exists:

- A person who stands in the doorway of a room is both in the room and not in the room.
- How many hairs does a man have to lose in order to be considered bald? Depending on which way the wind blows, he is sometimes considered bald and sometimes considered not bald.
- Is 42 a small or large number?

Human beings use vague ideas all the time. Our mindset and our concomitant human language are full of vague statements:

- Sometimes we say people in a doorway are in the room and sometimes we say they are not in the room.
- We call certain people with a few hairs bald and others not bald.
- If our bank account contains only $42, we say that 42 is a small number, but if we are talking about the number of diseases a person has, 42 is a large number.

Because vague ideas are outside the pristine world of science and mathematics, we cannot rely on some of the usual tools in addressing these ideas. Vagueness plays a major role in our discussions in chapter 3.

As a slight aside, special types of jokes are of interest for our discussion. We have seen that paradoxes are ways of showing that one has gone too far with reason. Violating a paradox means you stepped beyond the boundaries of reason and entered the land of the absurd. There are jokes that also play on the fact that we are taking reason too far. Such jokes take logic and reason to places where they were not intended. They start off with concepts that are well understood and then go farther or beyond their usual meaning. Consider the following:

- Woody Allen cheated on his metaphysics exam by looking into the soul of the boy sitting next to him.
- Steven Wright said he would kill for a Nobel Peace Prize.
- Groucho Marx didn't care to belong to any club that would have him as a member.

In all of these jokes, normal ideas are taken too far. Cheating on an exam, desiring a Nobel Peace Prize, or resigning your club membership in disgust are all common ideas. However, these great thinkers have taken these usual concepts where they do not belong: to the silly and ridiculous.

Even puns fall into this category. A pun is a joke where the meaning of a word or phrase is taken into an area where it was not intended:

- "Have you heard about the guy whose whole left side was cut off? He's all right now."
- "I'm reading a book about antigravity. It's impossible to put down."
- "Did you hear about the par-a-dox? . . . Doctor Shapiro and Doctor Miller."

Groan! (Sorry. The only thing worse than a pun is an analysis of a pun. Let us move on.)

I close this introduction with a few questions about the nature of reason and its limitations. Read the book with these questions in mind. I return to these issues in the last chapter and perhaps get closer to the answers using some of the ideas presented in the book.

I would be remiss in writing a book titled *The Outer Limits of Reason* without giving a definition of *reason*. After all, how can we say something is beyond the limits of reason if we do not define reason? What is a reasonable process to determine facts? Are there different levels of reason? How do we draw the line between alchemy and chemistry? Between astrology and astronomy? Why are some actions deemed reasonable and others not? Why does it make sense to check your blood pressure while it is ludicrous to check your horoscope? What thought processes are reasonable and will avoid contradictions?

The *Oxford English Dictionary* gives sixteen classes of definitions for the word *reason*. The definition closest to the one we want is the following: "The power of the mind to think and form valid judgments by a process of logic; the mental faculty which is used in adapting thought or action to some end; the guiding principle of the mind in the process of thinking. Freq. contrasted with *will, imagination, passion,* etc. Often personified." But this definition just raises more questions. What is a "valid judgment"? When is something a logical process as opposed to an illogical process? When is thinking part of the will and when is it reason? This definition is unsatisfying. Other purported definitions are not much better.

There is something self-referential in our entire enterprise. We are using reason to find limitations of reason. If reason is limited, how are we to use reason to discover those limitations? What are the limits to our limit-showing abilities?

Let's hold these questions in abeyance and return to them in chapter 10, when we conclude our explorations of the limits of reason.

Further Reading

Other books that discuss limitations of reason are Barrow 1999, Dewdney 2004, and Poundstone 1989. Sorensen 2003 is a wonderful history of paradoxes.

2 Language Paradoxes

What we cannot speak about we must pass over in silence.[1]
—Ludwig Wittgenstein (1889–1951), Proposition 7 of *Tractatus Logico-Philosophicus*

After all, Mr. Wittgenstein manages to say a good deal about what cannot be said.
—Bertrand Russell (1872–1970), introduction to Wittgenstein's *Tractatus Logico-Philosophicus*

Half the lies they tell about me aren't true.
—Yogi Berra[2]

Rather than jumping headfirst into the limitations of reason, let us start by just getting our toes wet and examining the limitations of language. Language is a tool used to describe the world in which we live. However, don't confuse the map with the territory! There is one major difference between the world we live in and language: whereas the real world is free of contradictions, the man-made linguistic descriptions of that world can have contradictions.

In section 2.1, we encounter the famous liar paradox and its many variants. These are relatively easy puzzles that will get us started. Section 2.2 contains a collection of self-referential paradoxes. I show that they all have the same form. In section 2.3 we meet several paradoxes involving descriptions of numbers.

2.1 Liar! Liar!

A linguistic paradox is a phrase or sentence that contradicts itself. A baby version of a linguistic paradox is an oxymoron (from the Greek *oxys* "sharp" and *moros* "stupid"—together they mean "pointedly foolish" or "pointedly dull"). These are phrases, usually consisting of two words, that contradict each other. Some examples are "original copies," "open secret," "clearly confused," "militant pacifist," "larger half," "alone together," and my favorite, "act naturally." Even though these phrases do not really make sense, we human beings have no problem using them in common everyday speech.

The classic example of a linguistic paradox is the famous *Epimenides paradox*. This dates back more than two and a half millennia to when Epimenides (600 BC), a philosopher and poet who lived in Crete, complained about his neighbors in a poem called *Cretica*. He wrote: "The Cretans, always liars, evil beasts, idle bellies!" This seems[3] paradoxical. If this statement is true, then since Epimenides is a Cretan, he is including himself as a liar and this line of the poem is false. In contrast, if it is false, then Epimenides is not a liar and the line is true.

There are many linguistic paradoxes similar to Epimenides' statement. The *liar paradox* is a simple sentence like

I am lying.

or

This sentence is false.

If these sentences are true, then they are false. Furthermore, if they are false, then they are true.

The liar paradox is found in many different forms. For example, we can denote a sentence L_1 and then say that L_1 asserts its own falsehood:

L_1: L_1 is false.

Again, if L_1 is true, then it is false. And if L_1 is false, then it is true. Other variations of the liar paradox have sentences that are not directly self-referential. Consider the following two sentences:

L_2: L_3 is false.

L_3: L_2 is true.

If L_2 is true, then L_3 is false, which would mean that "L_2 is true" is false and hence L_2 is false. In contrast, if L_2 is false, then L_3 is true and L_3 asserts that L_2 is true. Buzz! That's a contradiction.

It is important to note that just because sentences refer to themselves and their falsehoods does not mean there is a contradiction. Consider these two sentences:

L_4: L_5 is false.

L_5: L_4 is false.

Let's assume that L_4 is false. Then L_5 is true and L_4 is false. Similarly, if you start with the premise that L_4 is true, you get that L_5 is false, and hence L_4 is true. Neither assumption leads you to a contradiction.

There are many other forms of the liar paradox:

- <u>The only underlined sentence on this page is a total lie.</u>
- **The boldface sentence on this page is a blatant falsehood.**
- The sentence after the boldface sentence on this page is not true.

Are they true or false?

The liar paradox has been around for over 2,500 years and philosophers have devised many different ways of avoiding such contradictions. Some philosophers try to avoid these linguistic paradoxes by saying that the liar sentences are neither true nor false. After all, not every sentence is true or false. Questions such as "Your place or mine?" and commands such as "Go directly to jail!" are neither true nor false. One usually thinks of declarative sentences like "Snow is white" as either true or false, but the liar sentences show that there are some declarative sentences that are also neither true nor false.

There are those who say that the sentence "This sentence is false" is not even grammatically correct. After all, what does "This sentence" refer to? If it refers to something, we should be able to replace "This sentence" with whatever it refers to. Let's give it a try:

"This sentence is false" is false.

This is grammatically correct and it might be true or false. But it is not self-referential and not equivalent to the original liar sentence. This is similar to the sentence

"This sentence is false" has four words.

which is true, while

"This sentence is false" has five words.

is false. It would be nice to have a grammatically correct English sentence that is a self-referential paradox. W. V. O. Quine came up with a clever way around these problems. Consider the following *Quine's sentence*:

"Yields falsehood when preceded by its quotation"

yields falsehood when preceded by its quotation.

First notice that this is a legitimate English sentence. The subject is the phrase in quote marks and the verb is *yields*. Now, let us ask ourselves if it is true. If it is true, then when you attach the subject to the rest of the sentence, as we did, we get falsehood. So the sentence is false. In contrast, what if the sentence is false? That means that when you attach the subject to the sentence, you do not get a falsehood; rather, you get a true sentence. So if you assume that Quine's sentence is false, you derive that it is true. This is a grammatically correct English sentence that is self-contradictory.

Another potential solution to paradoxical sentences is to restrict language so as to avoid such sentences. Some have said that language should be stratified into different levels. They have declared that sentences cannot talk about other sentences of their own level or higher. For example, at the lowest level there will be sentences like "Grass is green" and "My pen is blue." The next level will be sentences about sentences on the lowest level. So we might have

"Grass is green" is an obvious sentence.

or

"My pen is blue" has four words in it.

One goes on to higher-level phrases like

"'My pen is blue' has four words in it" is a dumb fact.

By restricting the types of sentences, we will be avoiding sentences of the form

The sentence in italics on this page is grammatically correct.

This is a sentence dealing with itself and hence is a sentence on its own level. It is declared not kosher—that is, not a legitimate part of language. Every sentence is only permitted to talk about sentences that are "below" it. If a sentence does talk about a sentence that is on its own level, that sentence is proclaimed meaningless. This stratification will ensure that there are no self-references and hence no contradictions. With such restrictions in place, linguists are fairly certain that they have banned most paradoxical linguistic sentences. However, this solution is somewhat artificial. Common human language has always dealt with some type of self-reference without problem:

- Someone says, "Oh! I am groggy today and I do not know what I am talking about." Is he aware of saying this sentence?
- Carly Simon sings a song with the lyrics "You're so vain, you probably think this song is about you." But this song *is* about him!
- "Every rule has an exception except one rule: this one."

- "Never say 'never'!"
- "The only rule is that there is no rule."

In all of these cases—and many more—human language is violating the restriction of only dealing with sentences that are "below" it. In each case, a sentence discusses itself. And yet, somehow, all these examples are a legitimate part of human language.

Another possible solution to paradoxical sentences was mentioned in chapter 1, namely, human language is a product of the human mind and, as such, subject to contradictions. Human language is not a perfect system that is free of discrepancies (in contrast to perfect systems like mathematics, science, logic, and the physical universe). Rather, we should simply accept the fact that human language is faulty and has contradictions. This seems reasonable to me.

2.2 Self-Referential Paradoxes

The cause of the problem with the liar paradox is that language can be used to describe language. In particular, one can have a sentence that discusses its own truthfulness. This ability of language to describe language is a form of self-reference. Paradoxes that arise from such self-reference are the subject of this section. While these paradoxes are not linguistic paradoxes per se, they are similar to the liar paradox and will help us understand the true nature of self-reference.

The British philosopher Bertrand Russell described a delightful little paradox that has come to be known as the *barber paradox*. Imagine a small isolated village in the Austrian Alps that has only one barber. Some villagers shave themselves and some go to the barber. Everyone in the village abides by the following rule: all those who do not shave themselves must go to the only barber and all those who do shave themselves do not go to the barber. This seems like a pretty innocuous rule. After all, if they can save some money by shaving themselves, why go to the barber? And if they go to the barber, why shave themselves? Now, simply ask yourself:

Who shaves the barber?

He is a villager and so if he does not shave himself, must go to the barber. But he *is* the barber and so he shaves himself. If he does shave himself, then, since he is the barber, he goes to the barber and does not shave himself.[4]

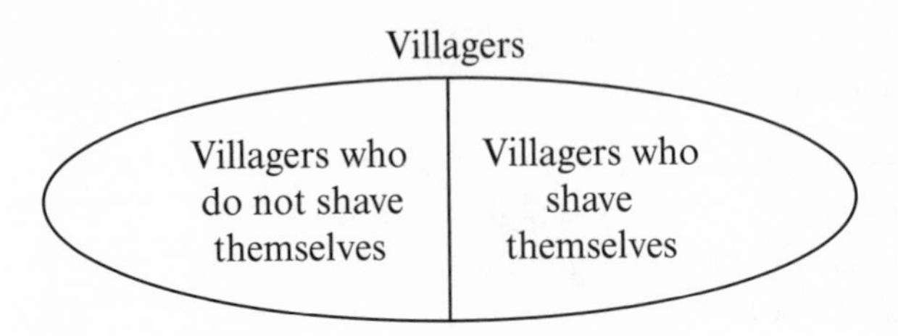

Figure 2.1
Which subset is the barber in?

We might envision the barber paradox with figure 2.1. We split the set of villagers into two parts and look to see if the barber is on the right or left.

In contrast to the liar paradox, the barber paradox has a simple solution: the village described simply does not exist. It cannot exist because there is a contradiction inherent in its description. Our description entails a contradiction with the barber. Since the real world cannot have contradictions, the village does not really exist. There are many other villages in the Austrian Alps, but they have different setups. They might have two barbers that shave each other; they might have a female barber that does not shave; they might have long-haired hippie types who do not go to any barber regardless of need. These descriptions of other villages are totally legitimate; no contradictions result from them. But the village Russell described cannot exist.

Another clever paradox deals with adjectives in English and is called the *heterological paradox* or *Grelling's paradox*. Consider the word *English*. *English* is an English word. In contrast, *French* is not a French word (it is an English word). Let us look at some other adjectives and see how they relate to themselves: *polysyllabic* is polysyllabic. *Monosyllabic* is not monosyllabic. *Pentasyllabic* (made of five syllables) is pentasyllabic. *Misspelled* is not misspelled. *Adjectival* is adjectival. *Female* is not female. *Awkwardnessfull* is awkwardnessfull. *Unpronounceable* is not unpronounceable. In effect, we have two groups of adjectives: those that describe themselves and those that do not. All adjectives that describe themselves are called *autological* (from the Greek *auto* meaning "self" or "one's own" and *logos* meaning "word," "speech," or "reason") or *homological*. In contrast, all adjectives that do not describe themselves are called *heterological* (from the Greek *heteros* meaning "other" or "different"). So we have that *English, polysyllabic, adjectival,* and so on are all autological. In contrast, *French, monosyl-*

labic, *unpronounceable*, and so forth are all heterological. With these two categories set up, we now pose the following question:

Is *heterological* heterological?

Let us say that *heterological* is heterological. Just as

English is English $\Rightarrow$ *English* is autological,

so too

heterological is heterological $\Rightarrow$ *heterological* is autological

and hence *heterological* is not heterological. In contrast, if we take the opposite view and say that *heterological* is not heterological, then just as we saw that

French is not French $\Rightarrow$ *French* is heterological,

so too

heterological is not heterological $\Rightarrow$ *heterological* is heterological.

We have come to the conclusion that *heterological* is heterological if and only if it is not heterological. Buzz! This is a contradiction and troublesome.

We can again envision this self-referential paradox as figure 2.2.

This paradox also seems to have a simple solution: there is no word *heterological*, or if the word does exist, it has no meaning. We saw that if one defines *heterological*, then we come to a contradiction. This is similar to saying that the village in the barber paradox does not exist.

However, we cannot simply solve all problems by waving our hand and declaring that the word *heterological* does not exist or has no meaning. The problem is too deeply rooted in the very nature of language. Rather than dealing with the word *heterological*, consider the related adjective phrase

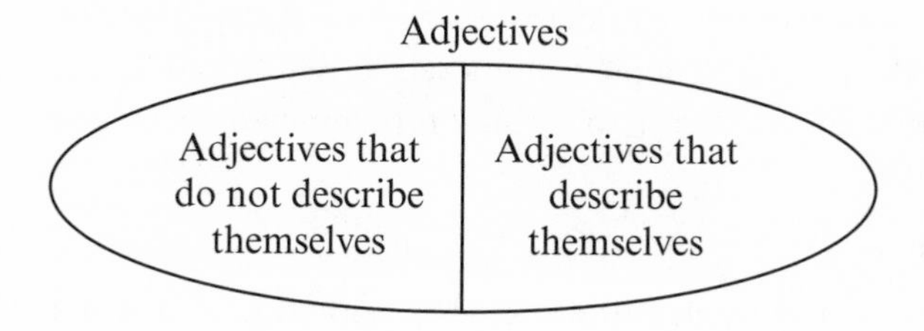

Figure 2.2
Which subset does *heterological* belong to?

"not true of itself." Simply ask if the phrase "not true of itself" is true of itself. It is true if and only if it is not true. Are we simply to posit that "not true of itself" is not a legitimate adjectival phrase? There are no problems with any of the words in the phrase. There is nothing about the phrase that is weird like the word *heterological*. Nevertheless, we come to a contradiction if we use it.

The *reference-book paradox* is very similar to the heterological paradox. A reference book is a book that lists books in different categories. There are many reference books that list books of many different types. There are reference books that list antique books, anthropology books, books about Norwegian fauna, and so on. Certain reference books list themselves. For example, if one were to publish a reference book of all books published, that reference book would contain itself. There are also certain reference books that would not list themselves. For example, a reference book on Norwegian fauna would not list itself. Consider the reference book that lists all reference books that do not list themselves. Now ask yourself the following simple question: Does this book list itself? With a little thought, it is easy to see that this book lists itself if and only if it does not list itself. We conclude that no such reference book with such a rule for its content can exist. (I leave to the reader the task of drawing a diagram similar to figures 2.1 and 2.2 for this paradox.)

Bertrand Russell used the barber paradox to explain a more serious paradox called *Russell's paradox*. This is more abstract than the other self-referential paradoxes we saw and is worth pondering. Consider different sets or collections of objects. Some sets just contain elements and some sets contain other sets. For example, one can look at a school as a set containing different grades, where each grade is the set of students in the grade. Some sets even contain copies of themselves. The set of all sets described in this book contains itself. The set of all sets with more than five elements contains itself. There are, of course, many sets that do not contain themselves. For instance, consider the set of all red apples. This does not contain itself since a red apple is not a set. Russell would like us to consider the set R of all sets that do not contain themselves. Now pose the following question:

Does R contain itself?

If R does contain itself, then, by definition of what belongs to R, it is not contained in R. If, on the other hand, R does not contain itself, then it satisfies the requirement of belonging to R and is contained in R. We have a contradiction. This can be visualized in figure 2.3.

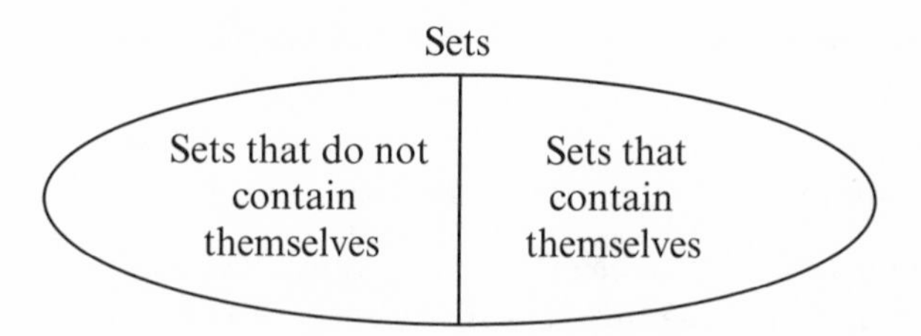

Figure 2.3
Which part contains R?

This paradox is usually "solved" by positing that the collection R does not exist—that is, that the collection of all sets that do not contain themselves is not a legitimate set. And if you do deal with this illegitimate collection, you are going beyond the bounds of reason. Why should one not deal with this collection R? It has a perfectly good description of what its members are. It certainly looks like a legitimate collection. Nevertheless, we must restrict ourselves in order to steer clear of contradictions. The obvious (and seemingly reasonable) notion that for every clearly stated description there is a collection of those things that satisfy that description is no longer obvious (or reasonable). For the clearly stated description of "red things," there is a nice collection of all red things. However, for the seemingly clear description of "all sets that do not contain themselves," there is no collection with this property. We must adjust our conception of what is obvious.[5]

Russell's paradox should be contrasted with the other paradoxes. There are simple solutions to the barber paradox and the reference-book paradox: those physical objects simply do not exist. And there is a simple solution to the heterological paradox: human language is full of contradictions and meaningless words. We are, however, up against a wall with Russell's paradox. It is hard to say that the set R simply does not exist. Why not? It is a well-defined idea. A collection is not a physical object, nor is it a human-made object. It is simply an idea. And yet this seemingly innocuous idea takes us out of the bounds of reason.

The liar paradox was summarized by one sentence:

This sentence is false.

It can also be summarized by the following description:

The sentence that denies itself.

Similarly, the other four self-referential paradoxes can be summarized by the following four descriptions:

- "The villager who shaves everyone who does not shave themselves."
- "The word that describes all words that do not describe themselves."
- "The reference book that lists all books that do not list themselves."
- "The set that contains all sets that do not contain themselves."

As you can see, all these descriptions have the exact same structure (as do figures 2.1 through 2.3). Every time there is self-reference, there are possibilities for contradictions. Such contradictions will have to be avoided and will require a limitation. We explore such limitations throughout the book.

Before moving on to the next section, there is an interesting result that demands further thought. One might think that every language paradox has some form of self-reference. That is, there must be some chain of reasoning that is circular and returns to where it started. This was the common belief until Stephen Yablo came up with a clever paradox called *Yablo's paradox*. Consider the following infinite sequence of sentences:

K_1 K_i is false for all $i > 1$

K_2 K_i is false for all $i > 2$

K_3 K_i is false for all $i > 3$

⋮ ⋮

K_m K_i is false for all $i > m$

K_{m+1} K_i is false for all $i > m + 1$

⋮ ⋮

K_n K_i is false for all $I > n$

⋮ ⋮

Every statement declares all the further statements to be false. Notice that no sentence ever references itself, nor is there any long chain that has some sentence referring back to itself. Nevertheless, this is a paradox in the sense that one cannot say that any sentence is either true or false. Imagine that for some m, we have that K_m is true. K_m says that all of K_{m+1}, K_{m+2}, K_{m+3}, . . . are false. Splitting this up, we have that K_{m+1} is false and all of K_{m+2}, K_{m+3}, . . . are false. However, K_{m+1} says that all of K_{m+2}, K_{m+3}, . . . are false, which makes K_{m+1} true. Hence, by assuming that K_m is true, we get a contradiction about the status of K_{m+1}. This can be viewed in figure 2.4.

In contrast, imagine that for any m, we assume K_m is false. That means that not all K_n for $n > m$ are false and there is at least one $n > m$ with K_n is true. But we saw that if any K_n is true, we get a contradiction as in figure 2.5.

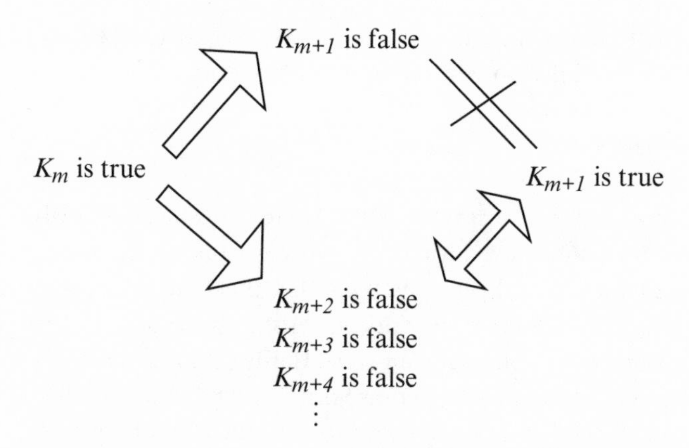

Figure 2.4
Yablo's paradox—assuming true

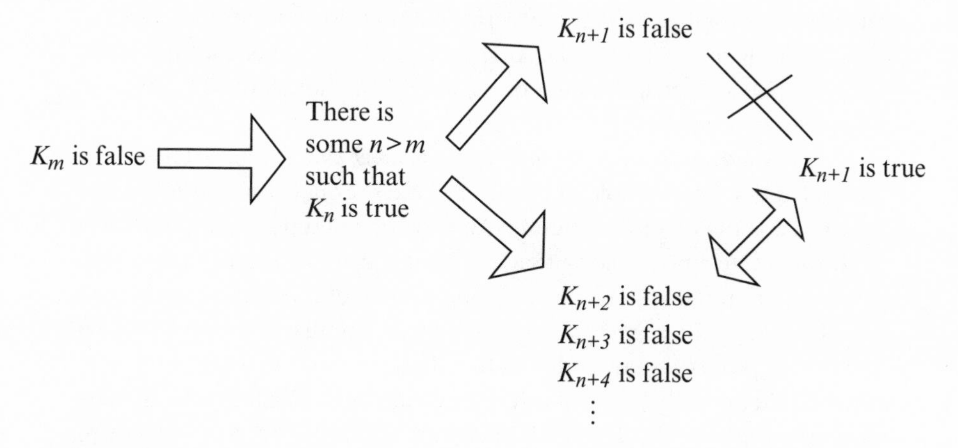

Figure 2.5
Yablo's paradox—assuming false

When we assume that any K_m is either true or false, we arrive at a contradiction. This is a contradiction without any self-reference.

2.3 Naming Numbers

Numbers are the most exact concepts we have. There is no haziness with the idea of 42. It is not a subjective idea where every person has their own concept of what 42 really is. And yet we will see that there are even problems with the description of numerical concepts. First a short story. In the early twentieth century, the mathematician G. H. Hardy (1877–1947) went to visit his friend and collaborator, the genius Srinivasa Ramanujan (1887–1920). Hardy writes: "I remember once going to see him when he was ill at Putney. I had ridden in taxi cab number 1729 and remarked that the number seemed to me rather a dull one, and that I hoped it was not an unfavorable omen. 'No,' he replied, 'it is a very interesting number; it is the smallest number expressible as the sum of two cubes in two different ways.'"[6] In detail, 1729 is equal to $1^3 + 12^3$ but it is also equal to $9^3 + 10^3$. Since 1729 is the smallest number for which this can be done, 1729 is an "interesting" number.[7]

This tale brings to light the *interesting-number paradox*. Let's take a tour through some small whole numbers. 1 is interesting because it is the first number. 2 is the first prime number. 3 is the first odd prime. 4 is a number with the interesting property that $2 \times 2 = 4 = 2 + 2$. 5 is a prime number. 6 is a perfect number—that is, a number whose sum of its factors is equal to itself (i.e., $6 = 1 \times 2 \times 3 = 1 + 2 + 3$, etc.). The first few numbers have interesting properties. Any number that does not have an interesting property should be called an "uninteresting number." What is the smallest uninteresting number? The smallest uninteresting number is an interesting number. We are in a quandary.

What went wrong here? The contradiction came about because we thought we could split all numbers into two groups: interesting numbers and uninteresting numbers. This is false. There is no way to define what an interesting number is. It is a vague term and we cannot say when a number is interesting and when it is uninteresting.[8] "Interesting" is a feeling that a person gets sometimes and hence is a subjective property. We cannot make a paradox out of such a subjective property.

A more serious and related paradox is called the *Berry paradox*. The key to understanding this paradox is that in general the more words one uses in a phrase, the larger the number one can describe. The largest number that can be described with one word is 90. 91 would demand more than one word. Two words can describe ninety trillion. Ninety trillion + 1 is the first number that demands more than two words. Three words can describe ninety trillion trillion. The next number (ninety trillion trillion + 1) would demand more than three words. Similarly, the more letters in a word, the larger the number you can describe. With three letters, you can describe the number 10 but not 11.

Let us stick to number of words. Call a phrase that describes numbers and has fewer than eleven words a *Berry phrase*. Now consider the following phrase:

the least number not expressible in fewer than eleven words.

This phrase has ten words and expresses a number, so it should be a Berry phrase. However, look at the number it purports to describe. The number is not supposed to be expressible in fewer than eleven words. Is this number expressible in eleven words or less? This is a real contradiction.

We may also talk about other measures of how complicated an expression is. Consider

the least number not expressible in fewer than fifty syllables.

This phrase has fewer than fifty syllables. Another phrase,

the least number not expressible in fewer than sixty letters,

has fifty-nine letters. Do these descriptions describe numbers or not? And if they do describe numbers, which ones? They describe a certain number if and only if they do not describe that number. But why not? Each certainly seems like a nice descriptive phrase.

Yet another interesting paradox about describing numbers is *Richard's paradox*. Certain English phrases describe real numbers between 0 and 1. For example,

- “pi minus 3” = 0.14159
- “the chance of getting a 3 when a die is thrown” = 1/6
- “pi divided by 4” = 0.785
- “the real number between 0 and 1 whose decimal expansion is 0.55555” = 0.55555

Call all such phrases *Richard phrases*. We are going to describe a paradoxical sentence. Rather than just stating the long sentence, let us work our way toward it. Consider the phrase the real number between 0 and 1 that is different from any Richard phrase.

If this described a number, it would be paradoxical since the phrase would describe a number and yet it would not be a Richard phrase. However, there are many real numbers that are different from all Richard phrases. Which one is it? The problem is that this phrase does not really describe an exact number. Let us try to be more exact. The set of Richard phrases are a subset of all English phrases, and as such, they can be ordered like names in a telephone book. We can first order all Richard phrases of one word, then the phrases of two words, and so on. With such an ordered list we can talk about the *n*th Richard sentence. Now consider

the real number between 0 and 1 whose *n*th digit is different from the *n*th digit of the *n*th Richard phrase.

This is just showing how the number described is different from all the Richard phrases, but it still does not describe an exact number. The number described by the forty-second Richard number might have an 8 as the forty-second digit. From this, we know that our phrase cannot have an 8 in the forty-second position. But should our number have a 9 or 6 in that position? Let us be exact:

the real number between 0 and 1 defined by its *n*th digit being 9 minus the *n*th digit of the *n*th Richard phrase.

That is, if the digit is a 5, this phrase will describe a 4. If the digit is an 8, this phrase will describe a 1. And if the digit is a 9, this phrase will describe a 0. This phrase is a legitimate English phrase that precisely describes a number between 0 and 1, yet it is different from every single Richard phrase. The phrase does describe a number if and only if it does not describe a number. What to do?[9]

These last two paradoxes can be seen as self-referential paradoxes. In a sense, they can be summarized by the following two descriptions:

- "the Berry phrase that is different from all Berry phrases"
- "the Richard phrase that is different from all Richard phrases"

From this point of view, they are simple extensions of the liar paradox. Self-reference is very common and we must be careful with it.

Further Reading

Many of the paradoxes can be found in places such as Quine 1966, Hofstadter 1979, 2007, Barrow 1999, and Poundstone 1989. Sorenson 2003 is a clear and well-written introduction to paradoxes. Chapter 5 of Sainsbury 2007 covers the liar paradox and other forms of self-reference. Chapter 3 of Paulos 1980 provides a humorous look at all self-referential paradoxes. Yablo's paradox is found in Yablo 1993.

A formal version of self-referential paradoxes can be found in Yanofsky 2003, which is derived from Lawvere 1969.

3 Philosophical Conundrums

Moreover, although these opinions appear to follow logically in a dialectical discussion, yet to believe them seems next door to madness when one considers the facts. For indeed no lunatic seems to be so far out of his senses.

—Aristotle (384–322 BC), *On Generation and Corruption*, 325a15

All are lunatics, but he who can analyze his delusion is called a philosopher.

—Ambrose Bierce, *The Collected Works of Ambrose Bierce*

It depends on what the meaning of the word "is" is.

—William Jefferson Clinton

Long before modern scientists took up the task of investigating the limits of reason, philosophers were analyzing the complexities of our world and our knowledge of it. In this chapter I explore some of the ancient and contemporary philosophical aspects of reason's limitations.

In section 3.1, I begin by discussing some very fundamental questions about concrete and abstract objects and the way we define them. In section 3.2, the very nature of space, time, and motion are analyzed using some of Zeno's paradoxes. The section ends with a short discussion of time-travel paradoxes. Section 3.3 is concerned with vagueness. Section 3.4 is centered on the very notion of knowing and having information. These sections are independent of each other and of the rest of the chapters. They can be read in any order.

3.1 Ships, People, and Other Objects

In ancient Greece, there was a legendary king named Theseus who supposedly founded the city of Athens. Since he fought many naval battles, the people of Athens dedicated a memorial in his honor by preserving his ship in the port.[1] This "ship of Theseus" stayed there for hundreds of years. As time went on, some of the wooden planks of Theseus' ship started rotting away. To keep the ship nice and complete, the rotting planks were replaced with new planks made of the same material. Here is the key question: If you replace one of the planks, is it still the same ship of Theseus? This question about a mythical ship is the poster child for one of the most interesting problems in all of philosophy, namely the *problem of identity*. What is a physical object? How do things stay the same even after they change? At what point does an object become different? When we talk about a certain object and say that "it changed," what exactly is "it"?

What happens if you change two of the ship's planks? Would that make it somehow less of the original ship than after one plank is changed? What if the ship consists of a hundred planks and forty-nine of the planks are changed? How about fifty-one changed planks? What about changing ninety-nine of the hundred planks? Is the single plank at the bottom of the ship enough to maintain the original lofty status of the ship? And what if all of the planks are changed? If the change is gradual, does the ship still maintain its status as the ship of Theseus? How gradual must the change be?

We are not answering these questions simply because there are no objective correct answers. Some maintain that changing one plank changes the ship and makes it no longer the ship of Theseus. Others claim that as long as there is at least one plank from the original, it is still the original. There are also those who maintain that the changed ship is always the same as the original ship because it has the form of the original. None of these different positions are wrong. However, there is no reason to say that any of them are correct either.

Let us continue asking more questions about our beleaguered boat. What happens if we switch the old wooden planks for more modern plastic planks? Then, as we change more and more of the planks, the ship will be made of a different material than the original. What happens if the people who replace the planks make mistakes in putting in the new planks and the ship has a slightly different form? Another question: Does it matter who is making all these changes to the ship—that is, whether one group of workers does it or another? If the ship is to be preserved for hundreds of years, then surely many different people will have to be making the changes. What if we make so many changes to the boat that it can no longer float out to sea? Can we still call it the ship of mighty Theseus if it cannot perform the same function as the original?[2]

Such questions go on indefinitely. I will restrain myself and discuss just one more scenario. Imagine that every time a plank is changed, rather than consigning the old planks to the scrap heap, we store them in a warehouse. After some time, all the old planks are assembled into a ship. This new construction is made to look exactly like the old ship with the planks in their original position. Question: Which ship has the right to call itself the ship of Theseus, the ship with the replaced planks or the ship constructed out of the old planks?

A common answer to some of these questions is that the ship remains the same because the changes are gradual. However, it is not clear why that should make a difference. How gradual must the changes be in order for the original ship to maintain its status? Is there a minimum speed limit for changes? To put the question of what is "gradual" in perspective, consider the case of Washington's ax. A certain museum wanted to preserve the ax of the founding father of the United States. The ax consists of two parts: a handle and a head. As time went on, the wooden handle would rot and the metal head would rust. When needed, each of these two parts was replaced. As the years passed, the head was changed four times and the handle was replaced three times. Is it still Washington's ax? Notice that here there is no question of the change being gradual. Every time a change is made, half the parts of the ax are replaced.

Our discussion is not limited to ships and axes. A tree is lush and green in the summer and bare and brown during the winter. Mountains rise and fall. Cars and computers get refurbished. Any physical object changes over time. This is the content of Heraclitus' famous dictum that you cannot step into the same river twice. For Heraclitus, the river changes at every instant.

Physical objects are not the only things that change. Businesses, institutions, and organizations are also dynamic entities that constantly change and evolve. Barings Bank was in existence from 1762 through 1995. In that time, the owners, workers, and customers all changed. The Brooklyn Dodgers have been around since 1883. Their players, managers, owners, and fans have definitely changed. What remains the same about a baseball team? After heartlessly betraying their city of birth, the Dodgers cannot even claim that they play in the same city as they originally did. In colleges, the students change every four years. Even the professors change over the years. The only real heart and soul of a college are the beloved secretaries. But, alas, even they change. Political parties are also not immune to change. The Democratic Party was founded in the 1790s to support states' rights over federal rights, the opposite of their current platform. Everything changes!

We are not only talking about change. Rather, we are discussing what it means for an object to be that object. What does it mean for a certain institution to be that institution? When we say that a certain object changes, we mean that it had a certain property beforehand and after the change it does not. In the beginning, the ship of Theseus had planks that Theseus himself touched. At the end, there were planks that he did not touch. That is a change in the properties of the ship. Our fundamental question is: What are the core properties of the ship of Theseus? We have shown that there are no clear answers to this question.

This discussion becomes far more interesting when we stop talking about ancient ships and start talking about human beings. Every person changes over time. We grow from infants to old people. What properties does a three-year-old have in common with their eighty-three-year-old self? These philosophical questions are called the *problems of personal identity*. What are the properties that make up a particular human being? We are not the same person we were several years ago. Nevertheless, we are still considered the same person.

Philosophers usually fall into one of several camps on this question. Some thinkers push the notion that a person is essentially their body. We each have different bodies and can say that every person is identified with their body. By postulating that a human being is their body, we are subject to the same insoluble questions that we faced with the ship of Theseus and other physical objects. Our bodies are in constant flux. Old cells die and new cells are constantly being born. In fact, most of the cells in our body are replaced every seven years. This leads to hundreds of questions that philosophers have posed over the centuries. Why should a person stay in jail after seven years? After all, "he" did not perform the crime. It was someone else. Should a person own anything after seven years? The old person bought it. In what sense is a person the same after having a limb amputated? Science fiction writers are adept at discussing challenging questions like cloning, mind transfers, identical twins, conjoined twins, and other interesting topics related to the notion that a person is the same as their body. When an ameba splits, which is the original and which is the daughter? When your body loses cells it loses atoms. These atoms can go on to belong to others. Similarly, other peoples' atoms can become part of your body. What about death? We usually think in terms of the end of a person's existence when they are dead even though the body is still there. Sometimes we use sentences like "She is buried there" as if "she" were still a person. And sometimes we use sentences like "His body is buried there" as if there is a difference between "him" and his body. In short, it is problematic to say that a human being is identified with their body.

Other thinkers favor the notion that a person is really their mental state or psyche. After all, human beings are not simply their bodies. A person is more than a physical object because there is thought. To such philosophers, a person is a continuous stream of consciousness—they are memories, intentions, thoughts, and desires. This leads us to ask other insoluble questions: What if a person has amnesia? Are they the same person? Doesn't a person's personality change over time? Who is the real you: the one who is madly in love with someone or the one who is bored with the same person two months later? Literally hundreds of questions can be posed about change in a person's thoughts, memories, and desires. Again, philosophers and science fiction writers have become quite adept at describing interesting scenarios that challenge our notion of a human being as a continuous stream of mental states. These scenarios are concerned with Alzheimer's disease, amnesia, personality changes, split-brain experiments, multiple personality disorders, computers as minds, and so on. There are also many questions along the lines of the mind-body problem. How much is the mind—that characterizes a human being—independent of the brain, which is a part of the body?

One of the more interesting challenges to the position that continuity of mental states characterizes a human being is the question of *transitivity of identity*. My mental states are essentially the same as they were ten years ago. That means I am the same person I was ten years ago. Furthermore, ten years ago, my mental states were essentially the same as they were ten years earlier. Hence the person I was ten years ago is the same as the person I was twenty years ago. However, at present, I do not have similar mental states to those I had twenty years ago. So how can it be that I am the same person I was ten years ago, and that person is the same as I was twenty years ago, but I am *not* the same as I was twenty years ago?

Yet another option is that everyone has a unique soul that determines who they are. Avoiding the questions of the definition or existence of a soul, let us concentrate instead on how this answers our question of the essential nature of a human being. Assuming the existence of a soul, what is the relationship between the soul and the body? What is the relationship between a soul and a person's actions, psyche, and personality? If there is no connection, then in what sense is one soul different from another soul?

How can you differentiate between souls if they have no influence over any part of you? What would the purpose of a soul be? If, on the other hand, a connection exists, then does the soul change when the body, actions, psyche, or personality changes? Is the soul in flux? If the soul does change, we are back to the same questions we had previously asked: Who is the real you? Are you the one with the soul prior to the change or are you the one with the changed soul?

Most people probably have an opinion representing some hybrid version of all three ideologies: a person is a composite of body, mind, and soul. Nevertheless, all schools of thought are somewhat problematic.

Rather than answering all the questions posed in this section, let us try to resolve the issues by meditating on why none of the questions have clearcut answers. Why is it that when we pose these questions to different people, we get so many different answers?

Examine the way people learn to recognize different objects, make definitions, and create distinctions. In the beginning, babies are bombarded with many different sensations and stimuli. As toddlers grow, they learn to recognize objects in the world. For example, when they see a shiny silver thing covered with brown gooey stuff coming toward them, they have to learn that it is applesauce on a spoon and that they should open their mouth. By learning to recognize that the physical stimulus of silver covered with brown gooey stuff is applesauce, they are able to handle life better. Human beings need to classify objects. We learn how to tell things apart and determine when they are the same. We learn that an object still exists even when it is out of sight ("object permanence"). Children learn after a while to recognize their mother. A few months later, they learn that even though she is wearing makeup—that is, even when she looks different—she is still the same person. Children have to learn that their mother is the same even when she is wearing perfume and smells totally different. Here toddlers are acting as philosophers and learning how to deal with different questions of personal identity. With all these skills, children are imposing order and structure on the complicated world they have entered. Before these skills are mastered, they are showered with an incomprehensible stream of stimuli and sensations. With these classification abilities the children can comprehend and start to control their environment. If they fail to learn the classification skills, they will be overburdened with external stimuli and unable to deal with their surroundings.

With enough sophistication, children also learn to classify abstract entities. For example, they might learn what it means to be a family. Their mother is a family member. Their father and siblings are also part of the family. What about first cousins? Second cousins? These are a little vague. Sometimes they are part of the family, and sometimes not. Children must learn what is a family and what is not. As they grow, they learn to classify even more abstract entities like numbers and political parties.

Not only do children learn to classify objects and people, they also learn to name them. They realize that they live in a society of other classifiers, and in order to communicate with these compulsive classifiers, they follow their example of giving names to objects. They first give the external stimuli their own names. As their communication skills progress, they learn to forgo their names and start to use other people's nomenclature for objects. They call brown gooey stuff "applesauce." They learn to call the woman who takes care of them "mom," regardless of her wearing makeup or not. By using the same names as others, children are showing society that they are conforming to the prevailing classification system and that their mental processes are similar to those of others. Society then rewards them by showering them with love and providing the protection they need.

The point is that classifying and naming are learned skills. Children do not learn exact definitions of things because they are never exposed to exact definitions. They learn to classify and name physical stimuli. Some notions are exact and unchanging. The concept of the number 4 is exact and has a clear definition. In contrast, many other notions lack sharp definitions. The first part of this section shows that even physical objects do not have sharp definitions.

With this in mind, we can discuss the many questions posed at the beginning of the section. Is the ship of Theseus the same after changing one plank? The proper response is that the definition we have for the ship is not clear enough to provide an answer to that question. *There is no exact definition of the ship of Theseus.* We only have what we learned—that is, the stimuli we were taught to associate with the ship.

The ship of Theseus does not really exist *as the ship of Theseus*. There is no exact definition of what is meant by the ship of Theseus. It exists as a collection of sensations but not as an object. Yes, if you kick it you will feel pain in your toes. When you look at it, you will see brown wood. If you lick it, you will taste stale wood and salt water. But these are all just sensations that one learns to associate with something we call the "ship of Theseus." Human beings combine these sensations and form the ship of Theseus. Of course, the ship exists as atoms. But it is made of atoms *as atoms*.[3] The atoms are not tagged as the ship's atoms. Rather, it is we who make those atoms into a whole entity called a ship. It is we who further demarcate this ship as somehow belonging to the mythical general Theseus. The first part of this section cited many examples demonstrating how the ship can lose and change atoms and still be the same ship. It's all in our mind. We are fortunate to live among other people who learned to give the same names to commonly occurring external stimuli. Each of us calls these similar stimuli the "ship of Theseus." Since we all agree with this naming convention, we do not commit each other to insane asylums. Nevertheless, the existence of the ship of Theseus is an illusion.

There are times where we are taught exact definitions and we can answer all questions based on the definition. For example, we are taught that driving more than 65 miles per hour is speeding. So if people drive 67 miles per hour, they are speeding; if they drive 64 miles an hour, they are not. We are very clear about this. However, for most physical objects, no objective definitions exist.

One can have a similar discussion on questions of aesthetics. Most people will agree that there are no correct answers when it comes to questions of taste. What is beautiful to one person is ugly to another. The present generation of art connoisseurs would spend millions of dollars for any sketch by Vincent van Gogh. In his own lifetime, Van Gogh was ignored and his paintings were not worth a pittance. Which generation has had the correct opinion about Van Gogh's work? There is no answer to this question because there is no such thing as objective aesthetics. It's a matter of taste. Similarly, whether changing a plank of a ship changes the ship cannot be given a definitive answer because there is no such thing as an objective ship of Theseus.

One can safely argue with what is posed here and claim that objects really do have an existence outside of the human mind and that what children are learning to do is classify and name those entities. They are learning to associate names of entities with physical stimuli. Weathered, rotting wood that looks like a ship in the port of Athens should be associated with the "ship of Theseus." This ideology might be called *extreme Platonism* (see figure 3.1). Classical Platonism is the belief that abstract entities have real existence outside of the human mind. The number 3 really exists. There is an exact idea when one refers to the U.S. government. An idea of a chair exists. However, classical Platonism takes no stand about concrete physical entities. In contrast, extreme Platonism is the belief that even a concrete physical object has some type of unchanging platonic entity associated with it. To someone who maintains this position, some platonic notion of "ship-of-Theseus-ness" exists and when a question is posed about a change to the ship of Theseus, all one has to do is somehow

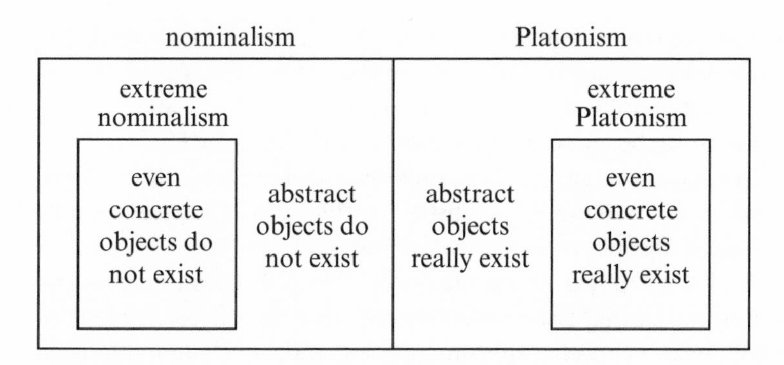

Figure 3.1
Different philosophical schools of thought

connect to the platonic notion and see if the changed ship still satisfies the definition. Extreme Platonism demands a fairly advanced metaphysics, and we cannot really say that as metaphysics, it is true or false. It is impossible to show that no such abstract entity exists. Nevertheless, as with all metaphysical notions, there is no real reason to posit such an existence.[4] If you claim that a name or a definition of an object is some type of "tag" on the object, then we can ask where the tag is. Why is it that people disagree so vociferously about the tag on the ship of Theseus?

In this chapter, I am promoting an idea that might be termed *extreme nominalism*. The philosophical position of classic nominalism is the belief that abstract entities really do not exist outside the human mind. To a nominalist, abstract ideas like the number 3, the idea of the U.S. government, and the idea of a chair or "chairness" do not really exist outside the minds of those who discuss them. Have you ever met a 3? Can you stub your toe against a 3? Can you point to the U.S. government? A classic nominalist would say that these entities only exist in the human mind. Since we share a similar education and social structures, we can banter about these different names and concepts with our neighbors. However, a classical nominalist does not have a position on the question of concrete physical entities.

Extreme nominalism takes nominalism a step further. It is the belief that even physical objects exist *as those physical objects* only in name. They do not have an external existence outside of a human mind. A particular chair is a chair because we call it a chair, not because it has properties of being a chair. The ship of Theseus is whatever people call the ship of Theseus. There are no exact, agreed-on definitions of the ship of Theseus.

I believe that extreme nominalism is correct because of the fact that there is so much disagreement about what constitutes a particular object. If there were exact definitions, presumably people would know about them. Another reason for believing in (extreme) nominalism is that any form of Platonism demands unnecessarily complicated metaphysics. Why do we need the supposed existence of an abstract entity or "tag" for every physical object? Such abstract entities serve no purpose.

From the view afforded by extreme nominalism, it becomes apparent that the reason we cannot answer questions about the ship of Theseus or changes to human beings has nothing to do with linguistic limitations. It is not that we lack the right words or definitions of these concepts. There is also no epistemological problem—that is, it is not a lack of knowledge of the exact definition of the real ship of Theseus. Nor is it a problem of having some type of deeper knowledge of the ship of Theseus beyond its physical stimuli.[5] Rather, we are dealing with a question of existence. In philosophical parlance this is an ontological problem. A real ship of Theseus need not exist.

It is interesting to note that with extreme nominalism, certain abstract objects, such as the number 42, have a clearer existence than physical objects such as ships. After all, we all agree about the many different properties of the number 42. If you take 42 and you subtract 1, you get 41 rather than 42. This is in stark contrast to subtracting planks from a ship.

I have shown that the ship of Theseus is part of our culturally constructed universe. There are other objects in this constructed universe such as Mickey Mouse and unicorns. In fact, more people know about Mickey Mouse than about Theseus' silly boat. Our friendly mouse is introduced to nearly every child, whereas only classics majors, philosophy majors, and privileged readers of this book know about Theseus. Furthermore, one can go to Disney World and actually see a physical manifestation of Mickey. You can even stub your toe against him (such actions are not recommended). In contrast, at present, we cannot find any trace of Theseus' ship in the port of Athens. We are left with the obvious question: In what way is the ship more existent than Mickey Mouse?

The resolution of the problems presented in this section is a challenge to the usual view of the universe. Most people believe that there are certain objects in the universe and that human minds call those objects by names. What I am illustrating here is that those objects do not really exist. What do exist are physical stimuli. Human beings classify and name those different stimuli as different objects. However, the classification is not always strict and vagueness prevails.[6]

3.2 Hangin' with Zeno and Gödel

Zeno of Elea (about 490–430 BC) was a great philosopher who was a student of Parmenides (early fifth century BC). Being a devoted student, Zeno promoted and protected his teacher from all criticism. Parmenides had the philosophical and mystical belief that the world was "one" and that change and motion were merely illusions that a person could see through with enough training. To demonstrate that Parmenides' ideas are correct, Zeno proposed several thought experiments or paradoxes that showed that it is illogical to actually believe that the world is a "plurality" and not "one," or that change and motion actually happen. In this section I will concentrate on four of those thought experiments that demonstrate that motion is an illusion. Since motion occurs within space and time, Zeno's paradoxes will challenge our intuition of these obvious concepts.

Unfortunately, most of Zeno's original writings have been lost. Our knowledge of the paradoxes largely comes from people who wanted to prove him wrong. Aristotle briefly sets up some of Zeno's ideas before knocking them down. Because Zeno's ideas were given short shrift, it is not always clear what his original intentions were. This should not deter us since our central interest is not what Zeno actually said; rather, we are more interested to know if something is wrong with our intuition and how it can be adjusted. These ideas should not be taken lightly. They have bothered philosophers for almost 2,500 years. Regardless of whether one agrees with Zeno or not, he cannot be ignored.

The first and easiest of Zeno's paradoxes of motion is the *dichotomy paradox*. Imagine an intelligent slacker waking up in the morning. He tries to get from his bed to the door in his room (see figure 3.2).

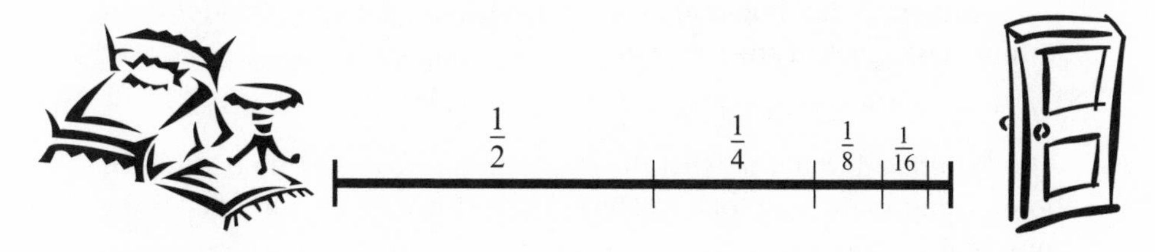

Figure 3.2
Zeno's dichotomy paradox

To get the whole way to the door, he must reach the halfway point. Once he reaches that point, he still must go a quarter of the way more.

From there he has an eighth of the way to go. At every point, he must still go halfway more. It seems that this slacker will never be able to reach the door. In other words, if he does want to get to the door, he will have to complete an infinite process. Since one cannot complete an infinite process in a finite amount of time, the slacker never gets to the door.

Our slacker can further justify his laziness with more logical reasoning. To reach the door, one has to go halfway. To reach the halfway point, he must first get to the quarter-way point, and before that the eighth-way point, etc. . . . Before any motion can be performed, he must perform half the motion. One needs to perform an infinite number of processes in order to get *anywhere*. An infinite number of processes demands an infinite amount of time. Who has an infinite amount of time? Why get out of bed at all?

Zeno's paradox is not only about movement but also about any task that has to be done. In order to complete a task, one must perform half the task first and go on from there. This shows that not only is movement an impossibility, but performing any task, indeed any change, within a time limit is unreasonable.

What are we to do with Zeno's little thought puzzle? After all, we do get to the end of our journey in a finite amount of time and when we do get out of bed in the morning, we can accomplish something. Following the theme of this book, Zeno's paradox has the form of a proof by contradiction. We are assuming something (that is wrong) and we are logically coming to a contradiction or an obvious falsehood. We came to the conclusion that there is no movement or change when, in fact, we see movement and change all the time. What exactly is our wrong assumption?

A mathematician might argue that there is no problem performing an infinite task. Look at the following infinite sum:

$$1/2 + 1/4 + 1/8 + 1/16 + 1/32 + \ldots .$$

The uninitiated would say that the ellipsis means that the sum is going on forever, and so the sum total will be infinite. However, the sum total is the nice finite number 1.[7]

There is a beautiful two-dimensional geometric way to see that this sum equals 1. Consider a square whose side length is 1, as in figure 3.3.

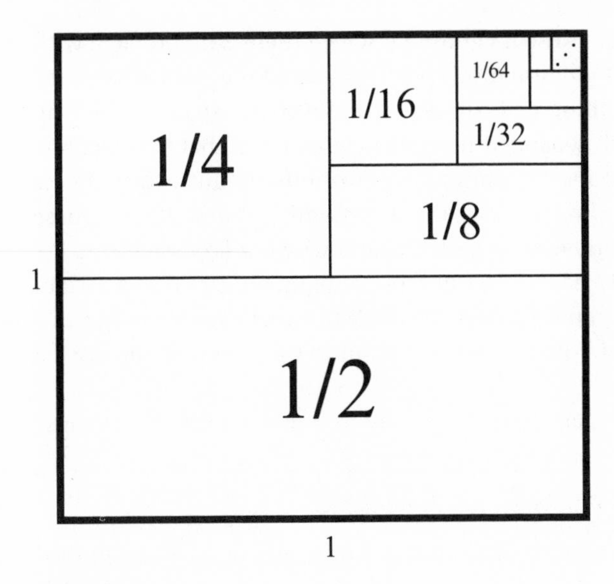

Figure 3.3
A two-dimensional infinite sum

One can see this square as made up of half of the square plus a quarter plus an eighth plus . . . Every remaining part can be further split in half. It is obvious that the area of the entire square is 1.

However, a mathematician would be somewhat disingenuous to claim that this solves Zeno's paradox of performing an infinite process in a finite amount of time. After all, the mathematician is not adding each of the infinite terms to a running sum. She is simply displaying the first few terms and then indicating with an ellipsis that there are an infinite number of terms. She is doing a trick that shows what the sum of all of them would be if she summed them. If one were to sit down and add all infinite terms, it would indeed take an infinite amount of time.

A better solution is to say that the problem with Zeno's reasoning is that he assumes that space is continuous. That means that space looks like the real-number line and is infinitely divisible—that is, between every two points lies an infinite number of points. Only with this assumption can one describe the dichotomy paradox. In contrast, imagine that we are watching the slacker go to the door in an old-fashioned television made up of millions of little pixels. Then as he is moving, he is crossing the pixels. He crosses half of the pixels and then he crosses half of the rest of the pixels. Eventually the TV slacker will be one pixel away from the door and then he will be at the door. There are no half pixels to cross. A pixel is either crossed or not crossed. On the TV screen there is no problem with the slacker getting to his destination and Zeno's paradox evaporates. Maybe we can say the same thing with the real world. Perhaps space is made up of discrete points each separated from its neighbor and that between any two points there is at most a finite number of other points. In that case we would not have to worry about the dichotomy paradox. If we assume such a discrete space, then we can understand why our lazy slacker makes it to the door: he only has a finite number of points to cross. At a certain point, the intervals could no longer be split into two. Objects move in this type of space by going from one discrete point to the next without passing between them.

In the language of chapter 1, we can say that this is a paradox because we are assuming that space is continuous:

Space is continuous $\Rightarrow$ movement is impossible.

Since there is definitely movement in this world, and our assumption led us to a false fact, we conclude that space is not continuous. Rather, it is discrete, or separated into little "space atoms."

Such ideas of discrete space are familiar to people who study quantum mechanics.[8] Physicists discuss something called *Planck's length*, which is equal to 1.6162×10^{-35} meters. Something smaller in length cannot be measured. To some extent, nothing smaller than that exists. Physicists assure us that objects go from one Planck's length to another. In high school chemistry it is taught that electrons fly in shells around a nucleus of an atom. When energy is added to an atom, the electrons make a "quantum leap" from one shell to the next. They do not pass in between the shells. Perhaps our lazy slacker also makes such quantum leaps and hence can finally reach the door.

Let us reconsider figure 3.3. The square is infinitely divvied up as illustrated. But this is only possible if we think of the square as a mathematical object. In mathematics every real number that represents a distance can be split into two, hence we can continue chopping forever. In contrast, let us think of the square as a piece of paper. We can start cutting paper into smaller and smaller pieces using finer and finer scissors. This will work for a while, but eventually we will reach the atomic level where no further cutting will be possible. This is true for any physical object made of atoms. We are forced to conclude that the square depicted in figure 3.3 is not a good model for the physics associated with the paper square. The real numbers can be infinitely divided but the paper cannot be. What Zeno is forcing us to do is to ask the question of whether space (which is not made of atoms) can be infinitely divvied up. If it can be, the slacker will not reach his goal. If it cannot be, there must be discrete "space atoms," and continuous real-number mathematics is not a proper model for space.[9]

We cannot, however, be so flippant about asserting that space is discrete and not continuous. The world certainly does not look discrete. Movement has the feel of being continuous. Much of mathematical physics is based on calculus, which assumes that the real world is infinitely divisible. Outside of some quantum theory and Zeno, the continuous real numbers make a good model for the physical world. We build rockets and bridges using mathematics that assumes that the world is continuous. Let us not be so quick to abandon it.[10]

Zeno's second paradox of motion is the story of *Achilles and the Tortoise.* Achilles was the ancient Greek version of the modern D.C. Comics character The Flash and was the fastest runner in town. One day he had a race with a slow Tortoise. To make the race more interesting (and because Achilles had a warm heart), Achilles gave the Tortoise a head start, as shown in the top line of figure 3.4.

Figure 3.4
Achilles not catching up to the Tortoise

The problem is, in order for Achilles to overtake the Tortoise, he must first pass the point where the Tortoise started (as in the second line in figure 3.4). At that point, the Tortoise has already moved further. Once again, in order for Achilles to overtake the Tortoise, he must get to the point where the Tortoise moved. At each point, Achilles is getting closer and closer to the pesky Tortoise, but he will never be able to reach him, let alone beat him.

Again there is a mathematical analogy to this. In calculus we say that the limit of $1/x$ as x goes to infinity is zero. That is, the larger x gets, the closer $1/x$ gets to zero. Since infinity is not a number, x can never get to infinity and $1/x$ can never get to zero. But the concept of a limit makes it meaningful. Similarly, the distance between Achilles and the Tortoise will never really be zero but the *limit* of the distance does get to zero. Again, we can find fault with this analogy. The concept of a mathematical limit is a type of trick. For no finite number will $1/x$ actually equal zero and at no time period will Achilles actually reach the Tortoise.

This paradox would also melt away if we assume that the racetrack is made out of discrete points. The fact that Achilles runs faster than the Tortoise simply means that he covers more of the discrete points in the same time. So eventually Achilles will overtake the Tortoise. Discrete space would answer the paradox, but again, we have to be careful. We should abandon the notion of continuous space with great trepidation since that mathematical model works so well in general physics.

In the third paradox, Zeno is not interested in determining whether a motion can be completed. Here he attacks the very idea of any motion whatsoever. In the *arrow paradox*, we are asked to think of an arrow flowing through space. At every instant in time, the arrow is in some particular position. If we think of time as a continuous sequence of "nows" that separates "pasts" from "futures," then for each "now," the arrow is in one particular position. At each point in time the arrow is in a definitive position and not moving. The question is, when does the arrow move? If it does not move at each of the "nows," when does it?

This paradox can also be solved if we introduce discrete ideas into the mix. But rather than saying that space is discrete, here we say that time is discrete. At each separate point in time there is no motion. But time leaps from one separated point to another and motion happens at that instant leap. In other words, say that time is discrete and not continuous. We do not see these magical leaps for the same reason we think we see continuous motion when we are watching a movie. In fact, a movie is made out of numerous discrete frames and there is no motion between them. Because the separate time points are so close to each other and there are so many, there is an illusion of continuity.

This paradox basically describes the following derivation:

Time is continuous $\Rightarrow$ movement is impossible.

Once more, since it is an obvious fact that movement is possible, we conclude that time is not continuous but is discrete.

There is also a mathematical analogy to this paradox. Consider the real-number line. Think of the real line as time. Each point of the real line corresponds to a "now." And yet each "now" has no thickness. In ninth grade you learned that the real line is made of an infinite number of points. Each point has zero length. So how can a finite line be made of points that do not have any thickness? Zero times anything is zero. Was your ninth-grade teacher lying to you? Does the real-number line make sense? Should we abandon it?

Again, there is a problem abandoning the notion of continuous time for discrete time. Modern physics and engineering are based on the fact that time is continuous. All the equations have a continuous-time variable usually denoted by t. And yet, as Zeno has shown us, the notion of continuous time is illogical.

The fourth and final paradox against motion is the *stadium paradox*. Zeno wants us to imagine three marching bands as in figure 3.5.

The A's are standing still, and behind them, the B's and C's are moving in opposite directions at the same speed. After some time, the marching bands will look as they do in figure 3.6.

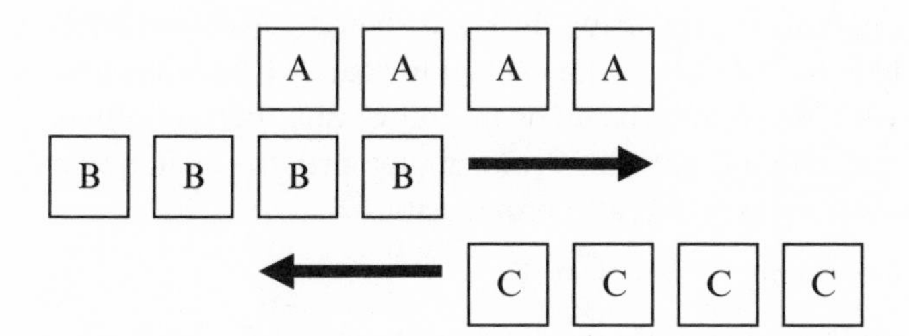

Figure 3.5
Three marching bands at starting time

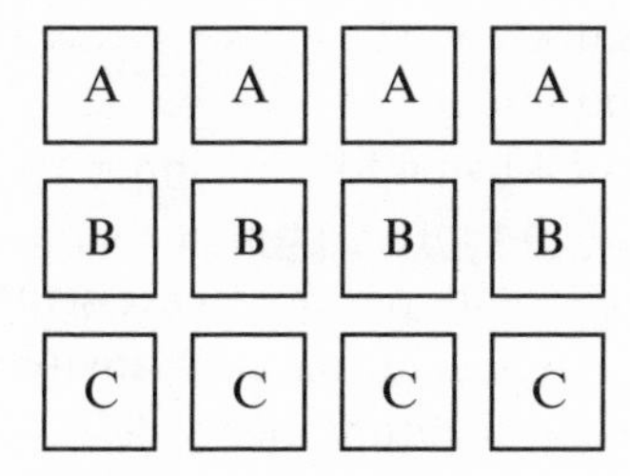

Figure 3.6
Three marching bands at ending time

Notice that in the same time period, the leading B passed two A's and four C's. Since the A's and C's are the same size, how can the B's pass different numbers of each? The obvious answer, and the reason that Aristotle dismisses this paradox out of hand, is that the A's are standing still while the C's are moving. There is a difference between velocity and relative velocity. We are used to such distinctions when driving or riding in a car and seeing how fast the houses pass as opposed to how fast the cars moving in the opposite direction pass.

Perhaps we should not be so dismissive of Parmenides' faithful student. There is really no way of ascertaining what Zeno's original intention was since we only have a brief discussion from Aristotle. Modern thinkers have postulated a scenario that is a bit more sophisticated. In the previous three paradoxes, we saw that our problems would evaporate if we thought of space and time as discrete or quantized. In this final paradox, let us assume that both space and time are discrete. Think of the members of the marching bands as having the smallest possible discrete size. At the same time, imagine that the B's are moving at the fastest possible speed. Two time clicks are needed for the B's to get from figure 3.5 to figure 3.6. At this speed, each of the B's crosses one box per time click. This is the smallest possible discrete time period for the B's to pass in front of both A's. How can it be that in this smallest time period, the B's passed twice as many C's? That means that the B's are passing the C's at an even faster rate. What would this look like to a member of the B marching band? It would appear to the B's that the C's are skipping boxes or are going faster than permitted. This reading of the stadium paradox demonstrates that the assumption that space and time are discrete is also problematic.

Space and time are discrete $\Rightarrow$ false fact.

We must conclude that space and time are not discrete.

Which is it? Are space and time continuous or discrete? On the one hand, the stadium paradox is pointing toward space and time being continuous. On the other hand, the first three paradoxes would be solved if we thought of space and time as discrete. The answer is that we simply do not know the nature of space and time.

This conflict is a microcosm of a battle in contemporary physics. The two great achievements of twentieth-century physics are relativity theory and quantum theory.[11] These two revolutionary sciences essentially describe most of the phenomena in the physical universe. Relativity theory deals with gravity and large objects, while quantum theory deals with the other forces and small objects. However, these two theories are in conflict with each other. One of the main reasons for their conflict is that relativity theory considers space and time to be continuous while quantum theory believes space and time to be discrete. For the most part, since the theories deal with different realms, the conflict does not bother us. Nonetheless, the conflict is apparent with certain phenomena such as black holes, which are termed the "edge of space." Since we cannot have conflicting physical theories, it must be that we do not know the final story. The jury is still out regarding the structure of space and time.

The most amazing aspect of Zeno's paradoxes is that they are 2,500 years old and they deal with such simple topics. What is the nature of space, time, and motion? It is doubtful that we have heard the last of our Elean friend.

Since we are discussing the relationship of space, time, and logic, let us talk about time-travel paradoxes. We first have to ask ourselves what it means to travel back[12] through time. What would it mean for me to go back to the Continental Congress held in Philadelphia in 1776, in order to witness the signing of the Declaration of Independence? If I am miraculously transported back there and see the signing, then the very fact that I am in the room on that hot day in July means that it is not the original Continental Congress. After all, I was not there during the original. In other words, if there were 150 people present at the original Continental Congress, when I go back there will be 151 people present. That is not the original. It is a major difference between what I was transported to and the original. What exactly am I being transported to? One thing is certain: not to the Continental Congress of 1776.[13] This conundrum shows how hard it is to understand the very basic concepts of time travel.

Be that as it may, let us imagine for a moment that we understood what traveling through time actually means, and furthermore, let us imagine that such a process was, in fact, possible. If time travel was possible, a time traveler might go back in time and shoot his bachelor grandfather, ensuring that the time traveler was never born. If he was never born, then he could not have shot his grandfather. Homicidal behavior is not necessary to achieve such paradoxical results. The time traveler might just ensure that his parents never have children,[14] or he might simply go back in time and make sure that he does not enter the time machine. These actions would entail a contradiction and hence cannot happen. The time traveler should not shoot his own grandfather (moral reasons notwithstanding) because if he shoots his own grandfather, he will not exist and will not be able to travel back in time to shoot his own grandfather. So by performing an action he is ensuring that the action cannot be performed. The event is self-referential. Usually, one event affects other events, but here an event affects itself. In the language of chapter 1, we are showing that

Time travel $\Rightarrow$ contradiction.

Since the universe does not permit contradictions, we must somehow avoid this paradox. Either time travel is impossible, or even if it was possible, one would still not be able to cause a contradiction by killing an earlier version of oneself. Which impossibility should we prefer?

Albert Einstein's theory of relativity tells us that the usual way that we conceive the universe makes time travel impossible. In 1949, Einstein's friend and Princeton neighbor, Kurt Gödel, did some moonlighting as a physicist and wrote a paper on relativity theory. Gödel constructed a mathematical way of looking at the universe in which time travel would be possible. In this "Gödel universe," it would be very hard, but not impossible, to travel back in time. Gödel, the greatest logician since Aristotle, was well aware of the logical problems of time travel. The mathematician and writer Rudy Rucker tells of an interview with Gödel in which Rucker asks about the time-travel paradoxes. The relevant passage is worth quoting: "Time-travel is possible, but no person will ever manage to kill his past self.' Gödel laughed his laugh then, and concluded, 'The *a priori* is greatly neglected. Logic is very powerful."[15] Gödel replies that the universe simply will not allow you to kill your past self. Just as the barber paradox shows that certain villages with strict rules cannot exist, so too the physical universe will not allow you to perform an action that will cause a contradiction.

This leads us to even more mind-blowing questions. What would happen if someone took a gun back in time to shoot an earlier version of himself? How will the universe stop him? Will he not have the free will to perform the dastardly deed? Will the gun fail to shoot? If the bullet fires and is properly aimed, will the bullet stop short of his body? It is indeed bewildering to live in a world that does not permit contradictions.

3.3 Bald Men, Heaps, and Vagueness

At what point does a man lose enough hair that he is considered bald? Do we have to be able to see his scalp? What if his hair is long but thin? Does that make a difference? When is someone considered tall? Is there a difference between a "pile" of toys and a "heap" of toys? Is that color red or maroon? All these questions are based on concepts that are somewhat vague. There does not seem to be universal agreement on when someone is bald and when someone is not bald. Nor is there a generally agreed on use of the terms *tall* and *short*. Even your interior decorator might have a hard time distinguishing dark red from maroon. In this section I explore the pervasive element of vagueness in our language and thought.

One of our core ways of describing limitations of reason is by finding contradictions. As I stressed in chapter 1, there are no contradictions in the physical universe. In contrast to the physical universe, in human language and thought there can be contradictions. Humans are not perfect beings. Our language and thought are rife with contradictory statements and beliefs. When we want to reason and talk about the physical world we must ensure that our language and thought do not have contradictions. There are, however, times when we are ostensibly thinking about or discussing the physical world and our meaning is not clear. This happens when there is vagueness. In contrast to contradictions where a statement is both true and false, a vague statement can be thought of as neither true nor false.

Vagueness is applied to terms that are not always perfectly defined. For example, a five-year-old is clearly a child. In contrast, a twenty-five-year-old is definitely not a child. At what point is a person no longer considered a child? There are *borderline cases* where someone is neither a child nor older than a child. Such terms with borderline cases are vague. Other terms with borderline cases are *tall*, *smart*, and *red*. Where does red end and maroon begin? How about scarlet, cardinal, crimson, cherry, puce, pink, ruby, and fuchsia?[16]

One must make a distinction between vague statements and *ambiguous statements*. An ambiguous statement is one in which the subject of the statement is unclear. For example, "Jack is above six feet" is ambiguous since you do not know which Jack is being discussed. Jack Baxter is above six feet, but Jack Miller is below six feet. However, this statement is not vague since six feet is an exact amount. Of course, we can make a statement that is both vague and ambiguous: "Jack is tall."

One must also make a distinction between vague statements and *relative statements*. "Jack Baxter is smart" might be true or not depending on who he is being compared to. If you are comparing Jack to the other people in his class, then he might very well be considered smart; however, the class might not be the smartest class. The truth of a relative statement can be determined by looking at the context of the statement. Who are we talking about? One can imagine the salutatorian at a Harvard University graduation legitimately being called stupid . . . by the valedictorian.

In both ambiguous cases and relative cases there is a lack of specificity. In other words, there is missing information. Usually, if one adds more information, then the statements can be clearly understood. If one identifies the subject of an ambiguous statement or the context of a relative statement, then we can determine if the statement is true or false. In contrast, vague statements usually cannot be tweaked by adding more information. There is no more information to add. When is a person considered bald? The answer is "blowin' in the wind." There is no real answer.

Vagueness is not necessarily a bad thing. Sometimes vagueness is a necessity. Biologists use vague characteristics to describe different species.[17] Many lawyers are employed to work with vagueness (and to obfuscate the truth). Diplomats are vague when they make treaties with foreign countries so that they are not caught by their own words. When a woman asks if a certain dress makes her look fat, it might be wise to be vague in your response.

Philosophers are usually split as to why there is vagueness. Some philosophers promote *ontological vagueness*—that is, the reason some terms do not have an exact meaning is that an exact meaning of these terms really does not exist. While there is an exact definition of "above six feet," there is no exact definition of "tall." In contrast, other philosophers promote *epistemic vagueness*. They believe there is an exact definition of vague terms but we simply do not know what it is.

Which is it? Ontological vagueness or epistemic vagueness? While everyone has an opinion, no one has the decidedly knockdown argument. Unfortunately it is an unanswerable metaphysical question. I humbly lean toward ontological vagueness.[18] For reasons elaborated in section 3.1, it is hard to believe that exact definitions of *tall*, *bald*, or *red* exist. Who determines these exact definitions? Are they to be found in Plato's attic? Is there some exact height that is considered tall? Is there an exact number of equally distributed hairs that make a person hirsute (not bald)? Is there an exact wavelength associated with red and not with cherry? I highly doubt it. Since we discounted an exact definition of the ship of Theseus, it stands to reason that we discount an exact definition of tallness, baldness, or redness.

One problem with vague terms is that the usual tools of logic and mathematics that we use to understand the world do not work for such terms. For example, one of the main rules of logic is that for any proposition P, it is always a fact that either P or not-P is true. So for example, "It is either colder than 32°F or it is not colder than 32°F now" is always true (and hence devoid of any content). This rule is called the *law of excluded middle*. That means that either a proposition is true or false but nothing in the middle. However, for vague predicates the law of excluded middle does not work. We all know people who are neither tall nor not tall. They are simply somewhere in the middle. There are men who are neither bald nor not bald . . . they are, like many men, heading toward baldness.

One of the main tools of logic is the law called *modus ponens*. This law says that if a statement P is true and the statement "P implies Q" is true, we can then derive that the statement Q is true. In symbols, we write this as

$$\frac{\begin{array}{l} P \\ P \rightarrow Q \end{array}}{Q}$$

For example, from the fact that "it is raining," and "if it is raining, then there are clouds in the sky," we can derive that "there are clouds in the sky." This basic law of logic is at the root of all reasoning. However, this law fails when we deal with vague terms. In the next few paragraphs I describe certain strange logical deductions that come about from the failure of modus ponens.

If a man does not have any hair on his head he is definitely bald. What if he has one hair on his head? Most people would say that a man with exactly one hair on his head is still considered bald. How about two hairs on his head? It is hard to believe that if a person with one hair is considered bald, one more little hair is now going to make him hairy. He must be considered bald. How about three hairs? There must be a rule that says that

If a man with 3 hairs is bald, then with 4 hairs he is also bald.

Again, we are only adding one little hair so this rule must be true. In fact we can generalize this rule to the following rule for all positive whole numbers n:

If a man with n hairs is bald, then with $n + 1$ hairs he is also bald.

Pressing on with our analysis, we can come to the conclusion that a man with 100,000 hairs or even 10 million hairs is still bald. But this is simply not true. A man with that much hair is not bald. This is called the *bald-man paradox*.

Such a paradox is an example of a type of argument that goes back to ancient Greek times and is called a *sorites paradox* (from the Greek word *soros* for "heap"). Eubulides of Miletus (fourth century BC) is usually credited with being the first to formulate this puzzle.[19] He asked how many grains of wheat form a heap. Is one grain of wheat considered a heap? Obviously not. How about adding one grain to it? Are two grains considered a heap? Still not. After all, we only added one grain. We can formulate the following law:

If n grains are not a heap, then $n + 1$ grains are also not a heap.

Following a similar analysis of the bald man, we come to the obviously wrong conclusion that no amount of grains form a heap. What went wrong?

Let us carefully analyze the argument given. We start with the obvious statement:

1 grain is not a heap.

We also use the n-grain rule for $n = 1$ to get

If 1 grain is not a heap, then 2 grains are also not a heap.

Combining these two rules using modus ponens, we get

2 grains are not a heap.

Furthermore, combining this with

If 2 grains are not a heap, then 3 grains are also not a heap.

gives us:

3 grains are not a heap.

Continuing on with this shows us that for any n, no matter how large,

n grains are not a heap.

This is obviously false.

We can also go the other way. Consider a heap with 10,000 grains of wheat. If we take off one little grain are we to come to the conclusion that 9,999 grains are not a heap? Obviously they are still a heap. A rule can be formulated:

If n grains are a heap, then $n - 1$ grains are also a heap.

Using this rule and applying the modus ponens rule many times, we arrive at an obviously false conclusion that a collection of 1 grain is also a heap. A similar argument can show that a man with 1 hair, or even no hairs, is not bald.

Another sorites-type paradox is the *small-number paradox* (also called *Wang's paradox*). 0 is a small number. If n is a small number, then so is $n + 1$. We conclude with the apparent false fact that any number is considered a small number. There are many other types of sorites paradoxes. Is a person tall if we add another centimeter to their height? Does a person become heavy if they add one more pound? Similarly, for any other vague terms like *rich, poor, short, clever,* and so on, one has an associated sorites-type paradox.

How is one to understand such paradoxes? Some philosophers say that the sorites paradoxes show us that there is something wrong with the logical rule of modus ponens. By following modus ponens we came to a false conclusion, so modus ponens cannot be trusted. This seems a little too harsh. The modus ponens rule works so perfectly in most logic, math, and reasoning. Why should we abandon it? Other philosophers (who believe that all vagueness is epistemic—i.e., they believe exact boundaries exist that we are not aware of) assume that the rule

If n grains are not a heap, then $n + 1$ grains are also not a heap.

is simply false. For them, there is some n for which n grains do not form a heap but $n + 1$ grains do form a heap. We mortals are not aware of which n this is but it nevertheless exists. For such philosophers modus ponens is true, but this implication is simply not valid and so cannot be used in a modus ponens argument. As noted above, to us it seems that vagueness is not an epistemic but an ontological problem. There are no exact boundaries and the implication from n to $n + 1$ grains is, in fact, always true.

Rather than saying that there is something wrong with the obvious rule of modus ponens, I prefer to say that this amazing rule is perfect but cannot always be applied. In particular, one should not use modus ponens with vague terms. Although modus ponens seems to work with the first few applications of the rule (i.e., that 2, 3, and 4 grains do not make a heap), for many more applications of the rule we come to obvious false conclusions. We must restrict ourselves to using modus ponens only with exact terms. We will not be able to use modus ponens with vague terms because that will take us beyond the bounds of reason.

It makes sense that these logical and mathematical tools do not work with vague terms since these tools were formulated with exact terms in mind. One needs exact terms to do science, logic, and mathematics. When we leave the domain of exact definitions—that is, when we talk about baldness, tallness, and redness—we are necessarily leaving the boundaries where logic and math can help us. Vagueness is beyond the boundaries of reason. While we all freely live and communicate with such terms on a daily basis we must, nevertheless, be careful about crossing the outer limits of reason.

As shown above, when it comes to vague statements, mathematicians and logicians are somewhat at a loss. Their usual tools in their toolbox do not work. However, since these vague terms are ubiquitous, we simply cannot ignore them. Researchers have developed a number of different methods to make sense of the vague world. Here I will highlight several of them.

Logic usually deals with terms that are either true or false. *Fuzzy logic* is a branch of logic that deals with terms that can have any intermediate value between true and false. Say that true is 1 and that false is 0. Rather than dealing with the two-element set {0,1}, fuzzy logic deals with the infinite interval [0,1] of all real numbers between 0 and 1. With this we can give different values in different cases. Telly Savalas and Yul Brynner are both totally bald and hence would have the value 0. People with full heads of hair would get a 1. People in the middle will get middle values. 0.1 means almost bald, while 0.5 is halfway there. Someone might get the value of 0.7235. With these different values set up, researchers have gone on to develop different operations similar to AND and OR to work in this logic.

Similar to fuzzy logic is a related field of logic called *three-valued logic*. Rather than saying that a statement is either true or false, say that a statement is true, false, or indeterminate. These branches of logic are used extensively in the field of artificial intelligence, which tries to make computers act more like human beings. If we are going to have computers interacting with human beings, then they are going to have to deal with vague terms like humans beings. These multivalued logics have been very successful in dealing with vague predicates.

Another method used to deal with vague terms is to restrict logic. Consider a man who is halfway between being bald and being hairy. Rather than saying he is neither bald nor not bald, say that he is both bald and not bald. In classical logic if a statement and its negation are both true, we have a contradiction and the system is inconsistent. The major problem with such a system is that anything can be proved within such a system—that is, from a falsehood one can derive anything. While most logicians avoid such systems, others, like Graham Priest, work with them. They attempt to extend the realm of logic to the vague by permitting certain types of contradictions. The belief that there are types of contradictions that are true is called *dialetheism*. The logics that they deal with are called *paraconsistent logics*. Essentially what they do is restrict the logic so that not every statement is derivable from a contradiction. With these restrictions in place, one can derive meaningful statements about vague terms. This direction of research has also progressed over the past few years.[20]

3.4 Knowing about Knowing

Imagine being a contestant on the television game show *Let's Make a Deal* with its host, Monty Hall. Monty presents you with three doors and tells you that behind two of the doors are goats and behind one of the doors is a brand-new fancy car. You are allowed to keep whatever is behind the door you choose. After selecting one of the doors but before you open the door to see what you won, Monty stops you and knowingly opens another door and shows you a goat. He now offers you the choice of staying with your original selection or of switching to the third unopened door. What should you do?

Your immediate reaction is that you might as well stay with your original choice. After all, when you started each door had one-third of a chance of having a car. Now that one of the doors is open, there is a fifty-fifty chance that the car is behind your original choice. What is to be gained by going to the other door?

This question was posed to Marilyn vos Savant, who wrote a special puzzle column in *Parade Magazine*. Vos Savant recommended that you switch doors. She said it was more likely for the car to be behind the other unopened door than the one you originally picked. If you thought that there was no reason to switch, do not feel bad: you are in good company. Her answer generated over 10,000 letters from readers telling her she was wrong. Included were over 1,000 letters from people who identified themselves as PhDs. The article and the letters made such a big impression that the story made it onto the front page of the *New York Times*.

To see why you should change doors, let us look at all the possibilities, as in figure 3.7.

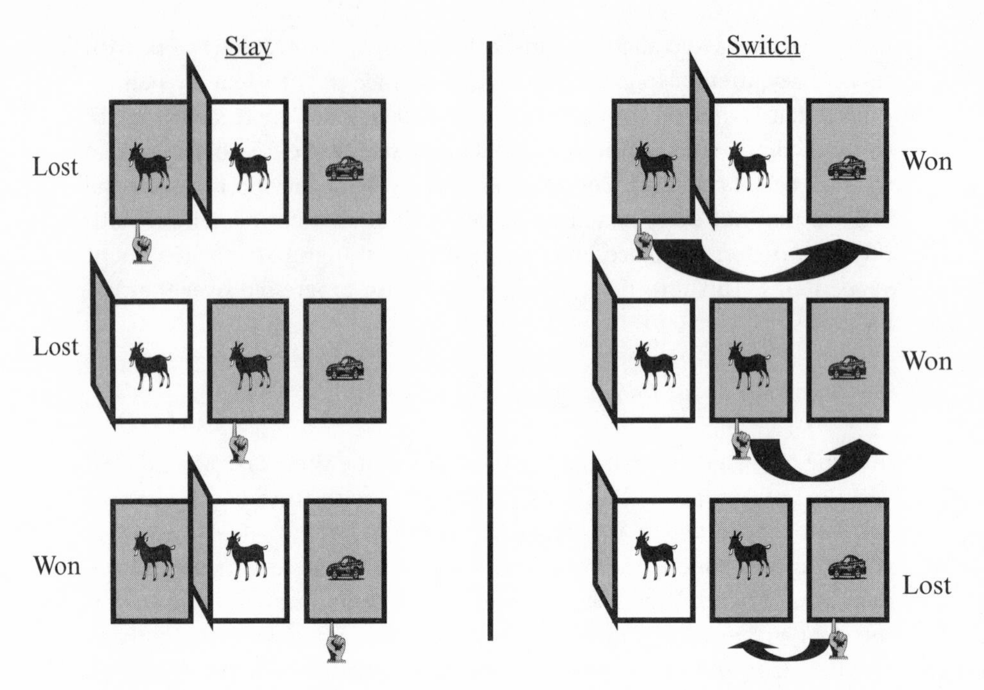

Figure 3.7
All possibilities for the Monty Hall problem

Assume Monty placed the car behind the third door. You have three possible doors to choose from. The three choices are depicted with the three rows. The left column shows what would happen if you stay with your original choice and the right column shows what would happen if you switch. Using the staying strategy gets you the car one out of three times, while the switching strategy has you winning two out of three times. You should indeed switch.

What's going on here? Why does switching help? The answer is that when Monty Hall opens the other door, he is giving you more information. Monty knows where the car is and is not going to open the door with the car. By avoiding the other door he is giving information that the other door was avoided. When he gives you information, the probabilities of what is behind each door change.

The way to see this more clearly is by imagining that Monty presents twenty-five doors to you and tells you that the car is behind one of the doors and there are goats behind the other twenty-four doors. You choose one of the doors and then Monty proceeds to open twenty-three other doors. Each door he opens reveals a goat, as in figure 3.8.

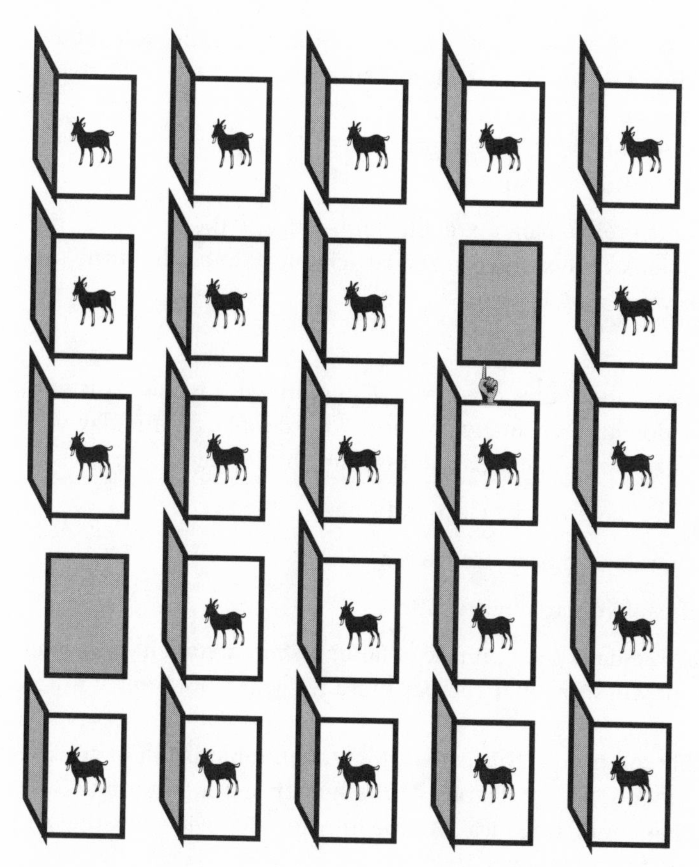

Figure 3.8
An extended version of the Monty Hall problem

Now there are two doors that Monty did not open: the one you chose and one other that he avoided. It could very well be that (a) the one you chose is the one with the car (a 1-out-of-25 chance) and that Monty is simply bluffing and hoping you change. Or it could be that (b) you picked a door with a goat behind it (a 24-out-of-25 chance) and since Monty knows where the car is, he is not going to open that door. It is obvious that you should switch. Here Monty is subtly giving you information about the whereabouts of the car *by not telling you where the car is.*

Here is an interesting scenario to think about. Imagine that Monty himself does not know where the car is. Then he will be randomly opening doors. He might accidentally open the door with the car in it and the game is over. But if he does not accidentally open the door with the car, then should you switch? Answer: nope! There is nothing to gain. You should switch only when you know that Monty knows, and he is subtly giving you the information.

This is just one of the strange aspects of knowledge and information that we explore in this section.

Probably the simplest paradox about knowledge is the cousin of the famous liar paradox that we met in the last chapter. Simply hold the following idea in your head:

This idea is false.

As with the liar paradox, this idea is false if and only if it is true. This self-referential paradox also has many variants. For example, on Tuesday you can have the idea that while today you cannot think straight,

Tomorrow, all my ideas will be clear and true.

Then, on Wednesday, you can realize that

All my thoughts of yesterday were false.

Question: Was Tuesday's thought true or false? A short argument following the implications will show that Tuesday's idea was true if and only if it was false.

One possible solution to this is that the human mind is full of contradictions. As I mentioned in chapter 1, the human mind is not a perfect machine and has conflicting ideas. A little introspection will show that we all believe ideas that contradict each other.

One of the more interesting paradoxes about knowledge is called the *surprise-test paradox*. A teacher announces that there will be a surprise test in the forthcoming week. The last day of class is Friday of that week. What day can the surprise test happen on? If the test is going to be on Friday, then after school on Thursday night the students will already know that the test is on Friday and it will not be a surprise test. So the test cannot happen on Friday. Since this was purely logical reasoning, everyone knows this. Can the test be on Thursday? After class on Wednesday night the students can deduce that since the test has not happened already and it cannot be on Friday, it must be on Thursday. But again, since they know that it must be on Thursday, it will no longer be a surprise test. So the test cannot occur on Thursday or Friday. We can continue reasoning in the same way and conclude that the test cannot happen on Wednesday, Tuesday, or Monday. When exactly will this surprise test occur?

Logic has shown us that a teacher cannot give a surprise test within a given time interval. This is a paradox because it goes against the obvious fact that teachers have been torturing students with surprise tests for millennia.

It is interesting to note that the paradox would not arise if the teacher just remained silent. The problems only arise because of the teacher announcing to the students that there will be a surprise test. The instant the students are told of the surprise test, they must hold the two contradictory thoughts simultaneously: there will be a surprise test and there cannot be a surprise test.

In 2006, Adam Brandenburger and Jerome Keisler published a groundbreaking paper about the very nature of reason and beliefs. When playing a game of chess, you must play rationally and take into account the position of the pieces on the board. You also must take into account that your opponent is rational. Realize that just as you are going to make a rational move, so too will your opponent see what move you make and similarly make a rational move. Your opponent also takes into account that you are rational and she knows that you know she is rational. This goes back and forth and happens anytime there are strategies involved (as in figure 3.9). There are, however, problems with such scenarios. The ability of beliefs to

Figure 3.9
Two people thinking about each other's strategies

deal with themselves will cause a self-referential paradox and hence a type of limitation.

A simple example has come to be known as the *Brandenburger-Keisler paradox*. It is a type of two-person liar paradox. Imagine Ann and Bob thinking about each other's thoughts. Now consider the situation described by these two lines:

Ann believes that Bob assumes that

Ann believes that Bob's assumption is wrong.

Pose the following question:

Does Ann believe that Bob's assumption is wrong?

If you answer yes, then you are agreeing with the second line. The first line says that Ann believes that this assumption is *correct* and not *wrong*. Hence the answer is no. Let us try the other way: the answer to the question is no. It is not the case that Ann believes that Bob's assumption is wrong. Therefore Ann believes Bob's assumption is correct. That is, the second line, which says *Ann believes that Bob's assumption is wrong,* is true. So the answer must have been yes. This is a contradiction.

Brandenburger and Keisler take such ideas and go much farther. Their revolutionary work proceeds to show that there will be limitations or "holes" in any type of game where two players reason about each other. That is, there will be situations where contradictions can happen.

Further Reading

Section 3.1

The section on the ship of Theseus, the problem of identity, and the problem of personal identity was mostly motivated by very passionate classroom discussions with my students at Brooklyn College and by reading too much David Hume. Unger 1979 comes to similar conclusions from a slightly different perspective.

Section 3.2

Zeno's motion paradoxes can be found in book VI of Aristotle's *Physics*. My discussion benefited greatly from the following publications: Grünbaum 1955, Huggett 2010, Makim 1998, Vlastos 1972, and especially Glazebrook 2001. There are also many wonderful papers in Salmon 1972. Chapter 1 of Sainbury 2007 has a nice exposition as well. Mazur 2007 is a popular history book on Zeno's paradoxes.

The discussion of Gödel's take on time travel can be found in Rucker 1982. In Yanofsky 2003, I show that the time-traveler paradoxes can be put into the same scheme as all other self-referential paradoxes.

Section 3.3

Sorensen 2001 is an important work on the general concept of vagueness. Chapter 2 of Sainsbury 2007 covers some of the same material. More on dialetheism and paraconsistent logic can be found in the works of Graham Priest, such as Priest 2003. Parikh 1994 is an interesting discussion of vague terms that is worth studying.

Section 3.4

The magazine article that made the Monty Hall problem famous was in *Parade Magazine*, September 9, 1990, 16. The front-page *New York Times* article (July 21, 1991) was by John Tierney: "Behind Monty Hall's Doors: Puzzle, Debate and Answer?"

You can read about the surprise-test paradox and many other epistemic paradoxes in Sorensen 2006. The Brandenburger-Keisler paradox and much more can be found in Brandenburger and Keisler 2006.

4 Infinity Puzzles

The last function of reason is to recognize that there is an infinity of things which are beyond it. It is but feeble if it does not see so far as to know this.[1]

—Blaise Pascal (1623–1662)

To infinity and beyond!

—Buzz Lightyear, *Toy Story* (1995)

There's an infinite number of monkeys outside who want to talk to us about this script for "Hamlet" they've worked out.

—Douglas Adams (1952–2001), *The Hitchhiker's Guide to the Galaxy*

Since ancient times, people have contemplated the infinite and its properties. For most of that time, our thoughts on the infinite were mired in strange ideas that could not withstand the rigor of exact reasoning. With such confusion, the medievals would endlessly discuss inane questions, like "How many angels can dance on the head of a pin?" In the late nineteenth century, Georg Cantor (1845–1918) and several associates were finally able to grab ahold of this slippery topic and make some progress. However, the new science of infinity has many counterintuitive concepts that are challenging to our intuition.

It is important to realize that ideas about infinity are not abstract scholastic thoughts that plague absentminded professors in the ivy-covered towers of academia. Rather, all of calculus is based on the modern notions of infinity mentioned in this chapter. Calculus, in turn, is the basis of all of the modern mathematics, physics, and engineering that make our advanced technological civilization possible. The reason the counterintuitive ideas of infinity are central to modern science is that they work. We cannot simply ignore them.

Section 4.1 is concerned with the basic language of sets. I restrict myself to the familiar world of finite sets and give a nice definition that determines when two sets are the same size. In section 4.2 I take this definition that works so well with finite sets and see what happens when we move to infinite sets. The strange world of infinity starts making life more interesting. The core of this chapter is section 4.3, where we encounter the different levels of infinity. Along the way, we will learn about a powerful proof technique called diagonalization. I close with section 4.4, where more advanced and philosophical topics are discussed.

4.1 Sets and Sizes

The ideas of infinity are expressed in the language of sets. A set is a collection of distinct objects. The objects can be anything and everything (including other sets). The objects in a set are called elements or members of the set. Sets can be denoted by braces (curly brackets) around their elements. So, the set

{a, b, c}

has three elements, which are the letters *a*, *b*, and *c*. We can talk about the set of students in a class, the set of red cars, the set of U.S. residents, the set of fractions, and so on.

There are different ways of denoting a set. We can list the elements of the set, such as

{dogs, cats, parrots, fish, snakes},

or we can describe the same set by giving a description:

{*x*: *x* is one of the five most popular household pets}.

This is read as "The set of all *x*, such that *x* is one of the five most popular household pets." Another example is

{3, 5, 7, 9, 11}.

This is the same set as

{*x*: *x* is an odd whole number greater than or equal to 3 and less than 12}.

Sometimes, when talking about infinite sets, I will use an ellipsis (. . .) to mean that the sequence continues. For example, the prime numbers can be written as

{2, 3, 5, 7, 11, 13, . . .}.

Capital letters will be used as names to describe certain sets:

$D = \{1, 3, 5, 7, 9, 11, 13, 15, \ldots\}$.

Two sets are equal if every element of one set is an element of the other set. So if

$F = \{x: x \text{ is a whole odd number}\}$

it is obvious that

$D = F$.

Certain sets will be subsets of other sets. It is obvious that the set of women in a class is a subset of the set of all students in the class. This is because every woman in the class is a student in the class. In general, given two sets, S and T, we say that S is a *subset* of T if every element of S is an element of T. Notice that a subset of T can be equal to the entire set T. S is a *proper subset* of T if S is a subset of T but is not equal to T. That is, a subset is not proper if it could be the same as the whole set. S is a proper subset of T if there is some element of T that is not an element of S. In terms of number of elements, S is a subset of T if S has fewer or the same number of elements as T. S is a proper subset of T if it has strictly fewer elements than T. This obvious fact about finite sets will be a sticking point when we meet infinite sets in the coming sections.

There is a special set that has no elements. This set is called the empty set and is denoted by $\varnothing$. For any set S, the following statement is always true:

Every element of $\varnothing$ is also an element of S.

After all, there are no elements in $\varnothing$. So $\varnothing$ is a subset of S.

For any set S, we will be interested in the set of all subsets of S, which is called the *powerset* of S and is denoted as $\wp(S)$. For example, if $S = \{a, b\}$, then

$\wp(S) = \{\varnothing, \{a\}, \{b\}, \{a, b\}\}$.

Notice that this set has four elements of which three are proper subsets of S. If there is a third element in S, such as $S = \{a, b, c\}$, then the powerset has the same subsets as before, namely $\varnothing$, $\{a\}$, $\{b\}$, $\{a, b\}$, but now each of those subsets can also contain c and so we have the subsets $\{c\}$, $\{a, c\}$, $\{b, c\}$, $\{a, b, c\}$. Hence we have

$\wp(\{a, b, c\}) = \{\varnothing, \{a\}, \{b\}, \{a, b\}, \{c\}, \{a, c\}, \{b, c\}, \{a, b, c\}\}$.

That is, by adding c to S we double the number of subsets. For a two-element set, the powerset has four elements. For a three-element set, the powerset has $2 \times 4 = 8$ elements. For a four-element set, the powerset has $2 \times 8 = 16$ elements. In general, the powerset of an n-element set has

$$\underbrace{2 \times 2 \times 2 \times \cdots \times 2}_{n} = 2^n$$

elements. So, usually the powerset of a set is much larger than the set.

When are two sets the same size? Consider the set

$S = \{a, b, c, d, e\}$

and the set of household pets

T = {dogs, cats, parrots, fish, snakes}.

It is obvious that these two sets are the same size: they both have five elements. However, let us examine this apparent fact in another way. Say that S and T are the same size because we can pair off every element of S with a unique element of T. That is, S and T are the same size because a correspondence exists that relates every element of S with a unique element of T and vice versa. For each element of S there is a unique mate in T and for each element of T there is a unique mate in S. This pairing or correspondence can be viewed as follows:

S	a	b	c	d	e
T	dogs	cats	parrots	fish	snakes

There might be other correspondences between the two sets—for example,

S	a	b	c	d	e
T	snakes	dogs	parrots	cats	fish

In fact, both S *and* T can be shown to correspond to the set

$\{1,2,3,4,5\}$.

With these correspondences we are confident that all of these sets have five elements.

This simple idea is at the core of this chapter. We say that two arbitrary sets, S and T, are *equinumerous* or the *same size* if such a correspondence exists between them. The sets will be deemed to have the same *cardinality*.

Examples of equinumerous sets abound:

• The set of human hearts in the world is the same size as the set of people in the world. (Note that this would not work with ears because people generally have two ears.)
• The set of states in the United States

{Alabama, Alaska, Arizona, . . . , Wisconsin, Wyoming}

can be put into correspondence with the set of state capitals in the United States

{Montgomery, Juneau, Phoenix, . . . , Madison, Cheyenne}

which can also be put in correspondence with the set

{1, 2, 3, . . . , 49, 50}.

• The set of ISBN codes can be put into correspondence with the set of published books.

The world of finite sets and correspondences between them is pretty straightforward. The definition stated for two sets to be of the same size is definitely reasonable. Now let us take a few small steps into the realm of the infinite.

4.2 Infinite Sets

David Hilbert (1862–1943), the greatest mathematician of his generation, told an interesting tale. Imagine owning a hotel with an infinite number of rooms. Let us call it "Hilbert's Hotel." Business is good and every one of the infinite rooms is occupied. Along comes a car with another guest who needs a room. You don't want to send the guest away on a cold blustery night, but all your rooms are full. What to do? Hilbert gave a suggestion: get on your hotel loudspeaker and have every one of the infinite guests move to the next room. So the guest in room 57 moves into room 58 and the guest in room 53,462 moves to room 53,463. In general, every guest in room n moves into room $n + 1$. This leaves room 1 open for the extremely grateful new guest.

Let's go on with the tale. Just as all the guests have gotten settled in their new rooms and are blissfully sleeping, a bus pulls up in front of Hilbert's Hotel with an infinite number of passengers, each demanding their own room. Every one of your infinite rooms is occupied and you are in desperate need of an infinite number of empty rooms. What is an honest hotel manager to do? Again, Hilbert has some good advice: get on your handy-dandy hotel loudspeaker and tell every guest to go to the room whose number is twice the number of their present number. That is, the guest in room 57 goes into room 114 and the guest in room 53,462 goes into room 106,924. In general, every guest in room n goes to room number $2n$. After this, all the even-numbered rooms will be occupied and all the odd-numbered rooms are empty and waiting for your tired bus passengers.

Notice that these tricks would not work for one of those boring standard hotels with a finite number of rooms. It's only in Hilbert's cool hotel with an infinite amount of rooms that we can move people without worrying about losing any guests. What Hilbert really shows in the first case is that there is a correspondence between the infinite set of rooms {1, 2, 3, 4, 5, . . .} and a proper subset of that set: {2, 3, 4, 5, . . .}. In the second case, Hilbert shows that there is a correspondence between the set of rooms {1, 2, 3, 4, 5, . . .} and the proper subset {2, 4, 6, 8, 10, . . .}. It's these strange counterintuitive correspondences that we are going to explore in this section.

Rather than telling stories about fictitious hotels, let us work with some real infinite sets. There are many examples of infinite sets, but these sets must contain more than physical objects because there are only a finite number of physical objects in the universe. Many other concepts like numbers can form infinite sets. There are many common infinite sets of numbers:

- Natural numbers, $N = \{0, 1, 2, 3, \ldots\}$
- Whole numbers or integers, $Z = \{\ldots, -3, -2, -1, 0, 1, 2, 3, \ldots\}$
- Rational numbers or fractions, $Q = \{m/n\text{: } m \text{ and } n \text{ in } Z \text{ and } n \text{ is not } 0\}$
- Real numbers, R

The natural numbers are a proper subset of all whole numbers. Every whole number n can be thought of as the fraction $n/1$ and so the whole numbers can be regarded as a proper subset of all rational numbers. Finally, the real numbers are the set of all numbers, even those that cannot be described as fractions. It has been known for over 2,500 years that there are certain numbers, such as $\sqrt{2}$, e, $-\pi$, and $\sqrt{5}$, that cannot be written as fractions. Such numbers are called "irrational"—that is, "not rational" or "not reasonable."[2] The real numbers, R, contain all the rational and irrational numbers. So the rational numbers are a proper subset of the real numbers.

Consider the set of even numbers

$$E = \{0, 2, 4, 6, \ldots\}.$$

Every even number is a natural number, so *E* is clearly a proper subset of *N*, the natural numbers. In fact, one could say that there are twice as many natural numbers as there are even numbers. After all, the natural numbers consist of the even numbers *and the odd numbers!* But we will not follow our intuition of sets and proper subsets. Rather, we will follow the definition of being equinumerous from the last section. A correspondence exists between the natural numbers and the even numbers: simply make every natural number, n, correspond to the even number $2n$:

N	0	1	2	3	4	5	. . .	n	. . .
E	0	2	4	6	8	10	. . .	2n	. . .

This shows that *N* and *E* are the same size. How can that be? How can there be just as many natural numbers as there are even numbers? What about the odd numbers? How can a part be the same size as a whole?

Where did we go wrong? The answer is that we did not go wrong. The same reasoning that was used for finite sets was used here. However, for finite sets, having the same size follows our intuition. After all, our intuition was developed by looking at the world and its finite sets. For infinite sets, logic dictates that our naive intuition is no longer valid and needs to be modified. Galileo Galilei (1564–1642) was the first to write of this strangeness about infinite sets. He showed that an infinite set can be equinumerous to a proper subset of itself. In the 400 years since Galileo, this new definition has worked its way into all of mathematics and physics. We cannot simply ignore it because it goes against our intuition. Rather, these definitions are crucial to the models we use in interpreting the universe and are used in scientific predictions. We must understand and accept the definition. Our intuition needs to be adjusted.

On to some more infinite sets! Consider the set *S* of square numbers—that is, numbers that are equal to the product of a whole number with itself:

$S = \{0, 1, 4, 9, 16, 25, \ldots\}$.

This set is an infinite set and it is also a proper subset of *N*, the natural numbers. There are far fewer square numbers than even numbers: in the first hundred natural numbers, there are only ten square numbers. However, *S* can also be put into a correspondence with the natural numbers:

N	0	1	2	3	4	5	. . .	n	. . .
S	0	1	4	9	16	25	. . .	n^2	. . .

This correspondence shows that the natural numbers and their proper subset of square numbers are equinumerous. They have the same cardinality. Can we assign a number to describe the amount of elements in these sets? Obviously no finite number will do. Cantor denoted the amount or cardinality of these infinite sets by the symbol $\aleph_0$, pronounced "aleph-null." Aleph is the first letter in the Hebrew alphabet. All sets that are equinumerous to *N*, the natural numbers, have cardinality $\aleph_0$ and are called *countably infinite* sets. It is important to realize that we cannot complete counting such infinite sets. Nevertheless, we can at least begin to count them. By looking at a correspondence with the natural numbers, we can say what the 0th element is, what the 1st element is, what the 2nd element is, and so on.

There are other sets with this cardinality. Remember that a number is prime if it is more than 1 and only divisible by 1 and itself. Consider the set of all prime numbers:

$P = \{ 2, 3, 5, 7, 11, 13, \ldots \}.$

Even though there are seemingly far fewer prime numbers than natural numbers, a correspondence still exists:

N	0	1	2	3	4	5	. . .	*n*	. . .
P	2	3	5	7	11	13	. . .	n^{th} prime	. . .

This shows that *P* is equinumerous to the natural numbers. How do we describe this correspondence? What is the 42nd prime number? No simple formula gives us this information. Nevertheless, even though it is not easy to describe a correspondence, all that we require is that a correspondence *exists* that shows that *P* is of cardinality $\aleph_0$.

So far, all the infinite sets we have discussed are proper subsets of the natural numbers. What about sets that are seemingly larger than the natural numbers? Are these sets, in fact, larger than the natural numbers? Consider the set of whole numbers

$Z = \{\ldots, -3, -2, -1, 0, 1, 2, 3, \ldots\}.$

The natural numbers are a proper subset of *Z*. However, *Z* also contains the negative whole numbers. How can we possibly make a correspondence between *N* and the set *Z* that contains positive and negative natural numbers? Luckily, very smart people like Cantor worked on this problem and were able to find a simple correspondence, as shown here:

N	0	1	2	3	4	5	. . .	2n–1	2n	. . .
Z	0	1	–1	2	–2	3	. . .	*n*	–*n*	. . .

By cleverly splitting *N* into even and odd numbers, we make the odd numbers of *N* correspond to the positive integers of *Z*, while the even numbers of *N* correspond to the negative integers of *Z*. Since we will never come to the end of the odds or evens, every number in *Z* will be paired. This is exactly what we did in Hilbert's hotel, where we made the even-numbered rooms for the infinite set of old guests and the odd-numbered rooms for the infinite set of new guests.

Let us look at a really large set. Consider the set $N \times N$ of ordered pairs of natural numbers. That is the set of pairs <*m*,*n*>, where *m* and *n* are natural numbers. For every *m*, there is a copy of natural numbers:

<*m*,0>, <*m*,1>, <*m*,2>, <*m*,3>, ...

Since there are infinitely many *m*, the set $N \times N$ has infinitely many copies of *N*. The integers, *Z*, had two copies of *N*, one for the positive numbers and one for the negative numbers. In contrast, this set $N \times N$ has *infinitely* many copies of *N*. Our intuition tells us that this set has far more elements than *N*. Our intuition is wrong! Cantor was a very intelligent man and was able to find a correspondence between *N* and $N \times N$. One can describe this correspondence as follows:

N	0	1	2	3	4	5	. . .
$N \times N$	<0,0>	<1,0>	<0,1>	<0,2>	<1,1>	<2,0>	. . .

To see this correspondence clearly, think of the natural numbers as a long line of numbers, as in figure 4.1.

0 ⟶ 1 ⟶ 2 ⟶ 3 ⟶ 4 ⟶ 5 ⟶ . . . *n* . . .

Figure 4.1
N as an infinitely long snake

Writing the set $N \times N$ as in figure 4.2, let the natural numbers "snake" their way through $N \times N$ in a zigzag pattern.

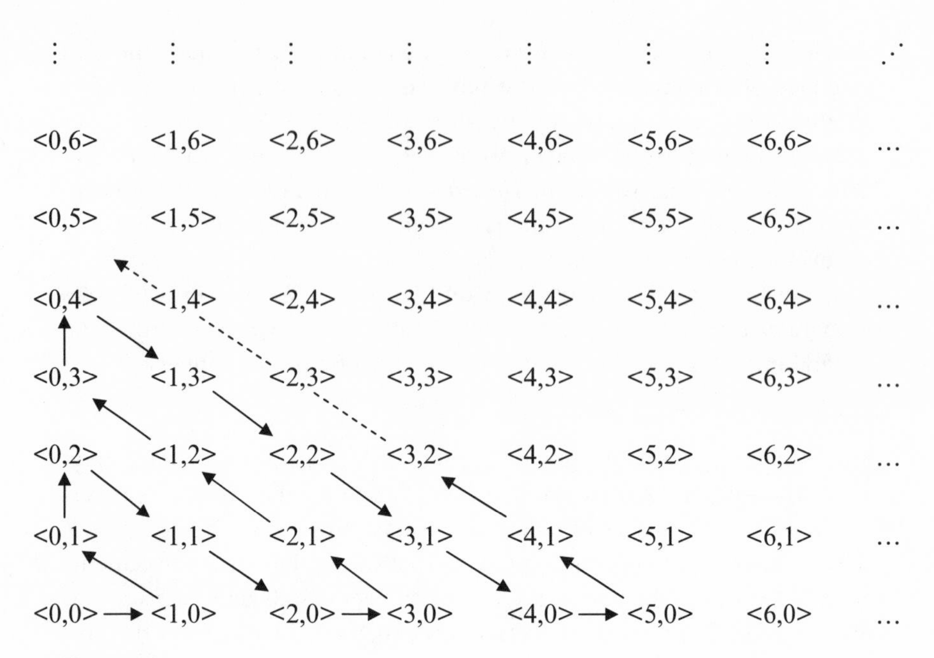

Figure 4.2
A correspondence between N and $N \times N$

Since these sets are infinite sets and we only have a finite amount of paper, we cannot display the entire correspondence. However, following this pattern, every ordered pair, including <303, 1227>, will eventually correspond to some natural number in N. For obvious reasons, this proof is sometimes called the *zigzag proof*. In summary, the set $N \times N$ is also equinumerous to N and has cardinality $\aleph_0$.

What about Q, the set of rational numbers? Surely there are more fractions than natural numbers! After all, every n in N is simply $n/1$ in Q. So Q has a copy of N inside it:

$0/1, 1/1, 2/1, 3/1, \ldots.$

But Q also has

$n/2, n/3, n/4, \ldots.$

Let us not forget negative fractions:

$-n/1, -n/2, -n/3, -n/4, \ldots.$

In addition, notice that between any two fractions, say 3/5 and 4/5, there is another fraction: 7/10. We can go on with this: between 3/5 and 7/10, there is 13/20. It would seem obvious that there are far more rational numbers than natural numbers. However, by now you should be expecting the unexpected. It might seem as though there are a lot more rational numbers than natural numbers, but, in fact, they are the same size. We will show this by exhibiting a correspondence between N and Q, as in figure 4.3.

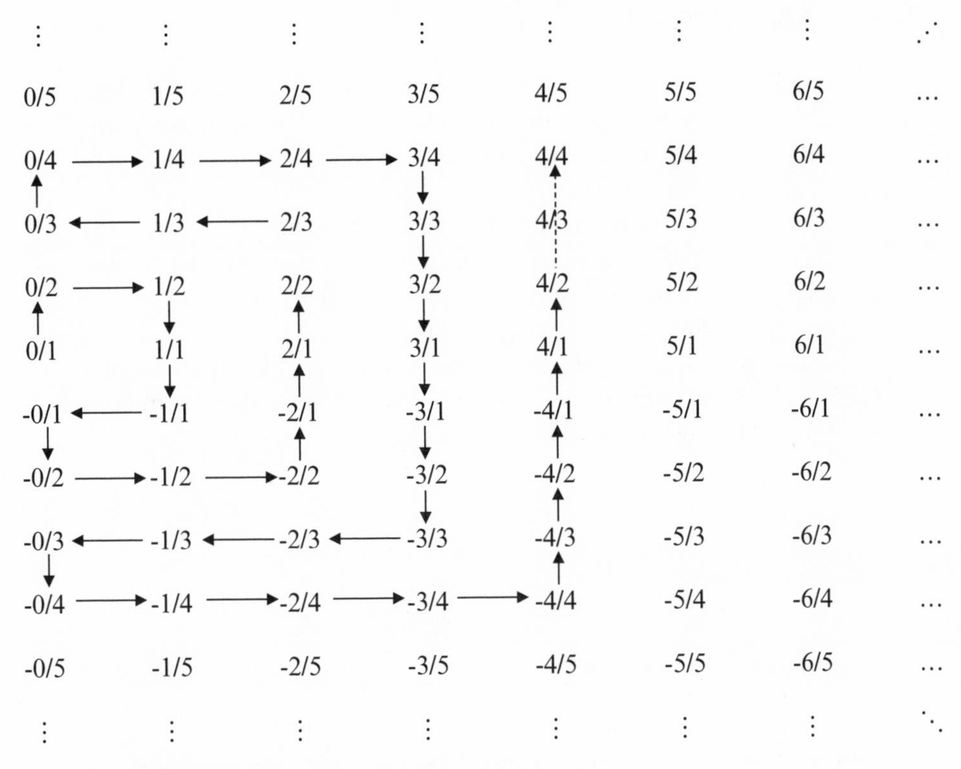

Figure 4.3
A correspondence between N and Q

Again, the natural numbers "snake" around the fractions and eventually hit every fraction. This proof is sometimes called the *necklace proof*. Can you see why?

However, there is a slight problem here. There are repetitions of rational numbers in our listing in figure 4.3. The rational number 4/7 has the same value as 8/14. So are we really making a correspondence with the set of rational numbers? In truth, we are doing something harder: we are making a correspondence between N and a set *larger* than rational numbers. There are, however, ways of making correspondences with only the rational numbers, simply by letting our snake skip over fractions that are already hit.

In conclusion, many sets that seem infinitely larger than the natural numbers are, in fact, equinumerous with the natural numbers. Is there any infinite set that is actually larger than the natural numbers?

4.3 Anything Larger?

While reading the last section, one can come to the conclusion that, with enough cleverness, every infinite set can be put into correspondence with the natural numbers. Cantor also thought this for a while . . . but then he looked at the real numbers.

Cantor considered the subset of the real numbers between 0 and 1. This subset is denoted (0,1)[3] and contains numbers like 0.43905346 . . . , 0.5, 0.373468 . . . , etc. He tried to find a correspondence between the natural numbers N and the set (0,1). He was looking for some type of trick similar to the zigzag or necklace proofs that worked so well in the last section. Maybe we can have the natural numbers "snake" their way through every point in (0,1), as depicted in figure 4.4.

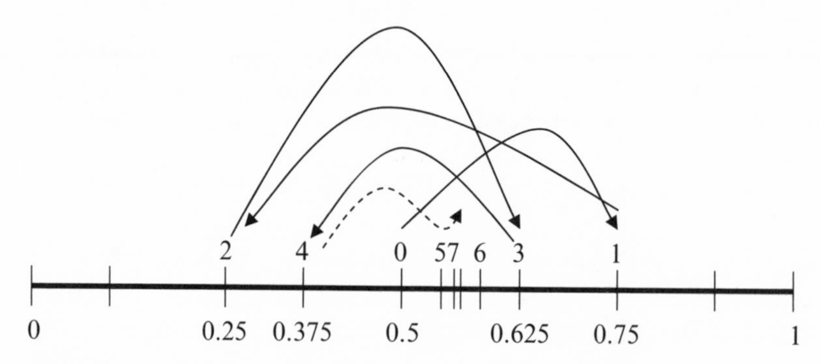

Figure 4.4
A (failed) attempt at a correspondence between N and (0,1)

The numbers on top of the line are the natural numbers and the numbers below are the real numbers that correspond to them. This correspondence could be written as follows:

N	(0,1)
0	0.500000 . . .
1	0.750000 . . .

2	0.250000 . . .
3	0.625000 . . .
4	0.375000 . . .
:	:

But this supposed correspondence did not work. In fact, every trick that Cantor tried in order to show that the natural numbers correspond to (0,1) failed. He always found that there were some elements of (0,1) that are missed in an attempted pairing.

Cantor failed to find a correspondence but he proved something far more interesting: that *no such correspondence could possibly exist.*[4] Rather than stating that he was not clever enough to find a correspondence, he showed that *no one,* with any amount of cleverness, will ever find such a correspondence because such a correspondence cannot exist. By showing that a correspondence between the natural numbers and the set (0,1) does not exist, Cantor demonstrated that the set (0,1) is strictly larger than the set N. The elegant and simple proof will be presented shortly in all its glory.

Infinite sets that are equinumerous to the natural numbers are called *countably infinite.* We can at least start counting such sets. The correspondence with the natural numbers will help us count these sets. We saw in the last section that the sets N, E, P, Z, $N \times N$, and Q are all countably infinite. Following that line of reasoning, infinite sets that cannot be put in correspondence with the natural numbers are called *uncountably infinite.* One cannot even begin to list the elements of an uncountably infinite set. We will prove that (0,1) is uncountably infinite. Sets that are uncountably infinite are vastly larger than sets that are countably infinite.

Cantor's result is the first proof that there are different types or levels of infinity. This is extremely counterintuitive. After all, who would have thought that there are different types of "going on forever and ever"? But in fact, there are. By simply following the logical definition of what it means for two sets to be the same size, Cantor came to this radical conclusion. Again—and this cannot be stressed enough—these distinctions between the different levels of infinities are used in modern calculus texts. With this knowledge in mind, engineers and physicists build bridges and rockets. It would be foolhardy to cross a modern suspension bridge if you knew that the engineer did not believe in Cantor's work. As counterintuitive as it seems, the different levels of infinity are fundamental to our understanding of the universe.

The actual proof is a proof by contradiction. In order for Cantor to show that no such correspondence between N and all of (0,1) exists, he assumes (wrongly) that such a correspondence does exist and arrives at a contradiction. Following the format introduced in chapter 1, we write:

There is a correspondence between N and (0,1) ⇒ contradiction.

The contradiction derived is a real number between 0 and 1 that is not paired off with any element of the proposed correspondence. Since we assumed that the correspondence pairs off *every* real number between 0 and 1, we have a contradiction.

Informally, the proof proceeds by describing a real number between 0 and 1 as

This real number is not in the given correspondence.

Or, to be more precise,

This real number is different from every other number in the given correspondence.

The proof is called a *diagonalization proof* and works as follows. Assume that there is some type of fancy correspondence between the natural numbers and every element of the set (0,1). We might illustrate such a correspondence as in figure 4.5.

On the left are the natural numbers and on the right we write what number in (0,1) it is paired with. With this alleged correspondence, we describe a number that is not on this list. The number will be a real number

(0,1)
Position

N		0	1	2	3	4	5	6	7	8	...
0	0.	5	0	3	0	3	2	0	0	0	...
1	0.	3	3	5	9	7	3	8	6	8	...
2	0.	2	5	9	4	1	1	7	8	3	...
3	0.	0	5	2	8	2	8	2	6	4	...
4	0.	5	0	0	0	0	0	0	0	0	...
5	0.	3	3	3	3	3	3	3	3	3	...
6	0.	9	9	1	1	2	3	0	4	1	...
7	0.	1	2	2	7	1	9	6	7	0	...
8	0.	1	0	5	4	1	7	3	5	6	...
⋮	⋮	⋮	⋮	⋮	⋮	⋮	⋮	⋮	⋮	⋮	⋱

Natural numbers

Figure 4.5
An alleged correspondence between N and (0,1), and its diagonal

between 0 and 1 and denoted as *D* (for "diagonal"). *D* is derived from the diagonal of the alleged correspondence presented in figure 4.5.

• The 0th digit of *D* will be 6 because that is one more than the 5 in the 0th position of the 0th number of the alleged correspondence.
• The 1st digit of *D* will be 4 because that is one more than the 3 in the 1st position of the 1st number of the alleged correspondence.
• The 2nd digit of *D* will be 0 because that is different from the 9 in the 2nd position of the 2nd number of the alleged correspondence.
• The 3rd digit of D will be . . .

And the description goes on. We can eliminate the unimportant part of figure 4.5 and see the description of the number *D* as in figure 4.6.

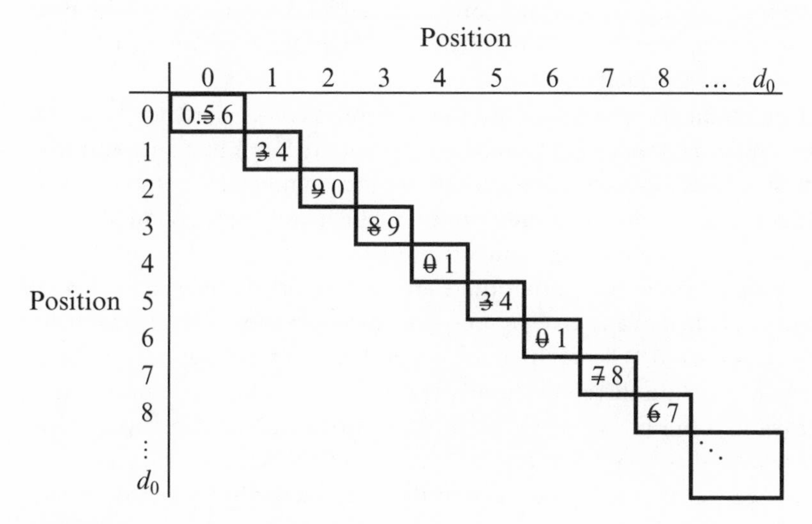

Figure 4.6
A description of the number *D* not in the alleged correspondence

So the number *D* is 0.640914187. . . . This number is clearly within (0,1). However, it is not in this correspondence. Let's look for it:

• Number *D* cannot correspond to *0* because number *0* in position *0* is a *5* and number *D* in position *0* is a *6*.
• Number *D* cannot correspond to *1* because number *1* in position *1* is a *3* and number *D* in position *1* is a *4*.
• Number *D* cannot correspond to *2* because number *2* in position *2* is a *9* and number *D* in position *2* is a *0*.
• Number *D* cannot correspond to *3* because number *3* in position *3* is an *8* and number *D* in position *3* is a *9*.

- etc.,
- Number D cannot correspond to d_0 because number d_0 in position d_0 is an x and number D in position d_0 is not x.
- etc., . . .

Since D is not in this correspondence, we conclude that the purported correspondence is not a correspondence at all. In fact, what we did was to describe D as different from every row in the purported correspondence. Notice that D is not the only number that fails to show up in the correspondence. To find other such numbers, all you have to do is systematically go through every row and change some digit. By doing this, we are finding numbers that cannot be the same as any row. In the above, we changed every row by altering every element along the diagonal, but we could have done it in other ways. Furthermore, we altered it by adding 1 to the digit (except to the digit 9, which we made a 0). There are, however, many other ways to alter the digits.

One might try to describe another potential correspondence in which the number D does occur. However, the same trick can be done with that correspondence as well. There will be another number, D', between 0 and 1 that should be in the correspondence but is not. Any potential correspondence will be missing elements of (0,1).[5]

We conclude by saying that there are vastly more numbers in (0,1) that are not matched in any given correspondence than those that are matched. That is, the set (0,1) is much, much larger than the set of natural numbers. Uncountably infinite sets are immense compared to countably infinite sets. Throughout this book, we will come back to this fact over and over. Many sets will be shown to be countably infinite, which is minuscule compared to a larger set that is uncountably infinite. In fact, when we "subtract" a countably infinite set from an uncountably infinite set, we are still left with an uncountably infinite set.

There are many other uncountable sets. Consider the powerset of the natural numbers, $\wp(N)$—that is, the set of all subsets of N. We saw that for finite sets with n elements, the size of $\wp(N)$ is 2^n. One might think that there is some trick for an infinite set that would show that a correspondence exists between a set and its powerset. This is not true. A correspondence between a set and its powerset cannot exist.

This is also shown with a proof by contradiction:

There is a correspondence between N and $\wp(N) \Rightarrow$ contradiction.

The contradiction is derived by describing a subset of natural numbers—that is, an element of $\wp(N)$—that is not in the proposed correspondence.

Since we assumed the correspondence pairs off every subset of natural numbers, we have a contradiction.

Informally the proof proceeds by describing a subset of natural numbers as

This subset of natural numbers is not in the given correspondence.

Or, more precisely,

This subset of natural numbers is different from every subset of natural numbers given in the correspondence.

Formally, this is again shown with a diagonalization proof. Assume that there is a correspondence between *N* and ℘(*N*)—that is, there is a way of pairing every natural number *n* and a subset of *N*. Rather than listing the elements of the subset, we will state yes or no depending if an element is in the subset. A description of a correspondence can be illustrated as in figure 4.7.

		Natural numbers									
		0	1	2	3	4	5	6	7	8	...
	0	Yes	No	No	No	Yes	No	Yes	Yes	Yes	...
	1	Yes	No	Yes	No	No	Yes	No	No	No	...
	2	No	No	Yes	No	No	Yes	No	No	No	...
	3	No	Yes	No	Yes	No	No	No	No	Yes	...
	4	Yes	No	Yes	Yes	No	No	Yes	No	Yes	...
Subsets of *N*	5	No	No	No	No	No	No	No	No	No	...
	6	No	No	No	No	No	Yes	No	No	No	...
	7	No	Yes	No	No	No	Yes	No	Yes	Yes	...
	8	No	No	No	No	No	No	No	No	No	...
	⋮	⋮	⋮	⋮	⋮	⋮	⋮	⋮	⋮	⋮	⋱

Figure 4.7
An alleged correspondence between *N* and ℘(*N*) and its diagonal

Let's look at a few of the subsets. The subset that corresponds to number 1 contains 0, does not contain 1, contains 2, contains 5, and so on. So this subset is

{0, 2, 5, . . .}

The subset that corresponds to number *7* is

{1, 5, 7, 8, . . .}.

This correspondence cannot contain all subsets of *N*. By looking at the diagonal, we can find a subset not on the list. Let's look at the opposite of what is on the diagonal, as in figure 4.8.

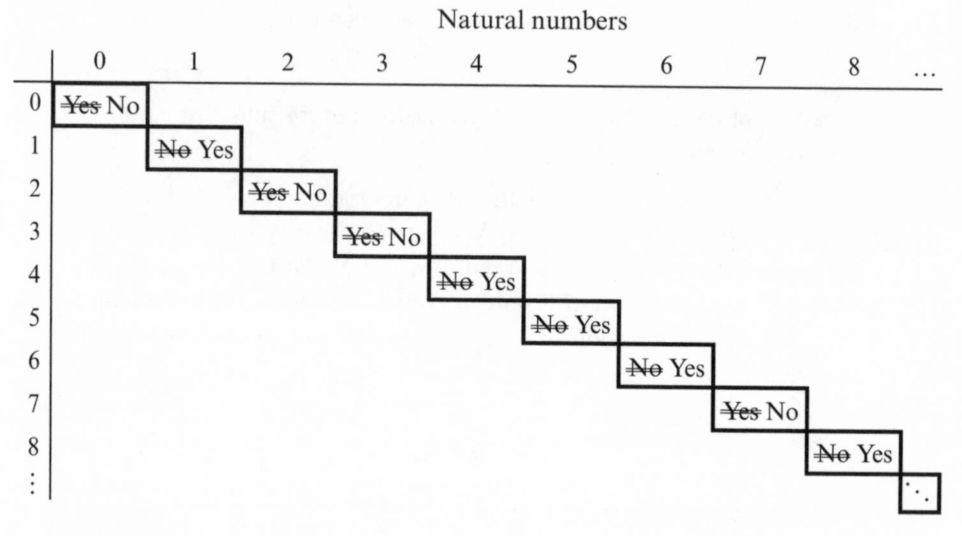

Figure 4.8
A subset of *N* that is not in the alleged correspondence

Subset *D* will

- not contain 0 because the subset that corresponds to number 0 contains 0;
- contain 1 because the subset that corresponds to number 1 does not contain 1;
- not contain 2 because the subset that corresponds to number 2 contains 2;
- not contain 3 because the subset that corresponds to number 3 contains 3;
- ⋮
- contain d_0 if and only if the subset that corresponds to number d_0 does not contain d_0;
- ⋮

We are really describing the subset of natural numbers:

$D = \{d$ in N: the subset that corresponds to d does not contain $d\}$.

The claim is that D does not correspond to any element of the natural number. If one were to say that subset D corresponds to number d_0, then see if the number d_0 is in D.

d_0 is in D if and only if d_0 is not in the subset that corresponds to d_0.

That is,

d_0 is in D if and only if d_0 is not in D.

This is a contradiction. We conclude that the subset D is different from the subset corresponding to number d_0. In fact, D is different from any subset in the proposed correspondence. Hence, our correspondence is missing at least one subset.

The natural numbers did not really play a major role in the last proof. We can generalize this proof and show that for any set, S, the powerset of S cannot be put into correspondence with S. That is, there are more subsets than elements of a set. This coincides nicely with our theme about the limitations of self-reference. The elements of the set S cannot "correspond," "describe," or "handle" all the membership properties of the set S.

The short proof goes as follows: imagine (falsely) that there is a correspondence between S and $\wp(S)$. Now consider the set

$D = \{d$ in S: the subset of S that corresponds to d does not contain $d\}$.

D is a subset of S and hence an element in $\wp(S)$, but D does not correspond to any element of S. In effect, D says

This subset is different from any subset in the correspondence.

If you (wrongly) claim that there is a d_0 in S that corresponds to D, then look at element d_0:

d_0 is in D if and only if d_0 is not in D.

This is a contradiction and we can conclude that D is not in the correspondence. Hence $\wp(S)$ is larger than S.

It was shown that both (0,1) and $\wp(N)$ are larger than N. In fact, a correspondence exists (which I will not describe) between these two sets showing that they have the same cardinality. Since the cardinality of the powerset of a set of size n is 2^n, and the cardinality of N is $\aleph_0$, cardinality of $\wp(N)$ is $2^{\aleph_0}$. Because this is also the cardinality of the *continuous* interval (0,1), it is also called the "cardinality of the continuum."

Why did we restrict ourselves to (0,1) only? What about the entire set, R, of real numbers? It would seem that the entire set of real numbers is much larger than (0,1). After all, the real numbers also contain the interval (1,2) and (2,3). Don't forget the negative intervals such as (-23, -18). The real numbers contain infinite copies of (0,1). However, following our definition of what it means for two sets to be the same size, we can show that (0,1) is equinumerous to *R*. The formal name for this proof is *proof by stereographic projection*, but I prefer the friendlier *sunshine proof*. The proof is essentially figure 4.9.

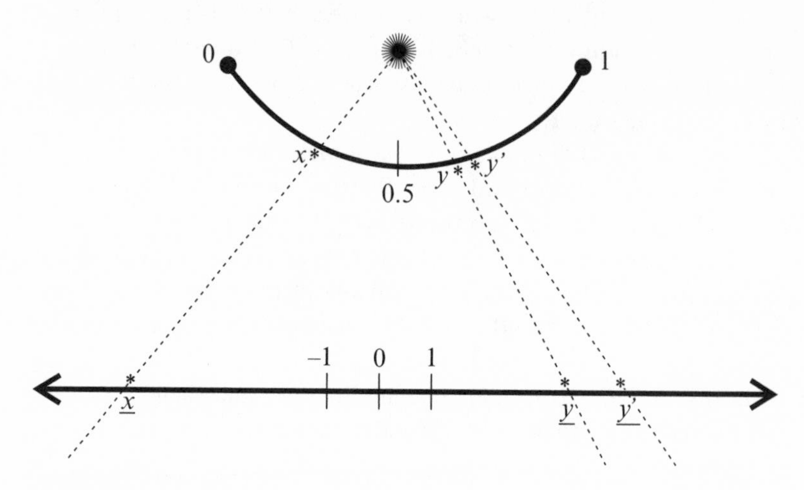

Figure 4.9
The correspondence between (0,1) and *R*

First take the sun, all bright and shiny, and put it on top of your picture. Then take the interval (0,1) and "bend" it around the sun. Now take the real-number line that represents the set *R* and put it on the bottom of the picture. Realize that the real line goes on to the right and left forever and ever. The description of the correspondence between (0,1) and *R* is as follows: for every point x in (0,1), draw a straight line from the sun through x and onto *R*. The point where this line crosses *R* will correspond to x and will be denoted as $\underline{x}$. To recognize that this is a good correspondence, all we have to do is realize that two different points y and y' in (0,1) will go to two different points $\underline{y}$ and $\underline{y}'$ in *R*. To show that every point $\underline{z}$ of *R* will be in this correspondence, all one has to do is to draw a straight line from the point $\underline{z}$ in *R* back to the sun. This line will pass through a unique point of (0,1). In conclusion, there is a correspondence between (0,1) and *R* and hence they are equinumerous.

We have shown that there are actually two ways of proving that an infinite set is of cardinality larger than $\aleph_0$. First, one can make a diagonal argument showing that no correspondence exists between the set and the natural numbers. Second, we can show that there is a correspondence between the set and another set that is already known to have cardinality larger than $\aleph_0$.

Several infinite sets with cardinality $\aleph_0$ have been described. We have also seen several infinite sets with cardinality $2^{\aleph_0}$. The obvious question is whether there is anything strictly larger than $2^{\aleph_0}$. The answer is yes. The powerset of a set is strictly larger than the set. From this we can see that the powerset of (0,1), denoted $\wp((0,1))$, will not be in correspondence with (0,1). That is, the set of subsets of the unit interval (0,1) will be larger than (0,1). This set will have cardinality $2^{2^{\aleph_0}}$ It is difficult to wrap one's mind around such a set. Try to write down some of the elements.

Of course, there is no reason to stop there. We can go on using the powerset function to describe sets of even higher cardinality. None of the sets in different levels of infinity can be put into correspondence with each other.

4.4 Knowable and Unknowable

The ideas of set theory described in the previous three sections are totally reasonable and rational. Unfortunately there is one little fatal flaw in them: set theory, as we have expressed it so far, is *inconsistent*. That means that with the language of set theory that we have used, we can formulate contradictions. This was first pointed out in a letter dated June 16, 1902, from Bertrand Russell (1872–1970) to Gottlob Frege (1848–1925) in which a simple contradiction in basic set theory was first illustrated. This contradiction is called Russell's paradox. Although we've already encountered this paradox in section 2.2, it is worth reminding ourselves what it is all about.

One can discuss many sets. Consider the set

$H = \{a, b, \{c, d\}\}$.

This set consists of three elements. Two of the elements are the letters a and b, and one of the elements is a set $\{c,d\}$. Now consider the set of

$J = \{a, b, J\}$.

This set also contains three elements but one of its elements is *itself*! Now consider the following set, which we call R for Russell:

R = The set of sets that do not contain themselves.

So the above H is an element of R but the above set J is not an element of R. Now ask the following simple question:

Is R an element of itself?

If R is an element of R, then according to the requirements of being an element of R, it must be that R is not an element of itself. In contrast, if R is not an element of itself, then it satisfies the requirement of being in R and so R is in R. Hence we have a contradiction, and we conclude that the naive form of set theory that we described is inconsistent. This was a tremendous blow to the researchers at the time. A contradiction within a system renders it useless.

This is a paradox. We made an assumption and were led to a contradiction. The subtle assumption we made is that for every description there is a set of elements that have the objects described. This works most but not all of the time. For example, if I think of the property of red, then I can form the set of all red things. With a description of pink Cadillacs there is a set of pink Cadillacs. But the description "does not contain itself" cannot correspond to a set of things that does not contain itself. This will lead to a contradiction. We must be careful.

To avoid contradictions such as Russell's paradox, researchers tried to formalize some of the notions of set theory and put some restrictions on the type of sets that can exist. This was done by developing a system of axioms that are self-evident and by using these axioms to generate theorems about sets.

One axiom system was formulated by Ernst Zermelo (1871–1956) and Abraham Fraenkel (1891–1965). This system, which came to be known as *Zermelo-Fraenkel set theory*, is the most important in the field.[6] The axioms of Zermelo-Fraenkel set theory are as follows:

1. *Axiom of extensionality* Two sets are the same if they have the same elements.
2. *Axiom of pairing* For any x and y, there exists a set $\{x,y\}$.
3. *Axiom of subset selection* (also called *axiom of restricted comprehension*) If X is a set, and φ is a property that describes certain elements of X, then a subset Y of X exists containing only those x in X that satisfy the property—that is,

$Y = \{x \text{ in } X \mid \varphi(x) \text{ is true}\}.$

(This almost says that if you have a property, say "redness," then you have a set of all things that are red. However, we need to restrict this axiom because otherwise there will be trouble with Russell's paradox by simply looking at the property of "not containing itself." We cannot talk of a subset of "everything." Rather, we can only talk of a subset of something. So for a property φ, we cannot say that

$$Y = \{x \mid \varphi(x) \text{ is true}\}$$

is a set. Rather, we must restrict this to a particular set X.)

4. *Axiom of union* The union of a set of sets is a set.
5. *Axiom of powerset* For any set X, the powerset of X is also a set.
6. *Axiom of infinity* There exist sets with infinitely many elements.
7. *Axiom of replacement* If F is a function—that is, a way of assigning elements from one set to another set—and X is a set, then $F(X)$, the set of values of F, is also a set

$$F(X) = \{F(x) | x \text{ in } X\}.$$

8. *Axiom of regularity* (also called the *axiom of foundation*) There is no infinite regression of a set that contains a set that contains a set . . . In technical terms, every nonempty set X contains a member Y such that X and Y are not the same sets.

An interesting philosophical question must be posed. Zermelo-Fraenkel set theory restricts us from discussing or accepting certain sets as legitimate. We can only consider certain collections as sets and are not permitted to consider other collections as sets. Does this mean that the other collections do not exist? Are they not also sets? Yes, it is good to steer clear of contradictions, and we like such error-free systems, but are we being truthful as to what really exists? Are we throwing away the baby with the bathwater?

The amazing fact about Zermelo-Fraenkel set theory is that the vast majority of modern mathematics can be formulated with sets and these few simple axioms. In a comprehensive encyclopedia of mathematics, we find the following observation: "Nowadays, it is known to be possible, logically speaking, that current mathematics, almost in its entirety, can be derived from a single source: the theory of sets."[7] In other words, most of mathematics can be seen to be built on the foundation of these few axioms. Most working mathematicians usually do not think about the axioms, nor do they care if their work can be put into the language of Zermelo-Fraenkel set theory. Nevertheless, with enough effort, their work can be stated within the language of Zermelo-Fraenkel set theory. From this important position, the axioms of Zermelo-Fraenkel set theory can be seen as the axioms of all of mathematics and hence the axioms of exact reasoning itself.

The obvious question is whether Zermelo-Fraenkel set theory is consistent. After all, one reason for putting set theory into axioms is to make sure that we steer clear of problems like Russell's paradox and other contradictions. It would be nice to know that no contradictions can be derived from these axioms. Concerning consistency, there is good news and there is bad news. The good news is that the Zermelo-Fraenkel set theory has been around for about a century and no one has derived any contradiction yet. Nor does it look like anyone will in the future. The bad news is that one of the consequences of Gödel's famous incompleteness theorems (which we will meet, in detail, in sections 9.4 and 9.5) is that the consistency of Zermelo-Fraenkel set theory is not provable within standard mathematics. And so we cannot be absolutely certain that Zermelo-Fraenkel set theory and all of the modern mathematics that it entails is consistent.[8]

Let's examine some topics about what can and cannot be proved using Zermelo-Fraenkel set theory. In section 4.2 we showed that there are many sets that are equinumerous to the natural numbers *N*. In Section 3 we showed that there are many sets that are equinumerous to the set (0,1) and that these sets are strictly larger than *N*. The obvious question arises: Are there any sets that are between *N* and (0,1)? That is, does there exist an infinite set *S* such that *N* is strictly smaller than *S* and, in turn, *S* is strictly smaller than (0,1)? What we are really asking is whether there is something between $\aleph_0$ and $2^{\aleph_0}$. This is a perfectly simple question. All we want to know is if a certain set of a particular size exists. We do not even care what the elements of the set are. Our only concern is the size of the set. Cantor was the first to ask this question in the 1880s. He believed the answer was no and formulated this conjecture as the "continuum hypothesis":

There does not exist a set whose size is strictly between *N* and (0,1).

Despite much effort, Cantor was unable to prove this conjecture. In 1900, David Hilbert gave a famous speech in which he listed twenty-three hard problems that were challenges to be solved in the twentieth century. The continuum hypothesis was number one.

In 1940, Kurt Gödel (1906–1978) showed that (assuming Zermelo-Fraenkel set theory is consistent) the continuum hypothesis is consistent with the axioms of Zermelo-Fraenkel set theory. This means that one cannot derive a contradiction from the axioms of Zermelo-Fraenkel set theory by simply adding an axiom stating that the continuum hypothesis is true. Another way of saying this is that there is a way of interpreting the axioms such that the continuum hypothesis is true and there *does not* exist a set of intermediate size.

In 1963, Paul Cohen (1934–2007), who once studied at Brooklyn College, gave a final answer to the eighty-year-old problem. He showed that (assuming Zermelo-Fraenkel set theory is consistent) the negation of the continuum hypothesis is consistent with the axioms of Zermelo-Fraenkel set theory. This means that one cannot derive a contradiction from the axioms of Zermelo-Fraenkel set theory by simply adding the axiom stating that the continuum hypothesis is false. Another way of saying this is that there is a way of interpreting the axioms such that the continuum hypothesis is false and there *does* exist a set that is strictly between N and (0,1).

With the results of Gödel and Cohen, one says that the continuum hypothesis is "independent" of the axioms of Zermelo-Fraenkel set theory. That means that the axioms cannot prove or disprove it. There is no way of answering these questions with the axioms of Zermelo-Fraenkel set theory, or any other equivalent set theory. With this independence from the axioms, one can go farther and ask if the continuum hypothesis is *really* true or false. Does there really exist a set that is intermediate between N and (0,1)?

The continuum hypothesis is just one of the many fascinating ideas in set theory. One of the more interesting statements in set theory is called the "axiom of choice." Let's start by looking at an easy example with finite sets. Consider the set of all U.S. citizens. They can be partitioned into fifty nonoverlapping sets that correspond to the fact that people live in fifty different states.[9] We may ask for a single set of citizens that has exactly one member or representative for each subset. The simplest way of forming this set is to choose the governor of each state as the representative of that state. We could also have chosen the senior senator from the state, or the oldest person in the state. We could have made many different choices. However, what if we are given a partition on an *infinite* set? Can we still choose one element of each subset? Things seem a little more complicated in the infinite case. Imagine being presented with an infinite set of pairs of shoes. (For some, this would result in an infinite amount of joy.) We may ask for one shoe from every one of the infinite pairs. This can simply be done by always choosing the left shoe in every pair. One may also choose the right shoe in every pair. However, what if we are presented with an infinite set of pairs of tube socks where each sock is identical to its mate? Can we still choose one from each pair? Which one? There is no way of describing the function. One might say no such choice is possible.

We will see that assuming one can always make such a choice leads to trouble.

The axiom of choice says that for any given set and any partition of that set into nonoverlapping subsets, a set can always be formed with one representative from every subset. This seems like a pretty innocuous requirement. The axiom of choice seems obviously true for finite sets but is slightly more problematic for infinite sets. Is it obvious that such a set can always be formed? Why should we not be able to form such a set? In 1963, Paul Cohen showed that not only is the continuum hypothesis independent of Zermelo-Fraenkel set theory, but the axiom of choice is also independent of those axioms. That is, we cannot prove it or disprove it with Zermelo-Fraenkel set theory.

Many mathematicians think that the axiom of choice is "self-evident" enough and should be added to the axioms of Zermelo-Fraenkel set theory to make a foundation for set theory and mathematics. They call this new axiom system *Zermelo-Fraenkel set theory with choice* or *ZFC*. It is the most popular foundational system in all of mathematics. Other mathematicians are more circumspect and worry about including the axiom of choice.

One of the major reasons for the suspicion about the axiom of choice is the *Banach-Tarski paradox*. This paradox says that with Zermelo-Fraenkel set theory and the axiom of choice (but not Zermelo-Fraenkel set theory alone) one can prove the following: given a three-dimensional ball of any size, one can chop the ball into five nonoverlapping pieces and put them together again into two balls *each of the same size as the original*. (See figure 4.10.)

The pieces that the original ball is chopped into are not your typical sane pieces. Rather, they will look like something that was done by Zeno

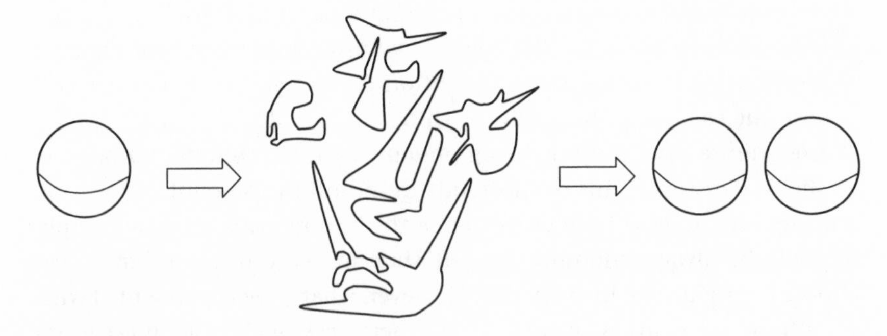

Figure 4.10
The Banach-Tarski paradox

while under the influence of psychedelic drugs. Each piece will be connected but very bizarre looking. Nevertheless, this fact is a provable consequence of the seemingly harmless axiom of choice. Another version of the paradox says that a sphere as small as a pea can be cut up into a finite set of different parts and then be put together to form a sphere the size of the sun. Many people say that since this paradox is the consequence of the axiom of choice, this axiom leads to an obvious false statement and should be excluded from what is reasonable. They want to banish the axiom of choice from mathematics. Others say that we should keep the axiom. They might appeal to Zeno's paradoxes that show that our concept of space is full of counterintuitive properties. Similarly, infinity has some strange mind-bending consequences. After all, as we saw in section 4.2, one infinite set can be put into correspondence with another set that is twice its size (even numbers with natural numbers, and natural numbers with positive and negative integers), so why can't an infinitely divisible ball have the same volume as two infinitely divisible balls?[10]

What are we to do with all these questions? Is Zermelo-Fraenkel set theory consistent or not? Is the continuum hypothesis true or false? Is the axiom of choice acceptable or unacceptable? These questions are independent of Zermelo-Fraenkel set theory, which is a basis of most of mathematics. So we cannot use mathematics to answer these questions. The answers to all of these simply stated questions are beyond contemporary mathematics, beyond rational thought, and perhaps even beyond us.

There are two broad philosophical schools of thought on how to deal with these questions. On the one hand there are Platonists or realists, who, following Plato, believe that in some sense sets really exist and that all these questions, as well as any other questions about mathematical objects, have definite answers. Mathematical objects and the theorems that describe the relationships of these objects are real and exist independently of human thought. Platonists believe that there are perfect ideal circles and that the ratio of the circumference to the diameters of these circles is π. If no human being ever existed and thought about numbers, π would still exist. To Platonists, whether or not the continuum hypothesis stands—that is, whether there really is a set between N and $(0,1)$—is something that is either true or false and human beings must venture to answer this question. Since the given axioms cannot answer the question, one must search for more or different axioms that would settle it once and for all. These axioms should somehow be self-evident, not cause contradictions, and not bring us to counterintuitive consequences.

In contrast to that school of thought, people at the other extreme are sometimes called nominalists, anti-Platonists, antirealists, formalists, or fictionalists.[11] They, in essence, do not believe that there is anything "out there." Mathematical objects are things that mathematicians talk about and that do not have any existence outside of language and human minds. The number 3 does not exist any more than other human-made fictional creations like Mickey Mouse or James Bond. To such philosophers, the reason the questions posed in this section do not have answers is that not enough about these mathematical objects has been described. To a nominalist, the mathematical objects do not really exist, only human descriptions exist. There are certain rules about mathematical objects just as there are certain rules about Mickey and Bond. One would never say that Mickey can be mean, because that would not conform to the fictional character we all grew up with. Similarly, we would never say that Bond was dressed like a slob. When it comes to mathematical objects, they seem more real because there are many more rules about them. So while it is conceivable, but improbable, that Bond would have his shirt untucked in a movie sequence, it is totally impossible for 3 + 2 to equal 6. Returning to our question about the continuum hypothesis: a nominalist would say that the language of sets has not been described well enough for one to make a judgment about the existence of such an intermediate set. Why look for new axioms that make the continuum hypothesis either true or false? There is no external reality to which our axioms must conform. Rather, we should study both systems: we should study Zermelo-Fraenkel set theory with the continuum hypothesis being true and Zermelo-Fraenkel set theory with the continuum hypothesis being false. Both systems are worthy of study. We have the independence, let us use it![12]

Some of the clashes between these two schools can be summarized by the answer to the following simple question: Are theorems of mathematics "discovered" or "invented"? The Platonists insist that the theorems and the objects that they are dealing with have always been there and always will be there. To them, a mathematician discovers what has always existed. In contrast, a nominalist would say that a mathematician invents a new theorem. The theorem has to conform to the rest of the known knowledge about some mathematical object, but he or she is nevertheless adding to the fictitious literature, as it were. Informal surveys have shown that the majority of mathematicians are, in fact, Platonists and feel like they are discovering while working. However, since this is a philosophical question and not a mathematical one, perhaps their opinion should not be taken as authoritative.

Do infinite sets exist? Does the number 3 exist? I have never seen an infinite set nor have I ever stubbed my toe against the number 3. We can talk about infinite sets, but I can talk about unicorns and Pinocchio as well. I can also tell a very long, logical, and plausible tale about Little Red Riding Hood even though she never existed outside a human mind. Can we claim the same nonexistence for the set of natural numbers with all its structure that is seemingly so real? Is the sequence 0, 1, 2, 3, . . . simply an invention of language and culture? Perhaps we are confused by the many possible meanings of the word *exist*? It is hard to imagine that these seemingly obvious ideas about natural numbers are simply a part of language and do not really exist in some true sense. Nevertheless, language, and the culture that governs the way we use language, seems sufficient to explain how we use numbers.

The strongest argument for Platonism is the amazing consistency of mathematics. For thousands of years, mathematicians have been working in isolation from each other and have come to similar, noncontradictory ideas. It seems that the only way this is possible is that they are all trying to describe something that is external to their mind.

The strongest arguments for nominalism are questions like the following: Who set up these Platonic ideals? Why are they there? For the past several hundred years, scientists have made steady progress by eliminating metaphysical presuppositions. Why keep any such metaphysics in mathematics and in set theory? A nominalist would counter a Platonist's proof by saying that the mathematicians are not all isolated from each other. Before they entered their lonely writer's garret, they were all aware of the rules for being a good mathematician. They knew that if they were to write anything that would cause a contradiction, they would lose their status as a mathematician. They were not isolated because they knew the language beforehand.

One issue that bothers both camps is the incredible usefulness of mathematics and set theory in the physical sciences. Why does the physical world somehow conform to the ideas of mathematicians and set theorists? The Platonists say that there is some type of (mysterious) connection between the Platonic realm of ideas and our physical world. They also posit some type of (mysterious) connection between the Platonic realm and our minds that permits us to discover these Platonic ideals. In contrast, nominalists say that the reason mathematics works so well is that mathematics was a language formed by humans with the intuition they received from the physical world. To them, it is not shocking that a system developed while observing the physical world should be suited to the physical world.[13]

Do not think that one can easily come to the "correct" choice between these two camps. The two giants who worked out the independence of the continuum hypothesis came to different conclusions. Gödel felt that we have to find new axioms that will somehow capture the Platonic world of sets. In contrast, Cohen felt that there is no real answer to the continuum hypothesis.[14]

These battles have raged for millennia with no apparent victor. To me, any argument that Platonists give can be answered better by the nominalists. However, I am aware that we, mere mortals, are not going to come to any firm conclusions.

Further Reading

Sections 4.1–4.3

The material in the first three sections can be found in many places. There are portions of popular history books such as chapter 24 of Kramer 1970 or section 15–4 of Eves 1976. There are also other nontechnical books like Kline 1980 and Rucker 1982. Two technical places to look are section 13.3 of Ross and Wright 2003 and section 3.4 of Truss 1998. Most of the ideas can also be found in any calculus text.

Joseph Dauben has written a fantastic biography of Georg Cantor. Although Cantor's work seems obvious to us now, in his lifetime, it was considered radical and unacceptable by many of his fellow mathematicians. Cantor suffered tremendously for holding fast to his ideas. Dauben 1979 describes Cantor's life and work in great detail.

The story of Hilbert's hotel can be found in Gamow 1988.

Section 4.4

For Russell's letter to Frege, see Van Heijenoort 1967, 124–125.

The Zermelo-Fraenkel set theory axioms were taken from chapter 1 of the authoritative Jech 1978 and chapter 2 of Devlin 1993.

Wapner 2007 is a very clear presentation of the Banach-Tarski paradox for the layperson.

One can read about the different types of Platonism in Wang 1996 and more about fictionalism in Balaguer 2008 and Eklund 2007.

Gödel presents his views on Platonism in his popular article, Gödel 1947. Cohen's anti-Platonism can be seen in the last section of Cohen 2005.

To learn more about the infinite and the paradoxes of set theory, see Rucker 1982 and Lavine 1994. Cohen 2005 also has many deep ideas related to our theme of the limits of formal reasoning.

There is a nice BBC Horizon documentary titled *To Infinity and Beyond* that covers some of these topics. For more information on the documentary, go to http://www.bbc.co.uk/programmes/b00qszch.

5 Computing Complexities

The guiding motto in the life of every natural philosopher should be, seek simplicity and distrust it.

—Alfred North Whitehead (1861–1947)

I do not like paradoxes, and I consider those who like them to be lacking in culture and intelligence.

—Rashid al-Daif[1]

A fool can throw a stone into the sea and ten wise men will not be able to retrieve it.

—Yiddish proverb

Computers are paragons of reason. They are heartless machines that relentlessly obey the laws of logic. There is nothing emotional or wishy-washy about computers: they do exactly as they are told by following the dictates of logic. It is from the perspective of computers as engines of reason that we examine these machines and determine what tasks they can and cannot perform. We all have an intuition as to what computers can do and what they can do with ease. They have no problem summing a list of numbers, nor do they have a problem sorting through myriad records or other simple tasks. However, it is surprising to learn that there are some easily stated problems that computers cannot satisfactorily solve. In this chapter I look at several problems that computers can theoretically solve, but that in practice would require an immense amount of time and resources. I also explain why these problems seem so hard and why researchers believe that they will never be solved easily.

In section 5.1 I present several examples of relatively easy problems. In encountering these easy problems, you'll become familiar with the language and notation used in the next two chapters. Section 5.2 covers five examples of hard problems. I demonstrate that although the problems are easy to state, they are not easy to solve. These problems will be shown to be related to each other in section 5.3. You'll also see why they seem to have no simple solutions. In section 5.4 I indicate how to cope with such hard problems. They cannot always be solved, but sometimes we can approximate their solutions. I close with section 5.5, in which I discuss even harder problems. In the next chapter, my focus will be on certain easily stated computer problems that simply cannot be solved, regardless of how much time or how many resources are used.

5.1 Some Easy Problems

Anyone who has worked with computers over the past ten years knows that they are getting faster and faster. What once took hours can now be done in seconds. What once took seconds is now done almost instantaneously. We have grown blasé about the processing speeds of personal computers. There are those who are not very tidy with their e-mail inboxes and have 10,000 messages awaiting deletion. By simply clicking a button, a modern-day pack rat can sort those 10,000 messages in seconds. This occurs so fast that we are not even aware that the computer is working. In this chapter, we will look at how much work is needed to sort and to perform other simple tasks. We will also see that some problems demand so much work that they simply cannot be performed in a feasible amount of time. First let us look more closely at some easy problems.

Addition and Multiplication

By the fourth grade, children learn the basic rules of how to add and multiply numbers that are several digits long. Students are taught the standard procedure that they must follow in order to obtain the correct answer. The term for a standard procedure or sequence of instructions is *algorithm*. The word is derived from the name Muhammad ibn Mūsā al-Khwārizmī (780–850), who wrote one of the first books that taught the Western world how to do algebra. We will describe and analyze various algorithms that solve different problems.

Let us start with something simple. To add two 7-digit numbers, one must add seven pairs of single digits (and keep track of numbers to carry) as in figure 5.1.

	6	7	3	9	2	7	5
+	7	6	1	0	6	7	8
1	4	3	4	9	9	5	3

Figure 5.1
Adding two 7-digit numbers

							6	7	3	9	2	7	5
						×	7	6	1	0	6	7	8
						5	3	9	1	4	2	0	0
					4	7	1	7	4	9	2	5	X
				4	0	4	3	5	6	5	0	X	X
				0	0	0	0	0	0	0	X	X	X
			6	7	3	9	2	7	5	X	X	X	X
	4	0	4	3	5	6	5	0	X	X	X	X	X
4	7	1	7	4	9	2	5	X	X	X	X	X	X
5	1	2	9	0	4	5	1	9	7	8	4	5	0

Figure 5.2
Multiplying two 7-digit numbers

Whenever we meet an algorithm, we must ask how many basic operations the algorithm requires. The number of operations needed depends on the size of the problem. For the two 7-digit numbers, seven operations are required. For two 42-digit numbers, 42 operations are required. In general, to add two n-digit numbers, one must do n additions of single digits.

In contrast to addition, consider multiplication. To multiply two 7-digit numbers, one must multiply every digit of the second number by every digit of the first number. Once this is done and everything is put in its correct column, the numbers must be summed as in figure 5.2.

Multiplication requires many more operations than addition. For each of the seven digits in the second number, we must perform seven multiplications, hence a total of $7 \times 7 = 7^2 = 49$ multiplications need to be done (ignore the number of additions that need to be done). In general, for two n-digit numbers, we need to multiply every one of the second number's n digits by every one of the first number's n digits. This gives us a total of n^2 multiplications. Usually n^2 is larger than n but—as we will see—it is still not very large.

How much time does it take to perform these tasks? The answer depends on the speed of your computer. The faster the computer, the less time will be needed. However, we will see that the speed of the computer is really not that important. Let's do some calculations. Say we want to multiply two 100-digit numbers. That will demand $100^2 = 10{,}000$ operations. Whether our computer can perform 100,000 or 1,000,000 operations per second, our task will be completed in less than half a second. It will become apparent that rather than looking at the speed of a computer, the important thing to look at is how many basic operations are needed.

We will see later that although multiplication is relatively easy, the opposite of multiplying two numbers—namely, factoring a single number into two numbers—is much harder.

Searching

Imagine that you have a cabinet full of files where each file contains information about a musical band. Assume that these files are not in any kind of order. Now, let us suppose that you are interested in finding the file for a particular band. This is similar to looking for a needle in a haystack. The only way to find the desired file is to examine them all, one at a time. Such a method of searching is called a *brute-force search algorithm*. If there are n files in the cabinet, how many files will you have to search? You might find the desired file after a few tries. However, in the worst-case scenario, you have to look through all n files to find the one you want, or to find that the desired file is not in the cabinet at all. So, a brute-force search of an unordered collection of n elements might demand n operations.

Consider a nicely alphabetized cabinet containing files relating to various musical bands. You might perform the same brute-force search on this ordered cabinet as you did on the unordered cabinet. If you are looking for the file on the group ABBA, then a brute-force search would work out well. However, if you are looking for the file on ZZTop, you would be wasting a lot of time. We are interested in an algorithm that performs the least amount of work in all cases. Let us try another algorithm called the *binary search algorithm*, which involves splitting groups of files into two. When searching for a file, rather than starting at the first file and then going forward, try the middle file and see if the desired one is before or after the middle file. Imagine that we are looking for Pink Floyd. We will look at the middle file, say Madonna, and since Pink Floyd is alphabetically after Madonna, we will ignore all the files from the beginning through the middle, and concentrate on the part of the cabinet that starts with Mariah Carey and ends with ZZTop. We might describe this search with the first column of table 5.1. We place a * next to the splitting file—that is, the file being compared.

Table 5.1
A binary search through an ordered list

First search	Second search	Third search	Fourth search
ABBA			
AC/DC			
Beatles			
⋮			
Led Zeppelin			
Madonna*			
Mariah Carey	Mariah Carey	Mariah Carey	
Metallica	Metallica	Metallica	
Neil Diamond	Neil Diamond	Neil Diamond*	
Paul McCartney	Paul McCartney	Paul McCartney	Paul McCartney
Pink Floyd	Pink Floyd	Pink Floyd	Pink Floyd*
Prince	Prince*		
Rod Stewart	Rod Stewart		
Tom Jones	Tom Jones		
U2	U2		
Van Morrison	Van Morrison		
ZZTop	ZZTop		

Our next guess will be the item that is halfway through this portion of the cabinet: this will be Prince. Since Pink Floyd comes before Prince, we will ignore all the files from Rod Stewart to ZZtop and concentrate on the files from Mariah Carey until Prince. We continue this method of splitting what remains until we find the file of our beloved Pink Floyd on the fourth guess.

How many comparisons do we have to perform? If we start with n files, after one check we will discard approximately $n/2$ of the files and focus on the other $n/2$ files. After another check, we will be concerned with only $n/4$ of the files. We continue in this manner until only one file remains and we see if it is the desired file, or we see if the file we want is not in our cabinet. For a value n, the number of times that n can be repeatedly split in half is described by the logarithm function that we write as $\log_2 n$. So the binary search algorithm works with $\log_2 n$ checks.

Let's look at a few examples of logarithms. For $n = 256$, we can make the following splits:

128, 64, 32, 16, 8, 4, 2, 1.

Since there are eight splits, we say that $\log_2 256 = 8$. For $n = 1{,}024$, we have

512, 256, 128, 64, 32, 16, 8, 4, 2, 1

and hence $\log_2 1{,}024 = 10$. And finally, for $n = 65{,}536$ we have

32768, 16384, 8192, 4096, 2048, 1024, 512, 256, 128, 64, 32, 16, 8, 4, 2, 1.

Since there are sixteen splits, we say that $\log_2 65{,}536 = 16$.

For $n = 1{,}000$, the brute-force search will perform, in the worst case, 1,000 checks while the binary search will do $\log_2 1{,}000$ checks. That is, about ten checks. This is a vast improvement.

For a particular problem, there might be several different algorithms that can be used to provide solutions. Sometimes certain algorithms should be used for certain types of inputs while other algorithms should be used for other types of inputs. We have seen that there is a difference if the cabinet is ordered or unordered. A brute-force search algorithm should be used for an unordered cabinet, while a binary search algorithm should be used on an ordered cabinet. For each problem, we will analyze different algorithms and try to find the one that is most efficient. The problem will be characterized by the most efficient algorithm that exists to solve it in the worst-case scenario. So, we say that searching an unordered list demands n operations and searching an ordered list demands $\log_2 n$ operations.

Sorting

In the previous example we saw that it pays to keep things tidy and sorted. How does one go about sorting a list of n elements? There are many different algorithms to perform the task at hand, but I present one of the simplest ones. The *selection-sort algorithm* works as follows:

1. Search through the elements and find the minimum element.
2. Swap that minimum element with the first element in the list.
3. Search the rest of the list to find its minimum.
4. Swap that element with the second element of the list.
5. Continue in this manner until all the elements are in their correct position.

We might envision this sort with the sequence of lists of three-letter words shown in figure 5.3.

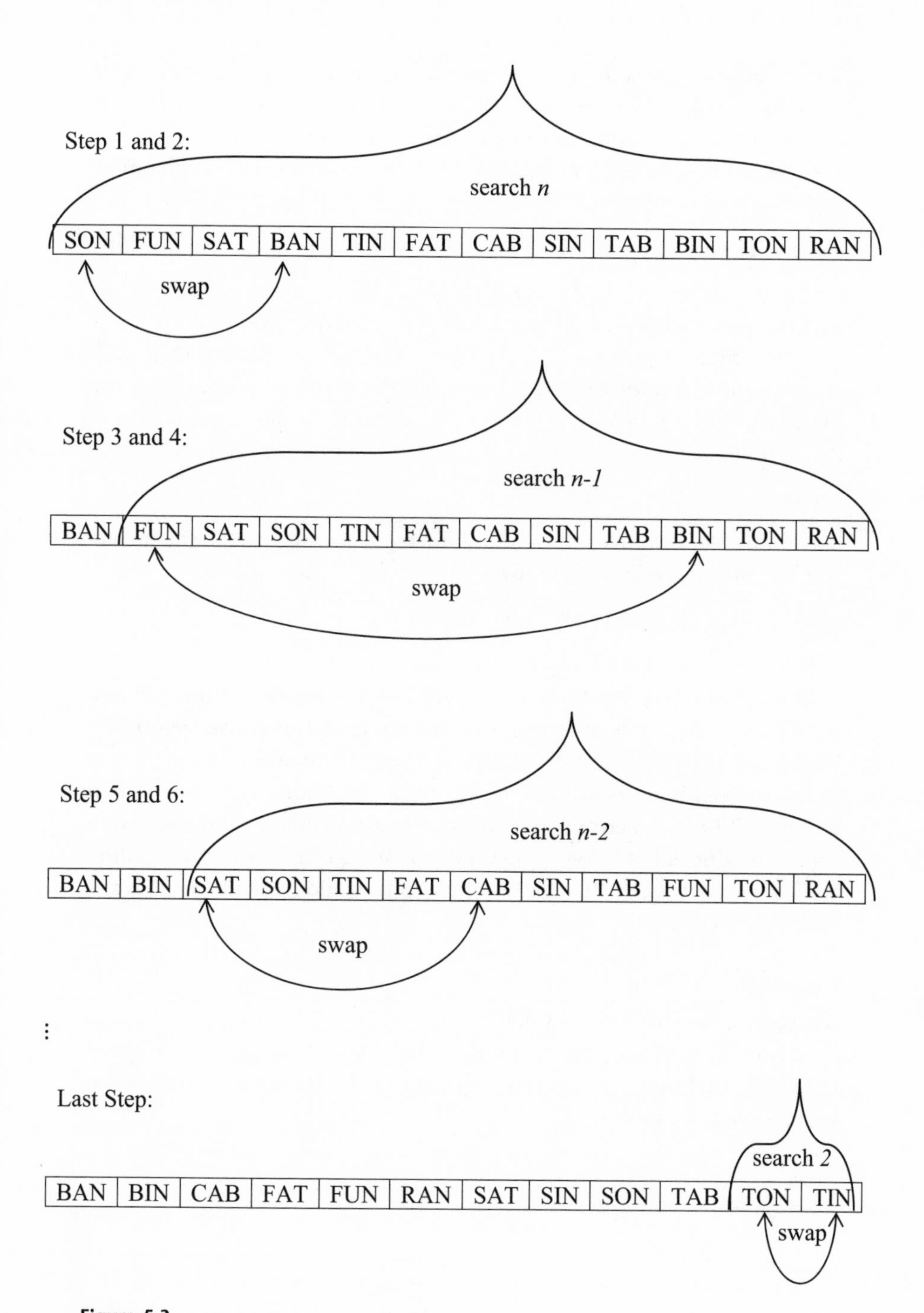

Figure 5.3
The steps of the selection sort

How much work was done? To obtain the minimum of all the elements we need to look at all n elements. To find the minimum of all the elements except the first one we need to look at n - 1 elements. Continuing in this way, we see that to perform the entire selection-sort algorithm, one must look at

$$n + (n - 1) + (n - 2) + \cdots + 2$$

elements. It turns out that this sum is about half of n^2 and so we will say that the selection-sort algorithm requires at most n^2 operations.

Other, more sophisticated sorts demand fewer operations. One such algorithm is called the *merge-sort algorithm*. We need not get into how the merge-sort algorithm works. Suffice it to say that for an input of size n, this sort algorithm demands

$$n \times \log_2 n$$

operations. For example, given an unordered list of size 1,000, the merge-sort algorithm would sort the list with

$$1{,}000 \times (\log_2 1{,}000) = 1{,}000 \times 10 = 10{,}000$$

operations.

You might think that there is not much of a difference between n^2 and $n \times \log_2 n$. Perhaps it is not worth the effort to have this fancier algorithm. Well, it is! Consider the task of sorting the four million numbers of the Brooklyn telephone book.[2] An n^2 algorithm that sorts the four million entries will have to perform 16,000,000,000,000 operations. In contrast, if the more efficient $n \times \log_2 n$ algorithm is used to sort the four million entries, approximately 88,000,000 operations would be required. This is a vast improvement.

Euler Cycle

Königsberg is a city in Russia that used to be part of Prussia. The city is on both sides of the Pregel River, and Kneiphof Island is in the middle of the river.[3] The main parts of the city and the island were connected by seven bridges, as in figure 5.4.

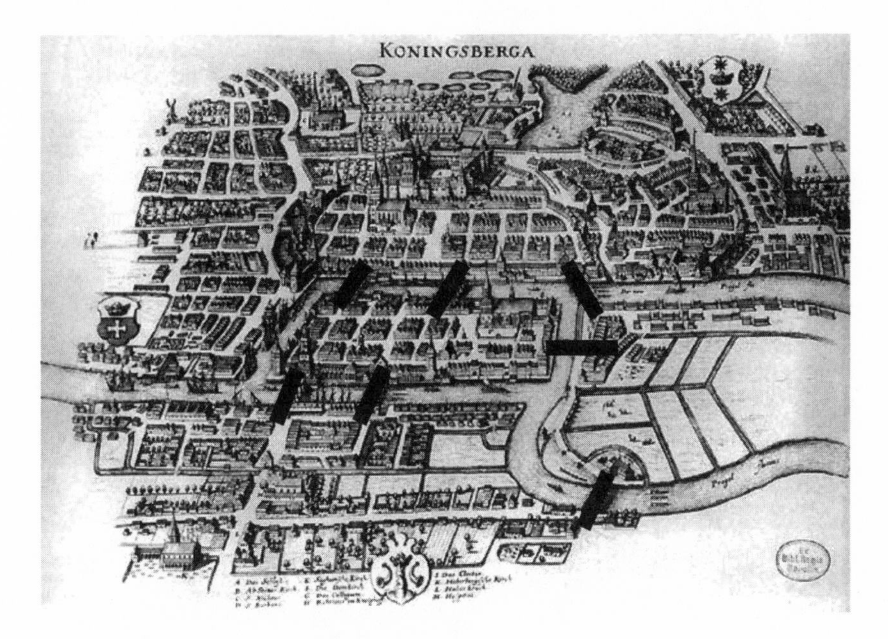

Figure 5.4
The seven bridges of Königsberg

Residents used to wonder if it was possible to take a leisurely stroll through the city in such a way that one starts and finishes in the same place and crosses each of the seven bridges exactly once. Try to find such a path.

In 1736, the question was posed to Leonhard Euler (1707–1783), one of the greatest mathematicians of all time. He realized that to find such a path through the seven bridges we do not care if the pedestrian stops for ice cream, wanders about, has a barbeque, or completes the trip in three days. What does matter is the way the land and the bridges are connected. He noticed that you might as well think of each landmass as a single point. It also does not matter how wide the bridges are or how old they are or how far away they are from each other. The only relevant information is which land masses the bridges connect. One might as well think of each bridge as a single line connecting the points. We can envision these points and edges superimposed on the map of Königsberg, as in figure 5.5.

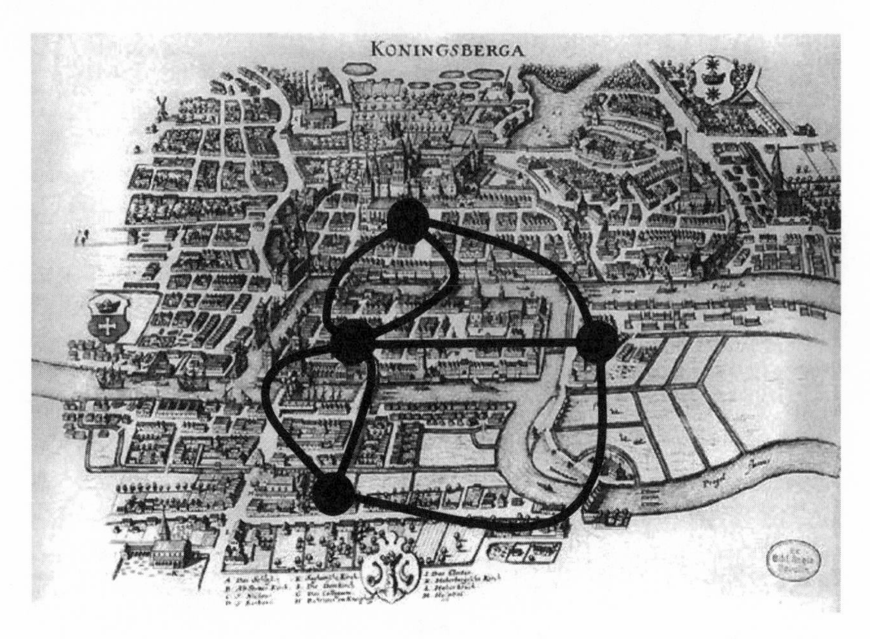

Figure 5.5
Königsberg and a graph that represents its bridges

With this insight, Euler essentially started the field of mathematics known as *graph theory*. A graph is a collection of vertices (or points) and edges (or lines) between the vertices.

A graph describes a relationship between things. Since "things" is general, graph theory has many diverse applications and has become one of the most important branches of mathematics. For example, graph theory can describe computer networks: the vertices would correspond to computers and an edge between two vertices would correspond to the connection between them. Another example is the World Wide Web: the vertices of a graph can represent web pages and an edge represents a link from one web page to another. A subway map is another common example of a graph.

Here is an interesting application of graph theory. Consider the graph whose vertices correspond to every person on earth today. Let there be an edge between any two people who are acquainted with each other. Researchers believe that in this graph, the maximum number of edges that one needs to traverse in order to connect any two vertices is six. That means that given any two people, there is a sequence of the first person who knows someone who knows someone who . . . who knows the second person. The sequence, in general, does not need to be longer than six. This is known as the "six degrees of separation."

Let us return to the Königsberg Bridge Problem. Once Euler ignored the unimportant part of the problem, he was easily able to see that no such path was possible. Euler reasoned that every time strollers enter a landmass or a vertex, they must also be able to leave the vertex. Of course, the strollers might come to that vertex again, but then they must leave again. Euler realized that a requirement for there to be a path is that every vertex must have an even number of edges touching it. If this requirement is met, then there is a path that starts and finishes at the same place and passes every edge exactly once. If this requirement is not met, then no such path exists. Since every vertex in figure 5.5 has an odd number of edges/bridges touching it, no such cycle is possible.

One need not be concerned with strollers in old Prussian cities to be interested in this problem. Given any graph, we call a path that starts and finishes at the same vertex a cycle. A cycle that passes every edge exactly once is called an Euler cycle. So for any given graph, we may ask if there exists an Euler cycle. This is the *Euler Cycle Problem.* We saw that the graph in figure 5.6 does not contain such a cycle.

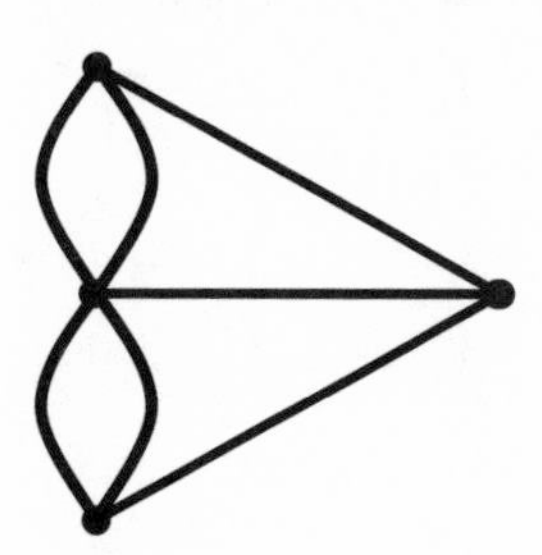

Figure 5.6
The graph of the Königsberg Bridge Problem

Notice every vertex has an odd number of edges touching it. In contrast, consider the graph in figure 5.7.

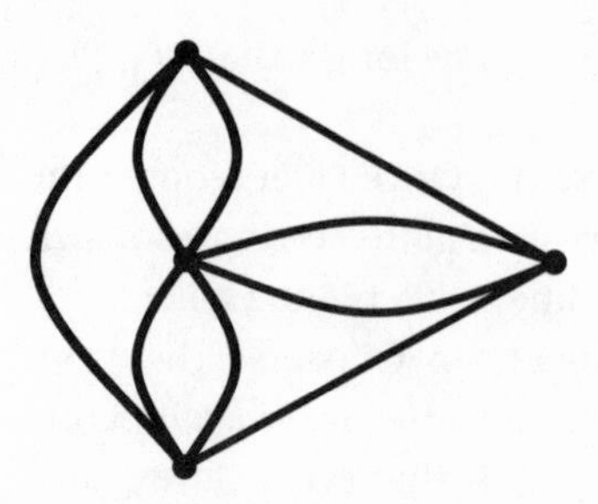

Figure 5.7
A modified Königsberg Bridge Problem

Every vertex touches an even number of edges and hence there does exist an Euler cycle. (Try to find one!)

We may drop the requirement that the paths start and end at the same vertex and pose the *Euler Path Problem*: Given a graph, does a path exist that crosses every edge exactly once but need not start and finish at the same vertex? With an understanding of Euler's requirement for an Euler cycle, the answer to the Euler path problem is obvious: there is such an Euler path *if at most two* vertices have an odd number of edges touching them and the rest of the vertices have an even number of edges touching them. The two vertices will be the starting and ending points of the desired path. Consider the graph in figure 5.8.

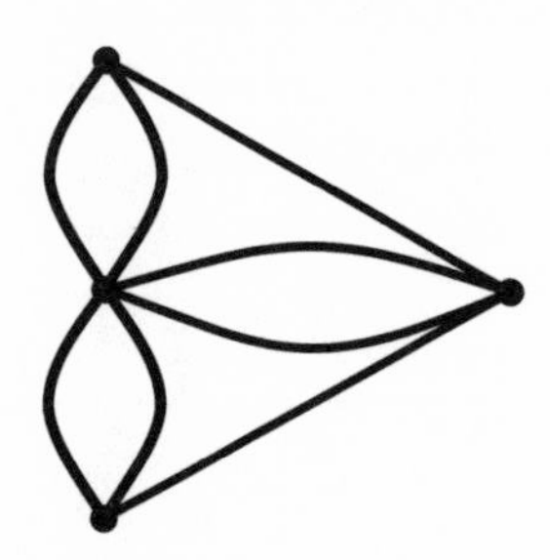

Figure 5.8
A graph that has an Euler path but no cycle

There is an Euler path from the top vertex to the bottom and vice versa, but no Euler cycle.

What does this have to do with computers and algorithms? Without the information presented in the previous few paragraphs, we might have thought that to determine whether a graph has an Euler cycle we would have to try all possible cycles and see if any cycle crosses every edge exactly once. For a large graph there is a tremendous number of cycles to check. Now, with Euler's trick in hand, we have a new method of determining if there is an Euler cycle. This trick has us simply verifying that every vertex has an even number of edges touching it, which can be done with relatively few operations. Euler's method has saved us a lot of work. We will always be looking for such tricks.

All problems presented in this section can be solved with algorithms that demand a polynomial number of operations such as n, n^2, or $n \times \log_2 n$ operations. Such problems are called *polynomial problems*. The collection of all polynomial problems is denoted as **P**. Most polynomial problems can be solved on a modern computer in a feasible or tractable number of operations. Such problems take a feasible amount of time to solve.

In contrast to the feasible problems discussed here, in the next section, we will look at infeasible or intractable problems—that is, problems that are unsolvable in a reasonable amount of time.

5.2 Some Hard Problems

To get a feel for how hard some problems are, let us introduce the following five problems, which are easy to describe.

The Traveling Salesman Problem

Consider a salesman who wants to drive to six specific cities across the United States. He would like to visit each city exactly once and when he is done with all of them, return to the starting city. There are many different routes that he can take. Our parsimonious traveler would like to travel the shortest route in order to save time and money. He looks up the distances between these cities and finds the information in table 5.2.

Table 5.2
Distances in miles between some major U.S. cities

	San Francisco	New York	Miami	Los Angeles	Denver
Chicago	2135	795	1380	2020	1000
Denver	1270	1780	2065	1025	
Los Angeles	385	2800	2740		
Miami	3115	1280			
New York	3055				

Since we have been learning about graphs, we might represent the same information with the graph in figure 5.9.

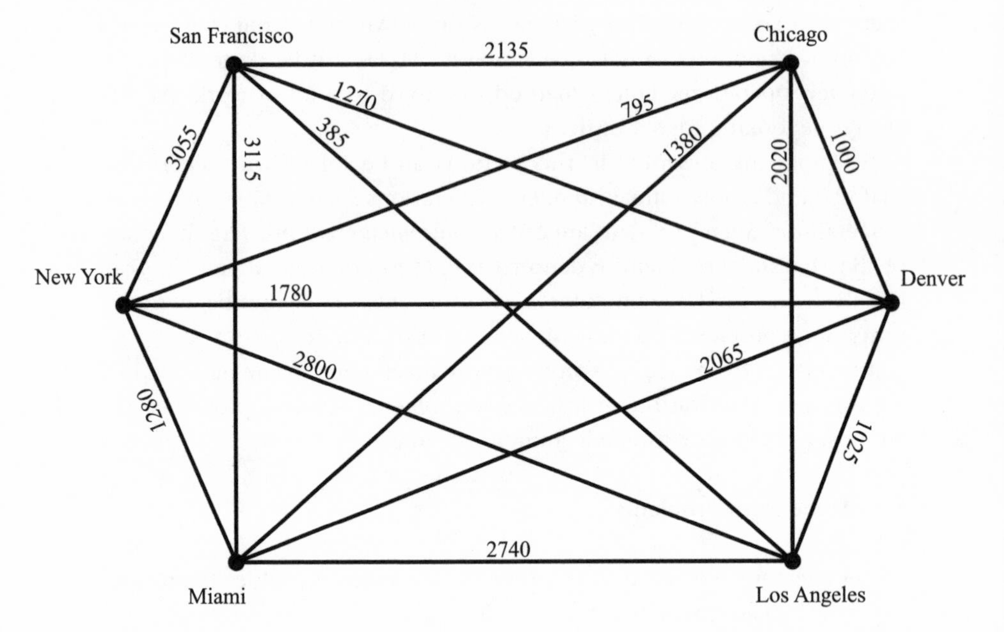

Figure 5.9
Complete weighted graph of cities

The number on each edge represents the distance between the cities. Such a graph is called a *weighted* graph because the edges are given weights. Since there is an edge between every two vertices, we also call this graph *complete*.

The question is which route through all the cities is the shortest? One method of going about this is to pick one possible route through all six cities, add up all the distances between the cities (don't forget to come back to the starting place), and see what that total is. Once that is done, try another possible route, sum up all of its distances, and see if it is shorter. To ensure that you have the shortest route, one would have to make a brute-force search through *all* possible routes.

How many possible routes are there to check? We have to search through all the possible orders or permutations of six cities. Let us count how many permutations there are. There are six possible cities where you can start. Once you are in the first city, since you do not want to return to that city, there are five possibilities for a second city. There are four possible cities for the next destination on your trip. Continuing this way, we find that the number of possible routes that you must search through is

$$6 \times 5 \times 4 \times 3 \times 2 \times 1 = 720.$$

That's a lot of work,[4] but with modern computers, going through all 720 possible routes can be done in microseconds.

Let us examine this problem from a somewhat more general perspective. Imagine that our traveler wanted to visit n cities. We can visualize this problem as a graph with n vertices. Between any two vertices there will be a weighted edge representing the distance between the two cities. So we are given a complete, weighted graph with n vertices and we are interested in the shortest route that hits every city (vertex) exactly once and returns to the starting city. Using the same reasoning as above: there are n possible first cities, $n - 1$ possible second cites, ... In conclusion, the number of possible routes that must be checked is

$$n \times (n-1) \times (n-2) \times \cdots \times 2 \times 1$$

We write this number as $n!$ and call this function "n factorial." The amazing fact about this function is that as n gets larger, $n!$ grows unbelievably larger.

Let us roll up our sleeves and do a little calculating. For $n = 100$, how many routes are there and how long would it take for a reasonable computer to search through all of those possible routes? We would need to calculate 100!. We can do this by sitting down and multiplying 100 times 99 times 98 times . . . times 3 times 2 times 1. Or, we can just cheat and type it into the fancy scientific calculator that comes with every computer. We obtain the following result:

100! = 93,326,215,443,944,152,681,699,238,856,266,700,490,715,968,
264,381,621,468,592,963,895,217,599,993,229,915,608,941,463,976,156,
518,286,253,697,920,827,223,758,251,185,210,916,864,000,000,000,000,
000,000,000,000.

Before your head starts spinning, let's look at this number for a minute. It consists of a 9 followed by 157 digits. A very large number called a *googol* is a 1 followed by 100 zeros.[5] Our number is much larger than a googol. It is an unimaginably large number.

For each of these potential routes, we would have to add up the distances between the 100 cities and compare the sum to the shortest distance that was already found. Imagine that we have a computer that can check a million routes in a second. Then dividing 100! by a million gives us

93,326,215,443,944,152,681,699,238,856,266,700,490,715,968,264,381, 621,468,592,963,895,217,599,993,229,915,608,941,463,976,156,518,286, 253,697,920,827,223,758,251,185,210,916,864,000,000,000,000,000,000 seconds.

To find out how many minutes this would take, we must further divide by 60 to get

1,555,436,924,065,735,878,028,320,647,604,445,008,178,599,471,073, 027,024,476,549,398,253,626,666,553,831,926,815,691,066,269,275,304, 770,894,965,347,120,395,970,853,086,848,614,400,000,000,000,000,000 minutes.

Dividing again by 60 will give us

25,923,948,734,428,931,300,472,010,793,407,416,802,976,657,851,217, 117,074,609,156,637,560,444,442,563,865,446,928,184,437,821,255, 079,514,916,089,118,673,266,180,884,780,810,240,000,000,000,000,000 hours.

A simple division by 24 gives us

1,080,164,530,601,205,470,853,000,449,725,309,033,457,360,743,800, 713,211,442,048,193,231,685,185,106,827,726,955,341,018,242,552, 294,979,788,170,379,944,719,424,203,532,533,760,000,000,000,000, 000 days.

Shall we continue? Dividing by 365 gives us

2,959,354,878,359,467,043,432,877,944,452,901,461,527,015,736,440, 310,168,334,378,611,593,658,041,388,569,114,946,139,776,006,992,588, 985,721,014,739,574,573,764,941,185,024,000,000,000,000,000 years.

Dividing by 100, we get

29,593,548,783,594,670,434,328,779,444,529,014,615,270,157,364,403, 101,683,343,786,115,936,580,413,885,691,149,461,397,760,069,925, 889,857,210,147,395,745,737,649,411,850,240,000,000,000,000 centuries.

That is 2.9×10^{142} centuries. That is a very long time. That is a *shockingly* long time!

A moment of meditation is in order. The Traveling Salesman Problem is a seemingly simple problem that can be explained to any elementary school student and for small inputs can be performed by everyone in seconds. But when the inputs get larger, the number of operations required is immense. Yes, a solution to the problem is possible, but there is no way we can say that a computer can "solve" this problem in a reasonable amount of time.

A problem in which we are given an input and asked to find the shortest or longest or best solution, is called an *optimization problem*. Many problems that computers solve are optimization problems. The Traveling Salesman Problem, like all the other problems that I will mention in the rest of this chapter, also comes in another form called a *decision problem*. A decision problem is one in which the computer is only required to give a yes or no answer. Such problems are not concerned with the shortest, longest, or best solutions. Rather, they are only concerned if a solution of a certain type exists. For example, an optimization problem would be concerned with determining the speed of the fastest runner in the United States. The related decision problem might be, "Can any runner in the United States run a mile in under 3.5 minutes?" Such a question demands a yes or no answer.

The traveling salesman problem in its decision-problem form is as follows: given a complete weighted graph and an integer *K*, tell whether or not there is a path through every vertex whose sum total of all the distances of the trip is less than or equal to *K*. If there is a route that is less than or equal to *K*, answer yes. If not, answer no. Notice that we are not required to find the route. It is worth mentioning that a decision problem has the potential to end sooner than an optimization problem. In a decision problem, if we find what we are looking for, we stop there and answer yes. By contrast, an optimization problem would have to examine all possible routes in its search for the best. Nevertheless, our major concern in this section and the next will be decision problems.

The traveling salesman problem must be seen as a limitation of the human ability to know. There is no way that one can possibly know the shortest route to take in a reasonably sized problem.

Hamiltonian Cycle Problem

This problem is also concerned with finding paths through a given graph. For a graph (we do not require that it be weighted or complete), we are to find a continuous path in the graph that hits every vertex exactly once and returns to the starting place. Such a cycle is called a *Hamiltonian cycle*. The decision version of the Hamiltonian Cycle Problem is to determine whether a Hamiltonian cycle exists for a given graph. There are many real-world applications of this problem. For example, a bus driver can be given a map of a city and told to pick up students at the given vertices. Can the driver make this trip without crossing any vertex twice? Consider the two graphs in figure 5.10.

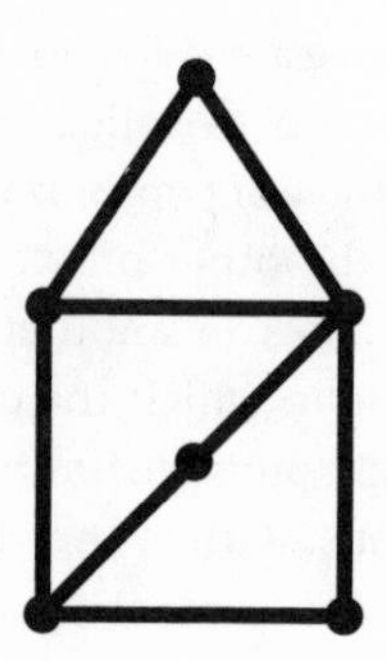

Figure 5.10
Instances of the Hamiltonian Cycle Problem

In the graph on the left, one can simply start at any vertex and go either clockwise or counterclockwise to obtain a Hamiltonian cycle. In contrast, try to find such a cycle in the graph on the right. (There is none.)

One way of solving this decision problem is to attempt a brute-force search through all possible permutations in a given graph. For every permutation, check to see whether it makes a connected path in the graph. For a graph of size *n,* there are *n*! possible permutations. As we saw with the traveling salesman problem, this is far from satisfactory. There is simply not enough time to solve this problem for any moderately large input. Can we do any better than an *n*! algorithm?

There is a certain similarity between the Hamiltonian cycle problem and the Euler Cycle Problem. In each case we are given a graph and are looking for a cycle. In the Hamiltonian Cycle Problem we are looking for a cycle that hits every vertex exactly once, whereas in the Euler Cycle Problem we are looking for one that hits every edge exactly once. In the last section, we saw that Euler taught us a cool trick to determine if an Euler cycle exists on a given graph: simply make sure the number of edges that touch every vertex is even. Is there a similar trick that tells whether a Hamiltonian cycle exists for a given graph? Unfortunately, your humble author does not know any such trick, and neither does anyone else. At this time, the only way to be absolutely certain of the existence of a Hamiltonian cycle in a given graph of *n* vertices is to perform a brute-force search through all *n*! possible cycles. So although the two problems are similar, one is easily solvable and the other seems to be very hard. Each problem presented to us must be examined carefully.

Just because we do not know of any better algorithm than a brute-force search does not mean that no such algorithm exists. Someone might come up with one at some point. However, in the next section, I show why most researchers believe that no such algorithm exists at all.

Set Partition Problem

Say we have a class of students whose ages are {18, 23, 27, 65, 22, 25, 19, 21}. We want to split them into two groups with equal years of "life experience." A little tinkering shows that the groups {18, 27, 65} and {23, 22, 25, 19, 21} will do; both groups have a sum of 110. The Set Partition Problem generalizes this example. Given a set of positive whole numbers, determine if they can be partitioned, or split, into two sets such that the sum of all numbers in one set is the same as the sum of all numbers in the other set.

Let us look at some other instances of this problem. How about the set {18, 23, 28, 65, 22, 25, 19, 21}? The numbers are almost exactly alike except that instead of a twenty-seven-year-old, we now have a twenty-eight-year-old. Some thought shows that this set cannot be split into two equal parts. In the last example, the sum of all the numbers was 220, and we were able to split them into two sets, each containing 110. In this example, the sum of all the numbers is 221. There is no way to divide the odd number 221 into two equal whole-number parts. So one way to determine that there is no solution is to add up all of the numbers and see if the result is odd. If odd, we are certain that no possible solution exists. But what if the sum is even?

Consider the set {30, 4, 32}. The sum of these three numbers is 66, which is even. However, it is obvious that there is no way to partition this set into two equal parts. So our trick of seeing if the sum of all the numbers is odd or even does not always work. The sum can be even but the numbers still cannot be partitioned into two equal parts.

How does one go about solving such a problem? We have to examine all possible splittings of the set. For each splitting, we need to sum one part and see if it is equal to half of the sum of all the elements. How many such splittings are there? For each element of a given set, we can either place the element in one set or in the other set. So we have two possibilities for the first element and two possibilities for the second element and . . . and two possibilities for *n*th element. This comes to

$$\underbrace{2 \times 2 \times 2 \times \cdots \times 2}_{n} = 2^n$$

possibilities. We say that this problem demands an exponential amount of work.

Let's say we have a hundred numbers that we want to partition into two groups. How many possible splittings would we have to examine? A calculation shows that there are

2^{100} = 1,267,650,600,228,229,401,496,703,205,376

splittings. If we had a computer that could check a million splittings per second, we would still need

1,267,650,600,228,229,401,496,703,205,376 / 1,000,000 = 1,267,650,600, 228,229,401,496,703 seconds.

Divide that by 60 to get 21,127,510,003,803,823,358,278 minutes. Divide again by 60 for 352,125,166,730,063,722,637 hours. That is 14,671,881, 947,085,988,443 days or 40,196,936,841,331,475 years or 401,969,368,413, 314 centuries. This is not exactly something for which we can simply wait!

One might ponder different, faster methods for solving this problem. However, no known method will work in all instances and in a shorter time. The only known method that will always find the exact solution in all cases is a brute-force search through all 2^n subsets.

Subset Sum Problem

Similar to the Set Partition Problem is the Subset Sum Problem. Imagine that you are packing your car for a long weekend in the country. There are many things that you would like to take but, alas, your car has finite capacity. You would like to fill your car to its exact capacity. This common situation is the informal version of the Subset Sum Problem: for a given set of whole numbers and a whole number C (for capacity), determine if there exists a subset of the set whose sum is exactly equal to C. For example, consider the set {34, 62, 85, 35, 18, 17, 52} and the capacity $C = 115$. If you play with the set for a long enough time, you will realize that the sum of the elements in the subset {62, 35, 18} equals 115. What would happen if we changed C to 114 or if we changed the given whole numbers?

In general, given a set of whole numbers and a whole number C, how should a computer go about solving this decision problem? As in the Set Partition Problem, we can sift through all subsets and for each one, sum up the elements to see if they sum to C. We have seen that for an n-element set, there are 2^n subsets. This will again lead to the same large number of operations and infeasible time situation as the Set Partition Problem for inputs of larger sizes.

Satisfiability Problem

This problem deals with simple logical statements. We assume some basic knowledge of elementary logic and the connectives $\wedge$ (and), $\vee$ (or), $\sim$ (not), and $\rightarrow$ (implies). Consider the logical statement

$$(p \vee \sim q) \rightarrow \sim (p \wedge q).$$

This statement is either true or false depending on which values are assigned to variables p and q. If, for example, we assign p = TRUE and q = FALSE, then the entire statement is true. If, however, we set p = TRUE and q = TRUE, then the entire statement is false. If a logical statement can be made true, then it is termed *satisfiable.* For a given statement in logic, the Satisfiability Problem asks whether there is a way to assign TRUE and FALSE to the variables such that the entire statement is true (i.e., satisfiable). We saw that the above logical statement is satisfiable. Consider the statement

$$a \wedge (a \rightarrow b) \wedge (b \rightarrow c) \wedge (\sim c).$$

For this statement to be true, we must have that a = TRUE and $(a \rightarrow b)$ = TRUE. By Modus Ponens, we have that b must be TRUE. Continuing with $(b \rightarrow c)$, we see that c needs to be TRUE. But c = FALSE because $\sim c$ must be TRUE. In conclusion, there is no way to make this statement logically true.

How does one go about solving the satisfiability problem? Most high school students know that to determine the value of a logical statement, one needs to construct a truth table. The decision version of this problem asks whether there is a row in the truth table of a logical statement that has the value TRUE.

One algorithm for solving this problem is to construct a truth table for a given statement and on inspection of all the rows, return a Yes or No depending on whether any row has the value TRUE. However, constructing such truth tables will require a lot of work. If there are two variables in the given logical statement, there will be four rows in the truth table. If there are three variables, there will be eight rows. In general, for a statement with n different variables, the truth table will have 2^n rows. This exponential growth in the amount of work done is a sign that this problem is as impractical as the other four discussed in this section.

Now that we've seen a few examples, let's come up with a nice definition. We will denote by **NP** the class of all decision problems that require 2^n, $n!$, or fewer operations to solve.[6] A problem in **NP** will be called an "**NP** problem." Since **P** is the class of problems that can be solved in a polynomial amount of operations, and polynomials grow more slowly than exponential or factorial functions, we have that **P** is a subset of **NP**. What this means is that every "easy" problem is an element of the class of all "hard and easy" problems.

One might have doubts that there really is a difference between polynomial functions and exponential or factorial functions. Table 5.3 should dispel any doubts.

Table 5.3
Several values for polynomial and nonpolynomial functions

n	$\log_2 n$	$n \log_2 n$	n^2	n^5	2^n	$n!$
1	0	0	1	1	2	1
2	1	2	4	32	4	2
5	2.32192	11.6096	25	3125	32	120
10	3.32192	33.2192	100	100000	1024	3628800
20	4.32192	86.4385	400	3200000	1048576	2.43×10^{18}
50	5.64385	282.192	2500	3.1×10^{8}	1.13×10^{15}	3.04×10^{64}
100	6.64385	664.385	10000	1×10^{10}	1.27×10^{30}	9.3×10^{157}
200	7.64385	1528.77	40000	3.2×10^{11}	1.61×10^{60}	7.8×10^{374}
500	8.96578	4482.89	250000	3.1×10^{13}	3.3×10^{150}	1.22×10^{1134}
1,000	9.96578	9965.78	1000000	1×10^{15}	1.1×10^{301}	4.02×10^{2567}
2,000	10.9657	21931.5	4000000	3.2×10^{16}	1.14×10^{602}	3.31×10^{5735}

The first few columns show typical polynomial functions and their values for various n's. The last two columns show the values for the exponential and factorial functions. Although, for small values of n, some polynomial functions are larger than the two right-hand columns, already for $n = 50$ the polynomial functions cannot keep up with the other columns. It is this tremendous growth that seems to mark a clear demarcation between polynomial and nonpolynomial functions.

Since NP problems were defined in the early 1970s, researchers have found literally thousands of such problems. NP problems occur in every aspect of business, industry, computer science, mathematics, physics, and so on. They are about scheduling, routing, graph theory, combinatorics, and many other topics. Lately, many problems in biology related to gene sequencing have been shown to be NP problems. As biologists have been learning to read people's DNA and use this information to customize medicines, they have come to realize that some tasks seem to demand too many operations and too much time.

Not all limitations are bad. We saw in section 5.1 that multiplying two numbers is a polynomial problem. By contrast, factoring a number is rather difficult. Consider multiplying 4,871 by 7,237. Any intelligent fourth grader can, in short order, determine the result of 35,251,427. Now consider the number 38,187,637. This number is a product of two prime numbers (that is, numbers that cannot be further divided). Can you find those prime factors? It would take a human being or a computer a long time to realize that the factors of this number are 7,193 and 5,309. The reason for this is that although multiplication is an n^2 problem, factoring is an NP problem. This asymmetry is used to send secret messages over the Internet. We will not get into the gory details of how this is accomplished, but there is an algorithm that uses large numbers to transmit messages in secret. The algorithm was described in 1978 by Ron Rivest, Adi Shamir, and Leonard Adleman and is called RSA. If eavesdroppers were listening in on a conversation, they would not be able to understand the message (or obtain the credit card number given in the conversation) without factoring a very large number. Since this seems to be a very hard problem, our security is safe. If the brilliant Lex Luthor ever found an algorithm that could easily factor numbers, RSA would be "broken" and much of the World Wide Web's security would be useless.

We have seen that all five of the problems highlighted in this section are essentially unsolvable for any large-sized inputs—that is, they cannot be solved in a reasonable amount of time with our present-day computers. One might try to ignore these limitations by expecting that as computers become faster and faster, NP problems will become easier and easier to solve. Alas, such hopes are futile. As we saw before, the Traveling Salesman Problem for a relatively small input size of $n = 100$ demands 2.9×10^{142} centuries. Even if our future computers are 10,000 times faster than they are now, our problem will still demand 2.9×10^{138} centuries. This remains an unacceptable amount of time. Conclusion: faster computers will not help us.

Similarly, as our technology and ability to multiprocess improve—that is, as our ability to use many processors working in parallel improves—we might shrug off the entire concept of an NP problem. One might think that such problems are hard for single processors performing one step at a time, but with many processors, the task might be accomplished in a reasonable period. This hope is also in vain. Even with 10,000 processors working in conjunction with each other, the time needed would not improve beyond 2.9×10^{138} centuries, which is still an unacceptable time requirement. Let us go further. Scientists estimate that there are 10^{80} atoms in the universe. Imagine that every single atom in the universe was a computer working on different parts of this problem. Splitting up the problem like this means that completing the entire problem will demand that all these computers work a measly 10^{62} centuries. And this is only for an input of size 100. What about larger inputs? The point is that such problems are not solvable in any machine within any reasonable amount of time.

You may have noticed a media buzz on the subject of quantum computers. These are, at present, hypothetical devices that use some of the strange and counterintuitive aspects of quantum mechanics[7] to improve our computational powers. One of the strangest notions of quantum mechanics is that subatomic objects can be in more than one place at the same time. While regular-sized objects are either in this or that position, subatomic objects have superposition—that is, they can be in many positions simultaneously. Regardless of any doubt you might have about such notions, superposition is an established fact of our universe. Quantum computer scientists try to use superposition to make better computers. A regular computer searches through a large number of possible solutions one at a time. In contrast, a quantum computer would put itself into a superposition of searching states and examining many possible solutions at one time. One might believe that the development of quantum computers will make NP problems easier to solve. Again, such hopes are dashed. It can be shown that when searching for a particular element in a list of size n, a quantum computer cannot do any better than $\sqrt{n}$ steps. So when searching through all 2^n possible solutions to an NP problem, a quantum computer will be able to do this task in $\sqrt{2^n}$ steps. However,

$$\sqrt{2^n} = 2^{\frac{n}{2}}$$

is still exponential. So, quantum computers are not the hoped-for solution to the NP problems.

In summary, all foreseeable future improvements to computer technology are essentially impotent in the face of these NP problems. The only way that these problems will be easily solved is to find nice polynomial algorithms for them. We will show in the next section why most researchers believe that there are no better algorithms for these problems. It looks as though they will remain problems that cannot be solved in a reasonable amount of time. These problems are not hard because we lack the technology to solve them. Rather they are hard because of the nature of the problems themselves. They are *inherently* hard and will probably remain on the outer limits of what we can solve.

5.3 They're All Connected

In the last section, we encountered several problems whose only known algorithms are brute-force searches, which take way too much time. At this point , no one has a polynomial algorithm to solve any of them. That is, no one has any easy trick to find a solution in a shorter amount of time. In this section, we will learn why most researchers believe that no such polynomial algorithms exist.

If you spend a little time working on the Set Partition Problem and the Subset Sum Problem, you will find that they are very similar and are, in fact, related. We can easily transform an instance of the Set Partition Problem into an instance of the Subset Sum Problem. The idea is that in the Set Partition Problem, you are really looking for a single subset of the elements that sum to half of the sum of all the elements in the set. For example, consider the set {12, 63, 13, 82, 42, 54, 24, 76, 22}. One way to determine if this set can be partitioned into two sets with equal sums of elements is to see if there is a subset that sums to half of the sum of all the elements. In this case, we would have to determine if there is a subset whose elements sum to

$$C = (12 + 63 + 13 + 82 + 42 + 54 + 24 + 76 + 22) / 2 = 388 / 2 = 194.$$

If such a subset exists, then the set can be partitioned into those elements in the subset and those elements not in the subset. If there is no such subset that sums to 194, then this instance of the Set Partition Problem yields a negative answer.

What we are saying here is that if we can solve the Subset Sum Problem, then we can most definitely solve the Set Partition Problem. In our notation:

solve Subset Sum Problem $\Rightarrow$ solve Set Partition Problem.

If you have an instance of the Set Partition Problem, just transform the problem into a Subset Sum Problem as we did above by setting C to be half of the sum of all elements in the set. To say that if you can solve one problem, then you can definitely solve a second problem means that the first problem is as hard or harder than the second. So we showed that the Subset Sum Problem is as hard or harder than the Set Partition Problem. In other words, the Set Partition Problem is as hard as or easier than the Subset Sum Problem. We write this as follows:

Set Partition Problem $\leq_P$ Subset Sum Problem.

Let us generalize what we just did for two arbitrary decision problems. Imagine that we have a Problem B that a computer can solve. This means we have a machine into which we input an instance of the problem and it will output either a Yes or No depending on the solution. We might represent this with figure 5.11. The input enters at the left and the machine calculates and outputs either a Yes or No.

Now imagine that we have a Problem A. If there is a way of transforming an instance of Problem A into an instance of Problem B, we can create a machine that decides Problem A by connecting the transformer to the Problem B decider as in figure 5.12.

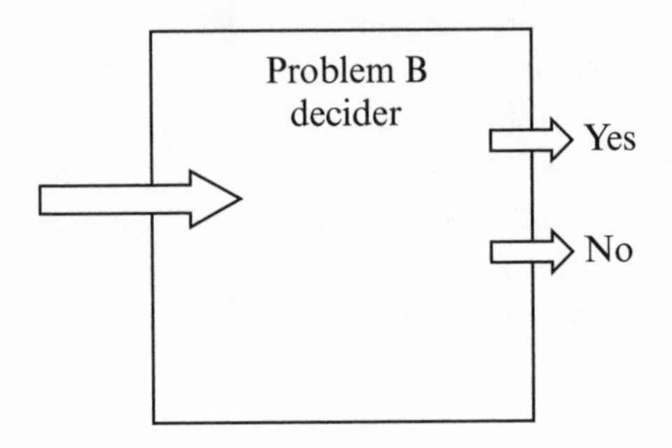

Figure 5.11
Machine to decide Problem B

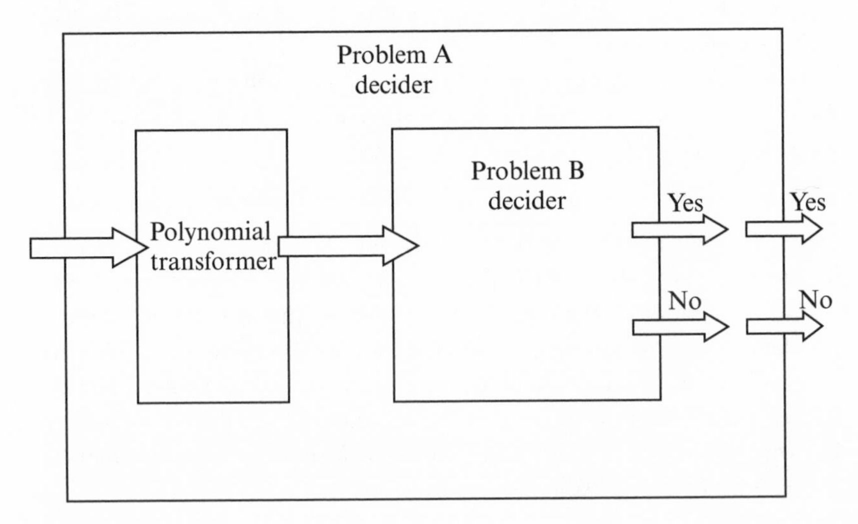

Figure 5.12
Machine to decide Problem A by using Problem B decider

An instance of Problem A enters on the left. The instance is transformed into an instance of Problem B and then goes into a Problem B decider to give an answer to both problems.

We don't want the transformer to arbitrarily change an instance of Problem A into an instance of Problem B. We require that the instances have the same answer. In other words, we will insist that the transformer take inputs with a "Yes" answer for Problem A to inputs that answer "Yes" for Problem B. If an input to Problem A gets a "No" answer from the Problem A decider, the transformer should output an instance that will have a "No" answer. We make one further requirement: this transformer should perform its task in a polynomial amount of operations. The need for this stipulation will quickly become apparent.

When we have such a relationship between Problem A and Problem B, we say that we have a "reduction of Problem A to Problem B" or "Problem A reduces to Problem B."

Let us look at another example of a reduction of one problem to another. The Hamiltonian Cycle Problem is reducible to the Traveling Salesman Problem. Consider the instances of the Hamiltonian Cycle Problem given in figure 5.10. We convert them into instances of the Traveling Salesman Problem in figure 5.13. Bear in mind that the Traveling Salesman Problem demands complete and weighted graphs. We complete the graphs by adding edges between those vertices that were not already connected. To differentiate among the types of edges, we will make the new edges gray.

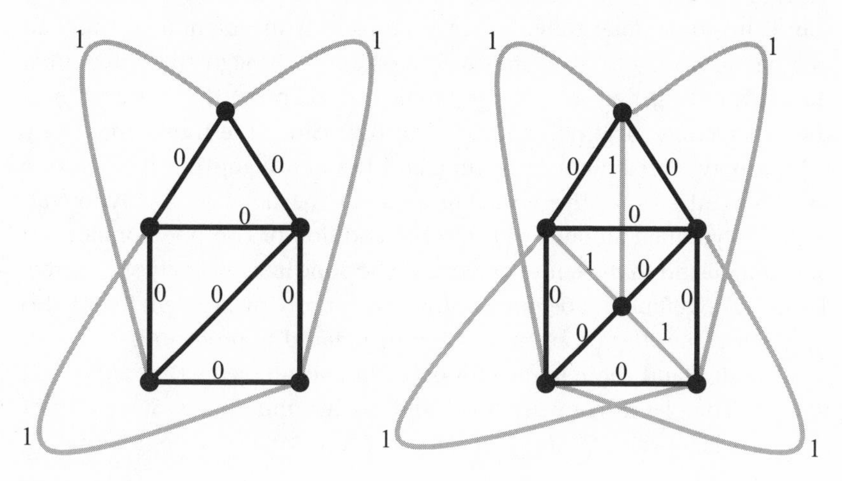

Figure 5.13
Hamiltonian Cycle instances transformed into Traveling Salesman instances

As for the weights, we say that all black edges have weight 0 and all gray edges have weight 1. And now for the final ingredient to make this an instance of the decision version of the Traveling Salesman Problem: we set the required integer K to 0. If there exists a traveling salesman cycle that has weight (less than or) equal to 0, none of the new gray edges were used in the cycle, so there was a Hamiltonian cycle in the original graph. The left graph in figure 5.13 contains such a traveling salesman cycle and the graph on the right does not. We have successfully reduced the Hamiltonian Cycle Problem to the Traveling Salesman Problem.

Why does it help us to know whether one hard problem can be reduced to another? After all, both problems are effectively unsolvable for any large *n*. In the next few pages, we will see many reasons why this concept of reduction is fundamental to the entire field of study.

Let us ponder this situation a bit. Imagine that NP Problem A can be reduced to NP Problem B, that is, Problem A $\leq_p$ Problem B. Now imagine that some supergenius comes out of nowhere with some fancy newfangled algorithm that can actually solve Problem B in polynomial time. This super-genius will no doubt earn many accolades for finally finding an easy way of solving Problem B. Those who waited for trillions of centuries while some exponential algorithm solves problem B can "break out" of that algorithm and run the new polynomial algorithm, getting a result in minutes.

But there is more. With a polynomial reduction, not only will the supergenius get credit for solving Problem B, but Problem A will also be solved in polynomial time. To solve Problem A in polynomial time, all one has to do is put an instance of Problem A through the polynomial transformer to get an instance of Problem B, then put that instance into the supergenius' new polynomial algorithm. Since the transformer only takes a polynomial amount of time and the new algorithm is also done in polynomial time, the entire process can be done in a-polynomial-plus-a-polynomial amount of time. The addition of two polynomials is a polynomial and so the entire process can be done in a relatively short time. So our supergenius will be praised not only for solving problem B, but also for solving *any other* problem that can be reduced to problem B.

To understand the relationship between two problems that are linked to each other, let us compare it to climbing two mountains. Since Mount Everest is taller than Mount McKinley, the following statement is true:

If you can climb Mount Everest, then you can definitely climb Mount McKinley.

That means that it is harder to climb Mount Everest than Mount McKinley. This is equivalent to saying:

If you cannot climb Mount McKinley, then you definitely cannot climb Mount Everest.

Similarly, when there is a polynomial reduction from Problem A to Problem B, the following statement is true:

If Problem B can be solved in polynomial time, then Problem A can also be solved in polynomial time."

That means that Problem B is harder than or as hard as Problem A. This statement is equivalent to:

"If Problem A cannot be solved in polynomial time, then Problem B also cannot be solved in polynomial time.

Until now, we have spoken about an NP problem that has another NP problem reducible to it. Now let's talk about an NP problem such that *every* NP problem is reducible to it. An NP problem such that every NP problem is reducible to it is called an *NP-Complete* problem. In a sense, NP-Complete problems are the hardest NP problems. All five problems presented in section 5.2 are, in fact, NP-complete problems and are reducible to one another.

Since any other NP problem is reducible to an NP-Complete one, if any NP-Complete problem can be solved in polynomial time, then all NP problems can be solved in polynomial time.

With the notion of NP-Complete problems, one can see why researchers believe that polynomial algorithms to solve such problems will never be found. Take any NP-Complete problem, say the Traveling Salesman Problem. For many years, people have looked in vain for a polynomial algorithm to solve this problem. This section has demonstrated the interconnectedness of the NP-Complete problems. If any person had found a polynomial algorithm for *any* NP-Complete problem, then the Traveling Salesman Problem would also have a polynomial solution. In a sense, researchers working on any NP-complete problem have also been working on the Traveling Salesman Problem. So we can say that thousands of people have been seeking a polynomial algorithm for our problem for many years and so far have found nothing. It seems likely that no such algorithm exists.

Why is it important to show that some problem is NP-Complete? Two vital reasons are usually given. First, by showing that a problem is NP-

Complete, we are demonstrating that finding an efficient algorithm for this problem is as hard as finding such an algorithm for other NP-Complete problems. Since no one has been able to find an efficient algorithm for any NP-Complete problem, we are showing that the problem is inherently hard. If you were given the job of writing a nice polynomial algorithm to solve a problem and you have a hard time completing the task, you will find yourself in trouble with your boss. If, however, you show the problem is NP-Complete, you can insist that not only are you unable to find a good algorithm, but no one else can either. This claim will save your job.

Another reason why the concept of NP-Complete problems is important is that once it is known that a problem cannot be easily solved, we are free to move on to find other algorithms that might help us *approximate* a solution to the problem. Algorithms that approximate solutions in a reasonable amount of time are introduced in section 5.4.

Once one has an NP-Complete problem, it is not hard to find others. If A is a known NP-Complete problem and you are interested in showing that B is an NP-Complete problem, you need only perform the following two tasks:

1. Show that B is NP.
2. Show that $A \leq_P B$—that is, that B is as hard as or harder than A.

From the fact that A is a known NP-Complete problem, and all NP problems are reducible to A, and from the fact that A is reducible to B, we know that all NP problems are reducible to B. This can be seen in the figure 5.14, in which we draw an arrow from one problem to another if there is a reduction between them.

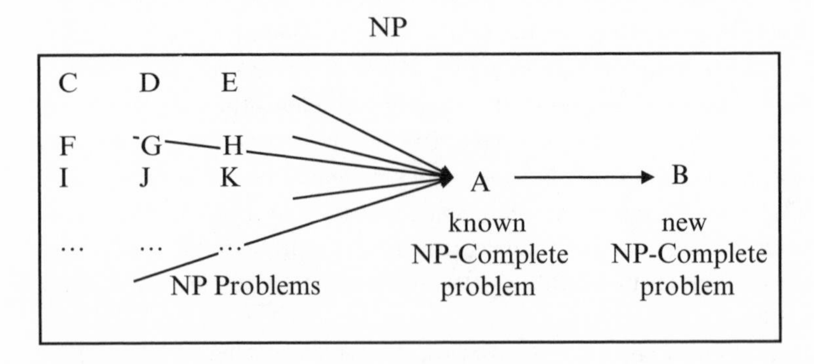

Figure 5.14
From one NP-Complete problem to another.

To recap: once we have one NP-Complete problem, we can easily find others. But how does one find the *first* NP-Complete problem? In the early 1970s, Stephen Cook of North America and Leonid Levin of Russia independently proved that the Satisfiability Problem is an NP-Complete problem. This theorem, which came to be known as the *Cook-Levin Theorem*, is one of the most amazing theorems in computer science. The theorem states that *all* NP problems can be reduced to the Satisfiability Problem. We have seen five diverse NP problems. There are currently thousands of known NP problems, some about graphs, some about numbers, some about DNA sequencing, some about scheduling tasks, and so on. These problems come in many different shapes and forms, and yet they are all reducible to the Satisfiability Problem. But wait . . . there's more! Cook and Levin did not demonstrate that every NP problem now in existence is reducible to the Satisfiability Problem; rather, they showed that *all* NP problems—even those not yet described—are reducible to the Satisfiability Problem.

The way Cook and Levin showed that the Satisfiability Problem is NP-Complete is clever and worth understanding. They started by asking what all NP problems have in common. By definition, every NP problem can be solved by a computer in at most an exponential or factorial number of operations. Now ask how such a computer works. What is the inner core of computers? The answer is simple: computers and their chips follow the laws of logic. Inside every computer there are literally billions of logical switches that perform the logical operations AND, OR, NOT, and IMPLY.[8] So since every NP problem can be solved by a logical computer, every NP problem can be reduced to the Satisfiability Problem. We kicked off this chapter with a discussion of how computers are engines of logic and reason. We now see that clearly. Every computer problem can be changed into a logical expression.

Before I end this section, let me tell you how to make a million dollars. At the turn of the millennium, in order to promote mathematics, the Clay Institute declared seven problems in mathematics to be the most important and hardest of their kind. Anyone who solves one of these "Millennium Problems" will receive a million dollars. One of these problems is the **P** =? **NP** problem or "is **P** equal to **NP**?" We saw at the end of section 5.2 that **P** is a subset of **NP**—that is, every easy problem can be solved in less than a large amount of time. But we can ask the reverse: Is **NP** a subset of **P**? Is it possible that every hard NP problem can be solved in a polynomial amount of time? If **NP** is a subset of **P**, then **P** = **NP**. If **NP** is not a subset

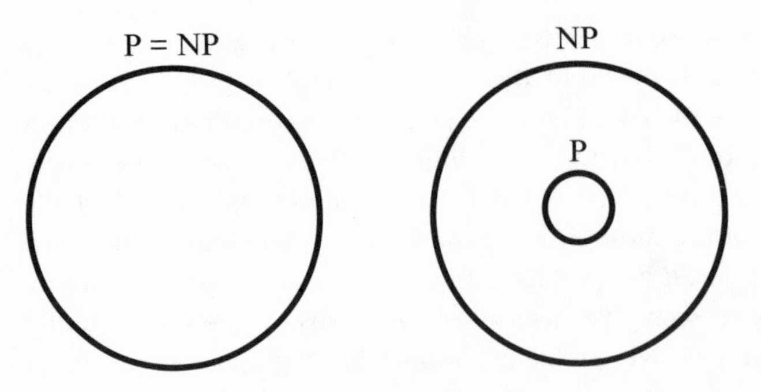

Figure 5.15
Two possibilities for the **P**-vs.-**NP** question

of **P**, then there is a problem in **NP** that is not in **P** and **P** ≠ **NP**. The two possibilities are depicted in figure 5.15. To win the prize, all you have to do is prove the answer one way or the other.

How should you go about claiming your award money? There are two possible directions to take. You can try to prove that **P** = **NP** or you can aim to show that **P** ≠ **NP**.[9] To show that **P** = **NP**, all you have to do is take one of your favorite NP-Complete problems and find a polynomial algorithm that solves it. As we have seen, if you do find such an algorithm, then all NP problems will be solvable in a polynomial amount of operations. It might seem strange to think that a problem that demands an exponential or factorial amount of operations can be done in a polynomial amount of operations. However, we saw something similar with the Euler Cycle Problem. Rather than look through all $n!$ possible cycles to see if any are Euler cycles, we used the trick of checking if the number of edges touching each vertex is even or not. Does a similar trick for the Hamiltonian Cycle Problem exist? For many years, the smartest people around have been looking for such a trick or algorithm and have not been successful. However, you might possess some deeper insight that they lack. Get to it!

On the other hand, you can try to show that **P** ≠ **NP**. One way to do this is to take an NP problem and show that no polynomial algorithm exists for it. It so happens that it is very hard to prove such a claim: there are a lot of algorithms out there. This has turned out to be one of the hardest problems in mathematics.[10] As a final hint, it should be noted that most researchers believe that **P** ≠ **NP**.

5.4 Almost Solving Hard Problems

These NP-Complete problems are not abstractions that computer scientists and mathematicians created. They come from real-world applications and are problems that need to be solved. Industry and computer professionals are always looking to find efficient solutions to these problems. Waiting several centuries for the answers is not acceptable.

To deal with these problems, computer scientists have invented helpful algorithms that will eliminate some of the pain of harder problems. Such algorithms are referred to as "approximation algorithms." These algorithms require a polynomial number of operations but do not always give the correct answer. Sometimes they are off by a little, sometimes by a lot.

Approximation algorithms are usually based on heuristics. These are recommendations that one learns through experience and are like "rules of thumb"—not always 100% correct but "close enough" to the solution.

Consider the traveling salesman problem, which is probably the most intuitive of the NP-Complete problems mentioned in section 5.2. Let's go back to the major-cities problem presented at the beginning of that section. Say the salesman starts in Los Angeles. From there, he does not think of the entire route that he needs to take; rather, he just determines the closest city on the list. The nearest city is San Francisco. Once he gets to San Francisco, again he looks for the nearest city he did not visit: Denver. And again, once he comes to each city, he simply looks for the nearest city he has not yet visited. This is an intuitive way of dealing with the problem. The traveler does not look at the "big picture." Rather he greedily picks the nearest city. This method is called the "nearest neighbor heuristic." At each vertex, simply go to the closest neighbor possible. This algorithm always works in a polynomial amount of time. However, it does not always find the correct solution.

While the nearest neighbor heuristic seems as though it will always find the correct solution, in actuality, it fails and it is easy to see why. Consider the complete weighted graph given in figure 5.16. Only the weights of neighboring edges are given, but the other weights can be calculated from them.

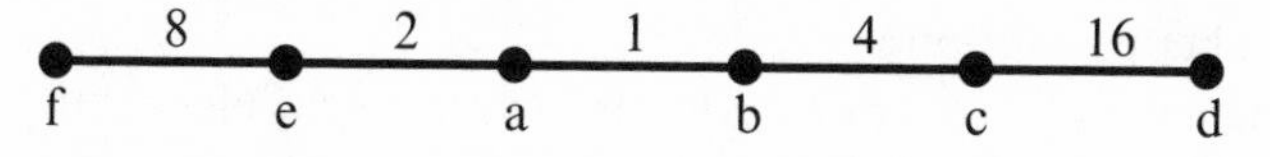

Figure 5.16
A counterexample for the nearest neighbor heuristics

Suppose we forced the traveler to start at vertex a. By the nearest neighbor heuristic, she would have to go to vertex b, which is only one mile away. From vertex b, the traveler would have two choices: either c, which is four miles away, or e, which is three miles away.[11] By our algorithm, she must choose e. Once the traveler is at e, she must go to c. By following the nearest neighbor heuristic, the traveler must make the following cycle:

$a \to b \to e \to c \to f \to d \to a.$

Our poor traveler would have to travel

$1 + 3 + 7 + 15 + 31 + 21 = 78$ miles.

As we mature, we learn that going through life taking shortcuts will not always get you to where you want to go in the shortest amount of time. Sometimes shortcuts will lead you astray. Consider this cycle, which does not follow the nearest neighbor heuristic:

$a \to b \to c \to d \to f \to e \to a.$

This cycle will need

$1 + 4 + 16 + 31 + 8 + 2 = 62$ miles.

Clearly, the nearest neighbor heuristic did not work very well for this complete weighted graph.

What about the Set Partition Problem? Here is a polynomial approximation algorithm that I call *extreme pairs*. Given a set of numbers, {24, 68, 61, 41, 35, 51, 58, 39, 49 54, 29, 23}, sort the elements as follows:

23, 24, 29, 35, 39, 41, 49, 51, 54, 58, 61, 68.

Take the pair of extreme elements, the minimum and maximum (23 and 68), and put them in one partition. Put the remaining minimum and the maximum (24 and 61) into the other partition. Continue in this manner until every element is in one of the partitions. This will leave us with

23, 68, 29, 58, 39, 51	24, 61, 35, 54, 41, 49

The left-hand partition sums to 268, while the right-hand partition sums to 264.[12] Is there a better solution?

It's worth taking a few minutes to see why the extreme-pairs algorithm works. Take the sequence of numbers in order as in figure 5.17.

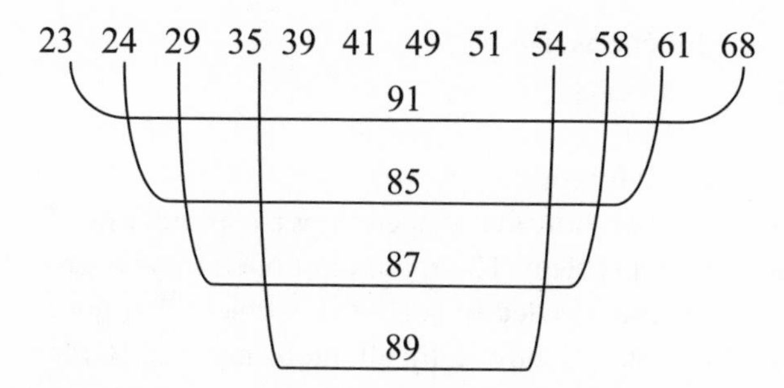

Figure 5.17
Sums of extreme pairs are almost the same.

The first number and the last number sum to 91. The second number and the next-to-last number sum to 85. These two sums are not the same but they are close enough. Continuing in the same way gets us numbers in the same ballpark. What this approximation algorithm does is send each of these similar sums to different partitions. It might not be the best solution, but it is better than waiting for 400 trillion centuries.

Every time a new NP-Complete problem is found, there is a search for good approximation algorithms to help solve the problem. As we said, NP problems arise all over and are important. Industry must find a way to deal with such problems. Approximation algorithms are not only formulated and described but are also compared with each other and analyzed. Which heuristic is better? Which algorithm gets you closer to the unobtainable real answer? Which algorithm works faster? Which of the algorithms gets the correct answer for more of the inputs? There is much work to be done.

5.5 Even Harder Problems

NP is not the end of the story. There are problems that demand even more operations than 2^n and $n!$. There are some (not-so-easy-to-describe) problems that require

$$2^{2^n}$$

operations called *superexponential problems*. Rather than spending time discussing such problems, let us look at how large this function is. For a small input of size $n = 10$, the exponential is $2^{10} = 1{,}024$. Using 1,024 as an exponent, we have 2^{1024}, which is larger than any imaginable number.

There are similar crazy functions like

$(n!)!$ or $2^{n!}$.

Try plugging in some small values for n.

Until now, we have focused on how many operations a computer needs to perform in order to solve a problem. The number of operations is proportional to the amount of time needed to solve the problem. However, there are other ways of measuring how difficult problems are. Harder problems not only demand a lot of time but also a lot of memory space to solve. When a computer solves a problem, it uses memory to store some calculations. The more space needed to store calculations, the harder the problem is.

As before, every algorithm has a function associated with it that describes how much memory space is needed in order to solve the problem. The larger the function, the more space is needed and the harder the problem.

An interesting class of problems consists of those that demand a polynomial amount of space. This class is denoted as **PSPACE**. It is known that **NP**, the set of problems that can be solved in exponential or factorial time, can be solved in polynomial space. In other words, **NP** is a subset of **PSPACE**.

There are many examples of problems that are in **PSPACE**, and some of them have to do with games. There are games with two players that have winning strategies—that is, there are surefire ways for one particular player to win the game. Consider the game of tic-tac-toe. It is known that the first player has a strategy to either force a draw or a win. Now consider a generalization of tic-tac-toe, which, rather than dealing with a three-by-three grid, deals with an n-by-n grid. Is there a winning strategy for one of the players in this game? The answer might depend on n. Determining whether there is a winning strategy for a given n-by-n game is a problem in **PSPACE**. Other types of generalized games like chess, checkers, connect-four, nim, go, and others are also known to be in **PSPACE**.

In conclusion, there are many different types of computer problems, ranging from easy to very hard. These classes of problems can be envisioned as in figure 5.18.

This diagram is a small part of a larger picture that we will meet in the next chapter when we encounter problems that cannot be solved in *any* amount of time and space.

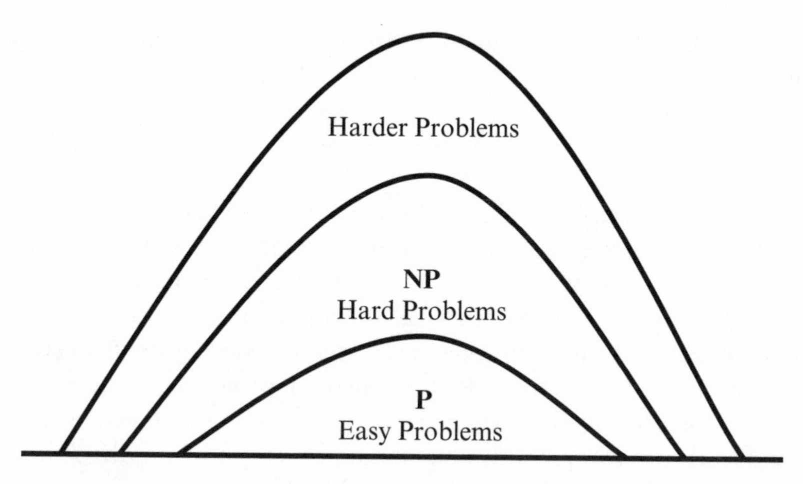

Figure 5.18
A hierarchy of solvable problems

Further Reading

Section 5.1

A popular history of algorithms can be found in Berlinski 2001. Harel 2003 is a popular text that covers everything done in this chapter. Pages 100–107 of Barrow 1999 also cover some of the same topics. There are many wonderful textbooks on algorithms; Baase 1988, Corman et al. 2002, and Dasgupta, Papadimitriou, and Vazirani 2006 are just a few.

The history and significance of the Königsberg Bridge Problem are discussed on pages 364–365 of Eves 1976 and on page 232 of Ross and Wright 2003. Ross and Wright is also a good place to learn more about basic graph theory.

For a more abstract view of algorithms and an exact definition, see Yanofsky 2011.

Section 5.2

All five of the NP problems can be found in Baase 1988, Corman et al. 2002, and chapter 9 of Papadimitriou 1994.

Section 5.3

For a general discussion of NP-Completeness see Garey and Johnson (1979), which remains an excellent book. The second half of the book lists over 300 NP-Complete problems. For a popular account, see chapter 9 of Poundstone 1989.

A nice short presentation of many different classes of problems and their relationship with each other can be found in chapter 8 of Yanofsky and Mannucci 2008. That wonderful textbook is also where you can learn a lot more on quantum computers and why they will not really help with NP problems.

The Cook-Levin theorem can be found in their original papers: Cook 1971 and Levin 1973. There is a very readable account with technical details in section 2.6 of Garey and Johnson 1979. There is also a more intuitive account that is closer to our presentation in section 34.3 of Corman et al. 2002. Stephen Cook, one of the founders of the field and one of the first to formulate the **P** =? **NP** question, wrote a very nice, readable paper introducing the problem. He also discusses some related results and explains why this problem is so important. This is one of the seven articles presenting the Millennium Problems of the Clay Institute.

Section 5.4

Approximation algorithms can be found in section 9.3 of Baase 1988, chapter 35 of Corman et al. 2002, chapter 6 of Garey and Johnson 1979, chapter 5 of Harel 2003, chapter 9 of Dasgupta 2006, and chapter 13 of Papadimitriou 1994.

Section 5.5

For more information about what is beyond **NP**, see chapter 6 of Garey and Johnson 1979, the end of chapter 3 of Harel 2003, and part V of Papadimitriou 1994.

6 Computing Impossibilities

I whispered, "I am too young,"
And then, "I am old enough";
Wherefore I threw a penny
To find out if I might love.
"Go and love, go and love, young man,
If the lady be young and fair."
Ah, penny, brown penny, brown penny,
I am looped in the loops of her hair.
O love is the crooked thing,
There is nobody wise enough
To find out all that is in it,
For he would be thinking of love
Till the stars had run away
And the shadows eaten the moon.
Ah, penny, brown penny, brown penny,
One cannot begin it too soon.
—William Butler Yeats (1865–1939), "Brown Penny"

The only way of finding the limits of the possible is by going beyond them into the impossible.
—Arthur C. Clarke (1917–2008)

In the end, we self-perceiving, self-inventing, locked-in mirages are little miracles of self-reference.
—Douglas R. Hofstadter, *I Am a Strange Loop*

Computers can do many wonderful things. However, there are many tasks that they cannot perform. Computers cannot determine whether a painting is beautiful; they do not "understand" moral issues; and they cannot fall in love. Such "human" processes are beyond computation.

Whether a painting is beautiful is subject to taste, and computers don't have taste. They also don't have a sense of ethics to deal with moral questions. All of these questions are subjective and computers don't do well with subjective questions. In this chapter, we will explore certain problems that have objective answers but that nevertheless cannot be solved by computers.

It is important to note that these tasks do not require a long time to compute (as in the last chapter); rather, they can *never* be done. Nor is this a problem related to our current level of computer technology. No computer, regardless of how fast and powerful, will ever be able to solve these problems.

Section 6.1 starts with a short discussion of programs, algorithms, and computers. A problem that cannot be solved by any computer, the *Halting Problem*, is introduced in section 6.2. I show that it is not solvable by any computer. This is not to be taken on faith. Rather, I carefully go through the proof that demonstrates any computer's inability to solve this problem. With the unsolvability of the Halting Problem in hand, I move on to section 6.3, where I show that many other problems are unsolvable. Section 6.4 describes a hierarchy or classification of unsolvable problems. I conclude with section 6.5, which addresses more philosophical questions like the relationship of brains, minds, and computers.

6.1 Algorithms, Computers, Machines, and Programs

We have all experienced computers "getting stuck" or entering into an "infinite loop." Try as we like, our computers get looped in the loops of their programs. Once in an infinite loop, they never leave. Microsoft Windows taunts us with the words "Not Responding."[1] Wouldn't it be nice to buy a piece of software with the reassurance that this could never happen? It would be great if there were some method to determine if a program will halt (or terminate) in a normal manner as opposed to entering into an infinite loop. Alas, no such method exists. This is one of the first problems shown to be unsolvable by a computer. Even though the question of whether a program will halt is objective and not subjective, there is no way a computer can solve it.

Before we begin, some terminology is in order. A *computer* is a physical *machine* that executes *algorithms*. *Programs* are exact descriptions of algorithms. When we say that there is no algorithm that solves a particular problem, we also mean that there is no program, computer, or machine that can perform that task. We are describing a limitation of mechanized processes and we will employ all four of these words interchangeably.

In my discussion of whether a program does or does not halt, rather than talking about all programs, I restrict myself to special types of programs that only deal with whole numbers. Before suspecting that I am trying to trick you by only looking at this restricted set of programs, you should keep two things in mind. First, I am going to show that even for this restricted set of programs, no computer will be able to determine if such programs halt or not. Certainly no computer will be able to determine this for *all* programs. Second, programs that deal with real numbers, graphics, robotics, and all the amazing machines that computers operate, can work by manipulating whole numbers. Such numbers encode different types of more complicated numbers and objects. So if we are going to show a limitation on programs that only deal with whole numbers, there will certainly be limitations on the more complicated types of programs.

The programs I work with are readable by anyone. No programming experience is needed. One simply analyzes the programs from the top down. There are different variables, like **x**, **y**, or **z**, which hold whole numbers. The statements in the programs are obvious. For example, there might be

```
x=y+1.
```

This means that the variable **x** is assigned the same value as **y** with 1 added to it. We can also do something like

```
x=x+1.
```

This means that the value of **x** should be incremented by 1. Since some programs demand inputs, we will write

```
x=?
```

when the computer should stop and ask the user for a number. Variable **x** will be given the value that the user entered. Some lines will have a label like **A**, **B**, or **C**. These labels will be used to control the execution of the programs. Programs that only go from top to bottom are not very interesting. There is a need for loops so that a program can perform actions repetitively. Labels allow us make loops by using the **goto** statement.

To get a good intuition about programs, let's look at a few examples:

```
x=?
x=x+1
x=x+1
x=x+1
print x
stop
```

What task does this program perform? If 15 is the input to this program, then the variable **x** would contain 15 because of the first line. The next three lines would each increment **x** by 1 with a combined effect of adding 3. The computer would then print the contents of **x**, namely 18. Once this is done, the computer would halt. If 56 were entered, the program would print 59. In general, the function this program computes is $f(x) = x + 3$.

This is fun! Let's try some more examples.

```
     x=?                          x=?
     y=10                         y=?
A    x=x+1                   B    x=x+1
     y=y-1                        y=y-1
     if y>0 goto A                if y>0 goto B
     print x                      print x
     stop                         stop
```

First focus on the left-hand program. If the user inputs 23, what happens? The variable **x** would have the value 23 and the variable **y** would have the value 10. The next two lines of the program work in tandem: **x** gets incremented to 24 and **y** decrements to 9. The next line is a conditional statement. Since **y** is 9 and hence more than 0, the computer would go back up to the line with the label **A**. From there, **x** would go up to 25 and **y** would go down to 8. This loop continues until **x** is 33 and **y** is 0. At this point, the conditional statement fails and the print statement is executed. The number 33 will be outputted since that is the value of **x**. If **x** started off as 108, then 118 would be printed. In general, this program computes the function $f(x) = x + 10$.

Now for the right-hand program. It is almost the same as the left-hand program, with a single difference. Rather than having **y=10**, there is **y=?**. This program demands two inputs, as opposed to one. Rather than counting down from 10, this program will count down from any starting value of **y**. It is not hard to see that this program will compute the function of two values $f(x,y) = x + y$. One does not need much convincing to see that programs written in this programming language can perform most of the tasks that calculators can perform. In fact, with enough encoding, programs like these can perform any task that any computer can perform.

Let us examine some more programs.

```
     x=?
A    x=x+1
     if x>10 goto A
     stop
```

```
     x=?
     y=x+15
A    x=x+1
C    if x> y goto B
     y=y-1
     if x<y goto A
B    x=x-1
     if x= y goto C
     stop
```

If one enters 5 in the left-hand program, the conditional statement will fail and this program will immediately terminate. In contrast, if one enters 15, the program will go into an infinite loop. In fact, entering any number less than or equal to 9 for **x** will halt the program, and for any number 10 or above, the program will go into an infinite loop.

What about the right-hand program? Can you determine when this program will halt and when it will get lost wandering around its loops? Neither can I! How about getting a computer to solve this problem . . .

6.2 To Halt or Not to Halt?

Let us formulate the Halting Problem. Given any program and an input for it, determine whether the program with that input will halt or go into an infinite loop. It is a decision problem—that is, it gives a Yes or No answer. Can a computer solve the Halting Problem?

The question was described by Alan M. Turing (1912–1954) and in 1936 he gave a definitive negative answer: there is no program that can solve the Halting Problem. There is no way for a computer to determine for any given program, and for any given input into that program, whether the program will halt on that input. If a computer can solve a decision problem, we say the problem is *decidable*. In contrast, if a computer cannot solve it, we say the problem is *undecidable*. The Halting Problem is undecidable.

A few minutes of meditation are in order. We first have to differentiate between problems being infeasible, which we focused on in the last chapter, and problems being undecidable. In the last chapter, problems could be solved, but for large inputs, an unreasonable amount of time is demanded. We are not saying that here. The Halting Problem cannot be solved in any amount of time. It is not a *hard* problem; it is an *impossible* problem.

Furthermore, observe that there is an objective answer to the question of whether a program for a given input will halt. It is not some subjective, wishy-washy idea like artistic taste or morality. Either the program will eventually halt or it will continue in an infinite loop, and yet computers cannot decide that objective question. Even human beings might have a hard time determining it (more about this in section 6.5). Nevertheless, there is a real, objective answer.

Another point to notice is that Turing did not say that he is incapable of writing such a program. Nor did he say that no one else is clever enough to find such a program. Rather, he proved that no such program could exist. The difficulty was not the lack of technology or ingenuity. Turing showed that no amount of technological innovation or ingenuity can ever solve this problem. On a deeper level, he was saying that neither computers nor any other physical devices obeying reason would be able to solve the Halting Problem.

You might think of an easy way to show that the Halting Problem is decidable: run the program with the given input and see whether it halts or goes into an infinite loop. Alas, this method will not work. If you run the program for ten minutes and it halts, you can safely say that this program for this input will halt. But what happens if the program is still running after ten minutes? That does not mean that the program is in an infinite loop because many programs require more than ten minutes to terminate. Perhaps you should run the program for twenty minutes. Again, if it halts, you are done. However, if it does not halt, what do we know? We are still not sure that it is in an infinite loop. How long should we run the program? For any given time limit, there are programs that need a little more time before they terminate.[2] The only way we can tell if a program is in an infinite loop is to wait an infinite amount of time and see if it is still running. But who has that much time?

One of the most amazing aspects of Turing's result is that it was demonstrated in 1936, long before any computer existed. Turing, a brilliant theoretical mathematician, demonstrated limits on the power of computers years before the practical engineers figured out how to create a computer. Theoreticians rock!

Now that we have meditated and absorbed why the undecidability of the Halting Problem is interesting and worthy of deeper understanding, let us go about actually proving it. The undecidability is proved using the fact that computers can talk about themselves. That is, computers have self-reference. Since programs can discuss programs, there is an element of self-reference and hence a limitation. By now, we have seen self-reference many times in our book—for example, with the liar paradox, where self-referential English sentences discuss themselves:

This sentence is false.

We showed that the sentence is true if and only if it is false. Here we will create a self-referential program that essentially says the following:

This program will give the wrong answer when asked if it will halt or go into an infinite loop.

As we will see, the existence of such a program will cause a contradiction. Such contradictions are not permitted in the real world of computers, hence we have a limitation: the Halting Problem cannot be solved. The proof is a type of proof by contradiction. Assume that the Halting Problem is not undecidable (i.e., that a computer *can* decide this problem); with this assumption we derive a contradiction:

The Halting Problem is decidable $\Rightarrow$ contradiction.

We have to conclude that our assumption must be incorrect.

Notice that the programs we talked about in the last section are very simple creatures. They are easy to describe and it turns out that we can encode such programs as whole numbers.[3] For any program, there is a unique number that corresponds to the program. For a given program the associated number will be called the "program number." We can also go the other way: from a number, we can find its program. For number *x*, we call the program associated with it "program *x*." The idea is that the programs deal with numbers and numbers represent the programs; therefore we will have programs deal with programs as in figure 6.1. This is the core of self-reference.

Let us imagine that the Halting Problem is decidable and we can actually create a program that decides it. That would mean that we could sit

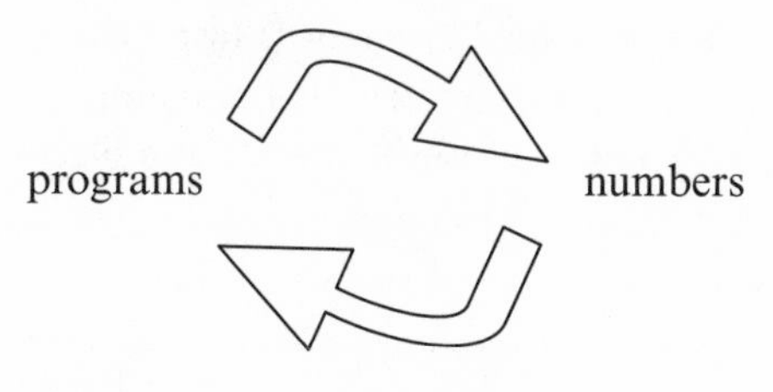

Figure 6.1
The self-reference of programs

down and write a computer program that works as follows: input a program number y and an input number x and the computer will output a Yes or No depending on whether program y with input x will halt. We write this function as

$$\text{Halt}(y,x) = \begin{cases} \text{Yes} & \text{program number } y \text{ halts on input } x \\ \text{No} & \text{program number } y \text{ loops on input } x. \end{cases}$$

The program that determines the halt(y,x) function is a black box that takes two inputs and outputs a Yes or No, as in figure 6.2.

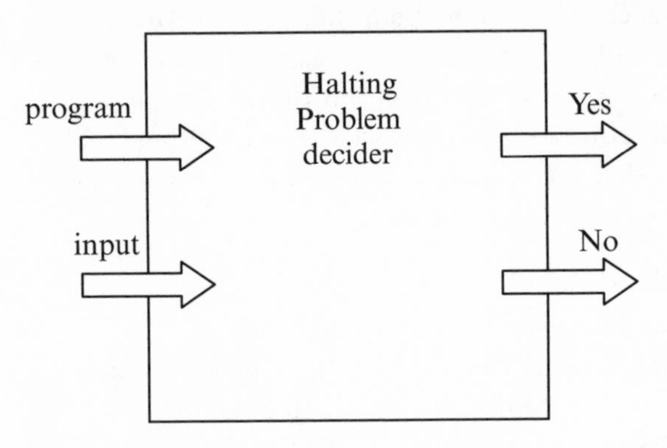

Figure 6.2
The (nonexistent) Halting Problem decider

We will denote the (nonexistent) program for this function as **Halt(y,x)**. Now that we have the ability to reference programs and numbers, let us use this presumed program as part of the following bigger program:

```
        x=?
A       if halt(x,x) = "Yes" then goto A
        print "No"
        stop
```

This program is very important and will be called Program D (for "diagonal"). Let's just look at this program as a diagram. The construction of Program D can be visualized in figure 6.3, which uses figure 6.2 as a black box within it.

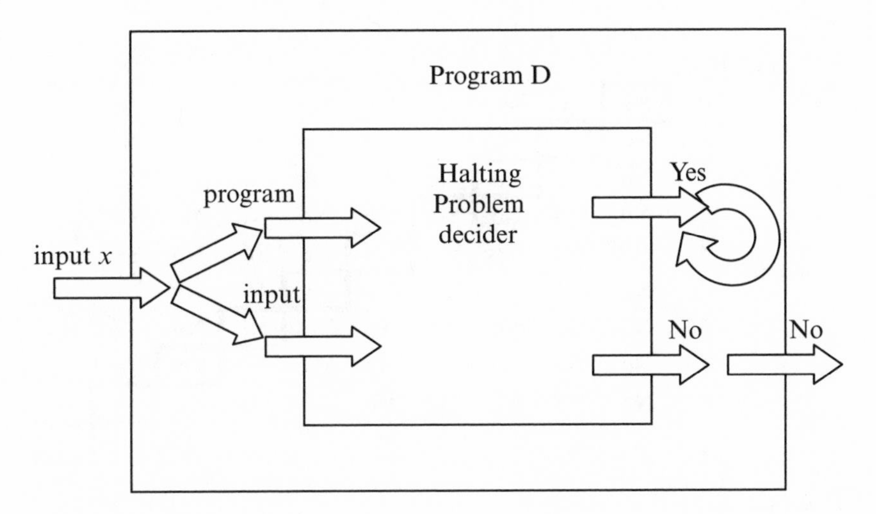

Figure 6.3
The (nonexistent) Program D

This program accepts one input, which becomes both the program number and the input (self-reference). If this program halts on its own input, it goes into an infinite loop. In contrast, if the program with the input goes into an infinite loop, it halts and prints a No. Essentially this program gives the wrong answer when asked about input x.

Formally, the action of Program D on input x is as follows:

Program D on input x halts if and only if program x on input x loops.

Hold on! Now comes the interesting part: if **Halt** is a real program, we have no worries using it as a part of another program and can create Program D. If Program D is a real program, then it has a number. We do not know exactly what the number is, but we might as well denote it as d_0. Now let us input the number d_0 into Program D. That is, let's see what Program D says about itself.

Program D on input d_0 halts if and only if program d_0 on input d_0 loops.

But program d_0 is exactly Program D, so we can restate this line as

Program D on input d_0 halts if and only if Program *D* on input d_0 loops.

This is a contradiction. We can say that

Program D gives the wrong answer about itself when asked if it will halt or go into an infinite loop.

Halt	Input 0	1	2	3	4	5	6	7	8	...
Programs 0	No	No	Yes	No	Yes	Yes	No	No	No	...
1	No	Yes	No	No	No	Yes	No	No	No	...
2	No	No	No	No	No	Yes	No	No	No	...
3	No	Yes	No	Yes	No	No	No	No	Yes	...
4	Yes	No	Yes	Yes	No	No	Yes	No	Yes	...
5	No	No	No	No	No	No	No	No	No	...
6	No	No	No	No	No	Yes	No	No	No	...
7	No	Yes	No	No	No	Yes	No	Yes	Yes	...
8	No	No	No	No	No	No	No	No	No	...
:	:	:	:	:	:	:	:	:	:	⋱

Figure 6.4
The (nonexistent) Halt Program and its diagonal

We are right back to the liar paradox. We have come to a point where a program halts if and only if it does not halt. Human language and human minds might have contradictions but computers do not. Something must be wrong. Only one assumption was made: that it is possible to write a program that decides the Halting Problem. That assumption caused us to reach a contradiction, hence it must have been wrong. It is impossible to solve the Halting Problem with a computer.

The above proof of the undecidability of the Halting Problem is a bit complicated and it will be helpful to look at it from another point of view. It might be easier to visualize it as a diagonalization proof. (If you have already glanced at chapter 4, this proof will be more palatable.) Assume for a moment that we have access to a program that decides the Halting Problem. We can write the output of such a program as an infinite matrix (see figure 6.4).

The numbers along the left-hand column are the program numbers. The numbers across the top row correspond to the inputs to the program. The Yes's or No's inside tell us whether that program with that input will halt. For example, program 7 with input 2 has a No. This means that if you input number 2 into program 7, the program will go into an infinite loop. In contrast to that, program 4 with input 8 will halt. We used this (supposed) program to create a new program by diagonalization. Program D only has one input and its value is determined by looking at the diagonal elements of the matrix in figure 6.4. The program accepts a number as input. On input x, Program D determines whether program x on input x halts and does the exact opposite, as in figure 6.5.

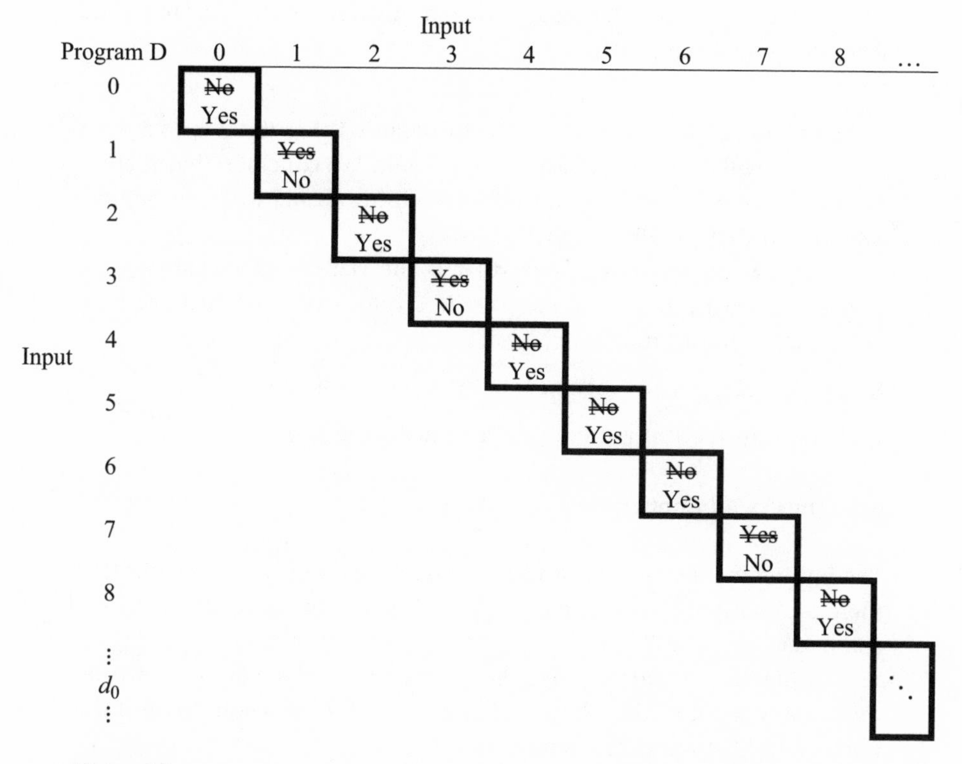

Figure 6.5
The (nonexistent) Program D

The numbers along the edges are the inputs to the program, and the Yes's and No's inside tell us if the program with those inputs will halt or loop.

Although this new (supposed) Program D is easily constructed from the **Halt** program, it can nevertheless be shown that Program D does not exist. If it did exist, it would have a number associated with it. But which number could it be?

- It cannot be program 0 because program 0 on input 0 goes into an infinite loop and Program D on input 0 halts.
- It cannot be program 1 because program 1 on input 1 halts and Program D on input 1 goes into an infinite loop.

• It cannot be program 2 because program 2 on input 2 goes into an infinite loop and Program D on input 2 halts.
• Etc.
• It cannot be program d_0 because program d_0 on input d_0 does "this" and Program D on input d_0 does "that."
• Etc.

In other words, Program D cannot exist because it has the wrong answer on what it will do when we ask it about itself. We conclude that there is no such Program D, so there must be something wrong with our original assumption that the **Halt** program exists.

Summing up, we have shown that if the **Halt** program exists, then Program D exists, and since Program D cannot exist, the **Halt** program cannot exist. In our notation:

Halt program exists $\Rightarrow$ Program D exists $\Rightarrow$ contradiction.

Which means the Halting Problem is undecidable.

6.3 They're All Connected

The Halting Problem is not the only undecidable problem. I will show that there are many other problems that cannot be solved or decided by a computer.

Consider the *Printing 42 Problem.* Given a program, determine whether there is any input to the program that will cause the program to print the number 42.[4] Let's look at a sample program:

```
     x=?
     y=3
A    z=10
B    x=x+1
     z=z-1
     if z>0 goto B
     y=y-1
     if y>0 goto A
     print x
     stop
```

This program has a loop within a loop. The inner loop adds the number in **z** to **x**. In the outer loop, **z** is set to 10 and this outer loop is done three times. In conclusion, this program adds $10 \times 3 = 30$ to **x**. So the only way this program will output 42 is if 12 is the input. We conclude that there is an input that will output 42. However, determining what this program does was fairly simple. One can give fairly complicated programs where it is hard to tell if the program will ever print out a 42.

The Printing 42 Problem is also undecidable. While we will not prove this result, we can get an intuition that this problem is harder than the Halting Problem. Our aim with the Halting Problem was to determine if a particular input would get the program to halt. With the Printing 42 Problem we are asking if there is *any* input that would halt and give an output of 42. We would have to search all inputs.

Let us consider the *Zero Program Problem*. Examine the following two programs:

```
     x=?                    x=?
A    x=x-1                  y=x-x
     if x>0 go to A         print y
     print x                stop
     stop
```

Regardless of what is entered, each of these programs will always print 0. There are millions of other programs that will always perform the same operation: regardless of what is entered, the program will always output a 0 and halt. Such programs are called *zero programs*. It would be nice to be able to tell when a program is a zero program. The Zero Program Problem asks whether a given program is a zero program. That is, we would like a computer that accepts a program and outputs a Yes or No, depending on whether that program always outputs a 0. Alas, this too is unsolvable. It would take us too far afield to prove this result. Suffice it to say that this problem demands to know that the program halts for *every* input. That is a lot harder than the Halting Problem.

It is important to emphasize the difference between the Printing-42 Problem and the Zero Program Problem. In the Printing 42 Problem we ask if there is *at least one input* that will make the program output 42. In contrast, the Zero Program Problem asks if *every input* makes the program output a 0. Either way, both problems are undecidable.

Once we have shown that a particular problem is unsolvable, it is not hard to show that another problem is as well. The method used is *reducing one problem to another*, or a *reduction*.[5] Suppose there are two decision problems: Problem A and Problem B. Furthermore, assume that there is a method of transforming an instance of Problem A into an instance of Problem B such that an instance of Problem A that has a Yes answer will go to an instance of Problem B with the same answer, and similarly for No answers. (We do not impose the requirement as we did in the last chapter that the transformation be performed in a polynomial number of operations. Here we have no interest in how long such a transformation takes, just whether it can be done.) We might envision this transformation as in figure 6.6.

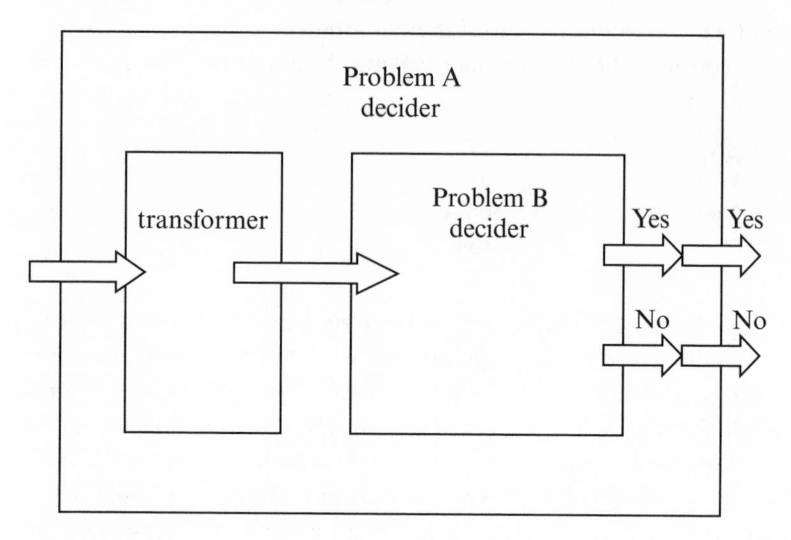

Figure 6.6
Transforming one problem into another

If such a transformation is possible, then

If Problem B is decidable, then Problem A is also decidable.

To decide Problem A, simply place an instance of Problem A through the transformer and see the results of the associated instance of Problem B. What if Problem A is undecidable? In that case, it must be the case that Problem B is also undecidable.

If Problem A is undecidable, then Problem B is also undecidable.

With such a transformation, we can say that Problem B is as hard as or harder than Problem A and write it as:

Problem A ≤ Problem B.

In the last section, I showed that the Halting Problem is undecidable. It is possible to describe transformations that show that

Halting Problem ≤ Printing 42 Problem

and

Halting Problem ≤ Zero Program Problem.

We conclude that these two problems are also undecidable.

Let's move on and show that we can reduce the Zero Program Problem to another problem. Consider the following two programs:

```
x=?                    x=?
y=3x+2                 z= x
print y                z=z+x
stop                   z=z+x
                       t=z+2
                       print t
                       stop
```

Although these programs look different and have different variables, it is easy to see that regardless of input, the same output is produced. Programs that perform the same task are called *equivalent programs*. It would be nice to be able to recognize when two programs are equivalent. The *Equivalent Program Problem* asks for a computer that accepts two programs and determines whether or not those two programs are equivalent. As you've probably guessed by now, the Equivalent Program Problem is unsolvable. This is actually easy to see by making the following reduction:

Zero Program Problem ≤ Equivalent Program Problem.

I will show that

Equivalent Program Problem is decidable ⇒ Zero Program Problem is decidable.

We showed before (but did not prove) that

Zero Program Problem is decidable ⇒ contradiction.

Putting these two together, we can piggyback on the previous result and show that the Equivalent Program Problem is undecidable.

Before we demonstrate this reduction, consider the short program

```
x=?
print 0
stop
```

This program always outputs 0, regardless of the input value. Let's call this program Z (for zero).

Assume (falsely) that you have a way of determining when two programs are equivalent. Suppose that you want to determine whether a program is the zero program.

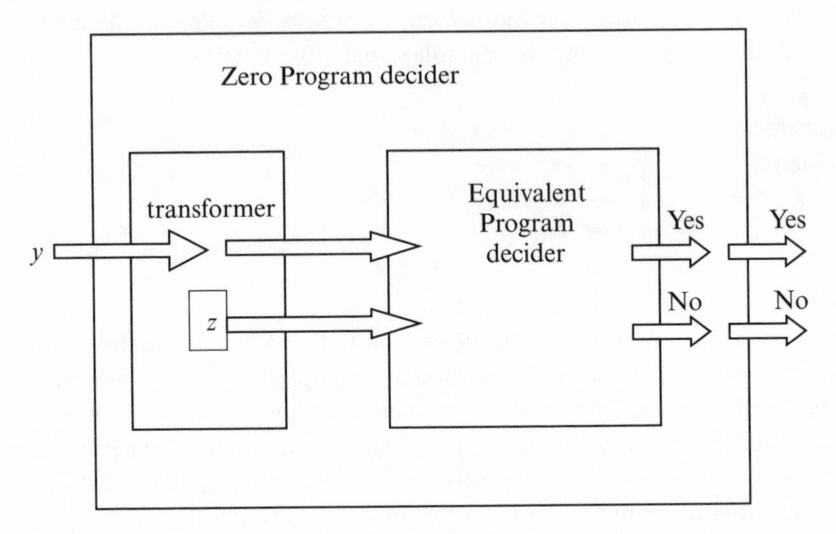

Figure 6.7
The reduction of the Zero-Program Problem to the Equivalent-Program Problem

Simply send the program and the program Z into the imagined equivalent-program decider as in figure 6.7. If the program is equivalent to Z—then the equivalent-program decider would tell us so. Using this procedure, we have created a zero-program decider. But we know that no such decider is possible. In conclusion, the assumption that there is an equivalent-program decider is untenable.

As you may have noticed, all the problems that we have so far shown to be undecidable have to do with determining different properties about programs. In other words, we have shown that there are no programs to determine certain properties about programs. This follows our theme that there are limitations when there is self-reference. In 1951, Henry Rice proved the granddaddy of all such theorems. In what has come to be known as Rice's theorem, it was shown that there is *no interesting property* about programs that can be determined by a program. This rather sophisticated result is proved by showing that for any interesting property P,

Halting Problem ≤ Property P Problem.

Since the Halting Problem is undecidable, the Property P Problem is also undecidable.

What about computers working in other areas like mathematics and physics? Are there other objective areas in which computers have limitations? In section 9.3 we will see that there are many other problems in mathematics for which computers are not up to the task.

A reader might be unmoved by this chapter. After all, what was shown is that there are several easily stated problems that a computer cannot solve. One might be led to believe that there are a few strange pathological problems that computers cannot solve, in contrast to the many that they can. Let us think about this more carefully.

Consider the problem of determining whether a number belongs to a set. This problem is easy for the following sets:

- The set of odd numbers
- The set of even numbers
- The set of prime numbers
- The set of numbers that can be written as a sum of five square numbers

For every one of these sets, a program can be written that accepts a whole number as input and—depending on whether the input is in that set—outputs a yes or no answer.

However, there are other types of sets. This chapter has shown that there are certain sets of whole numbers for which no computer program can be written that can decide if they are in the set or not. For example,

- The set of numbers of programs that will output 42 for some input
- The set of numbers of programs that always output a 0

So, certain sets of whole numbers are decidable and others are not.

Now let us do some counting. We saw in section 4.3 that there are uncountably infinite subsets of whole numbers. How many of those sets are decidable and undecidable. I mentioned in the beginning of section 6.2 that for every program, there is a unique whole number that represents it. Hence, there are only a countably infinite number of programs. Since every decidable set needs a program to decide it, there are only a countably infinite number of decidable sets of whole numbers (see figure 6.8). All the other sets of whole numbers are undecidable, so there are an uncountably infinite number of undecidable sets and only a countably infinite number of sets that are decidable. Chapter 4 showed the immense difference between countably and uncountably infinite.

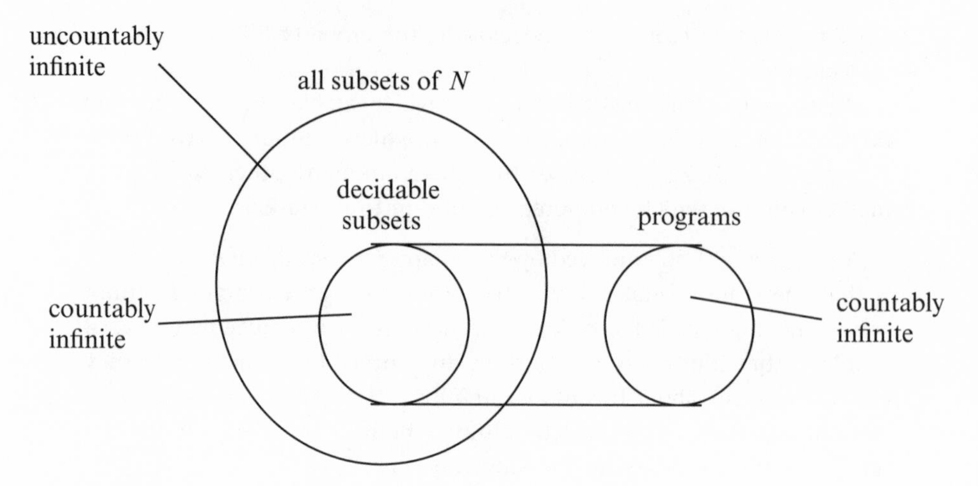

Figure 6.8
Properties and the programs that decide some of them

At the beginning of this chapter, we asked what tasks a computer can perform and what tasks are beyond a computer's ability. We discovered that computers can only do a minuscule amount. In fact, the vast majority of tasks cannot be performed by computers and are beyond the boundaries of reason.

6.4 A Hierarchy of the Unknown

So far, this chapter has shown that many problems are beyond the capabilities of any computer to solve. The obvious question is what is beyond the computability barrier. Turing asked this question in his original paper on the Halting Problem, and he gave an ingenious answer that provides a structure for the unsolvable problems.

We cannot solve the Halting Problem. However, imagine for a moment that we are able to solve it. The process that solves the Halting Problem could not be an ordinary computer, for we have shown that no ordinary computer can do it. Rather, there would have to be something nonmechanical about it, something "spooky." In ancient Greece (and in almost every other society), there were certain special people through whom the gods communicated. Such people were called *oracles* (from the Latin word "to talk"). On being asked a question, these oracles would fall into a trance and then supposedly give the answers from the gods.

Turing used the concept of an oracle to help classify unsolvable problems. Imagine that there was some type of oracle that could solve the

Halting Problem. Use this "spooky" halt oracle in a computer. During a computation, let the computer ask the oracle questions about whether certain programs halt. We might envision this as in figure 6.9.

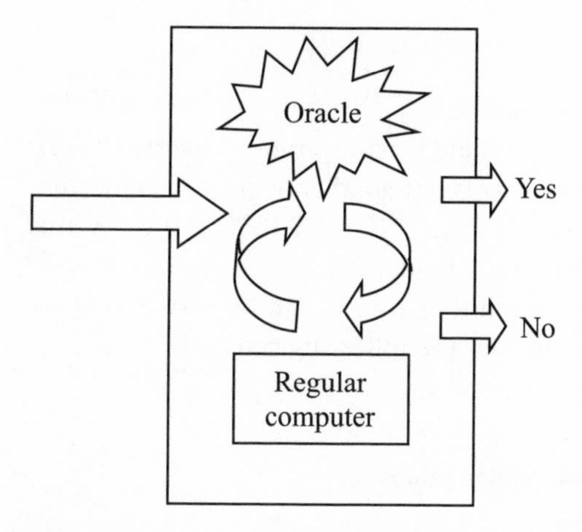

Figure 6.9
A computer that can ask questions from an oracle

Input enters on the left and the computer performs some regular computations. Along the way, the regular computer could ask the halt oracle a Yes or No question and depending on the answer, perform different computations. The computer might ask the oracle several questions as the computation continues. This extra feature of being able to ask questions might be added to our programming language by adding commands such as

ask Halt Oracle about z
if Oracle answers Yes goto A
if Oracle answers No goto B

Such commands might happen several times in a program and might happen for different values of **z**. A computer with this power would not only be able to solve the Halting Problem but many other problems that we saw were not computable by regular computers.

With the halt oracle, even problems outside the realm of computer science can be solved. One of the hardest open problems about numbers is called the *Goldbach conjecture*. The problem is to prove or disprove a conjecture about numbers that dates back to the middle of the eighteenth century:

Every positive even number greater than 2 is a sum of two primes.

This statement is easily shown to be true for small numbers:

- 4 = 2 + 2
- 18 = 5 + 13
- 220 = 23 + 197
- 8206 = 59 + 8147.

In fact, mathematicians have shown that this conjecture is true for all even numbers less than 10^{17}. However, that is not good enough. The conjecture says that it is true for *every* even number. After more than 250 years, this easy-to-state problem still taunts mathematicians.

Determining whether this conjecture is true would be very easy if we had access to the halt oracle. Consider the following program:

```
        x=2
A       x=x+2
        if x is the sum of two primes goto A
        stop
```

This simple program searches for a counterexample to the Goldbach conjecture. If no such counterexample exists, the program will go on forever. In contrast, if a counterexample does exist, then the program will halt. So, all you have to do is ask the halt oracle if this program will halt. Notice that saying there is no counterexample to the Goldbach conjecture will not earn you much fame. The problem is to *prove* that the conjecture is true.

There are many other problems in mathematics that could be solved if we had access to the mythical halt oracle. We will meet some of these problems in section 9.3.[6]

There are other possible oracles besides the halt oracle. For any oracle X, all programs that can ask questions of that oracle will be called *X oracle programs*. In particular, programs that pose questions to the halt oracle will be called *halt oracle programs*. Many unsolvable problems would be solved with halt oracle programs. The question then arises: Can halt oracle programs solve all unsolvable problems? Turing showed that they cannot. Just as we demonstrated that there is a unique number for every regular program, so too is there a unique number for every halt oracle program. With these numbers in hand, we can ask whether a particular halt-oracle program with a given number will halt. This decision problem is called the *Halting Problem for Halt Oracle Programs*. Using arguments similar to those in section 6.2, it can be shown that the Halting Problem for halt oracle programs cannot be solved by any halt oracle program. This problem is another unsolvable problem, but it is not computable even with access to the halt oracle.

Turing wasn't finished. Imagine that we had an oracle that can solve the Halting Problem for halt oracle programs. Call such an oracle the *halt′ oracle*. With this oracle we can solve many more problems. Any program that employs this oracle is called a *halt oracle program*. And, once again, we may pose the question of whether all problems can be solved by halt′ oracle programs. As you probably guessed by now, the answer is no. It can be shown that no halt′ oracle program can solve the Halting Problem for halt′ oracle programs. For that, one would need a *halt″ oracle*. And this goes on and on . . .

A hierarchy of unsolvable problems has been described. One can say that certain unsolvable problems are harder than others and certain problems are easier than others. Computer scientists have been able to characterize certain problems as halt′-computable problems, but not as halt″-computable problems. They have described different problems in each part of the hierarchy. We have been concerned with the limits of reason and here we have some clear structure of what is beyond the limits of reason.

Figure 5.18 in chapter 5 can now be extended to include problems that are not computable to form figure 6.10.

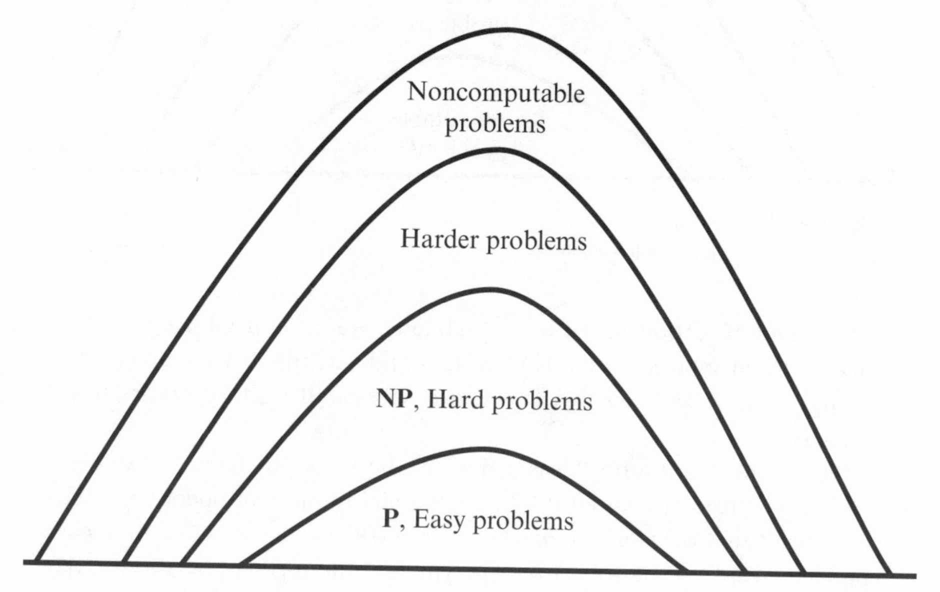

Figure 6.10
A hierarchy of problems

The set of easy, hard, and harder problems can all be considered one set that we called "computable problems." This section shows that the set of noncomputable problems has a hierarchical structure. Putting all this together gives us figure 6.11.

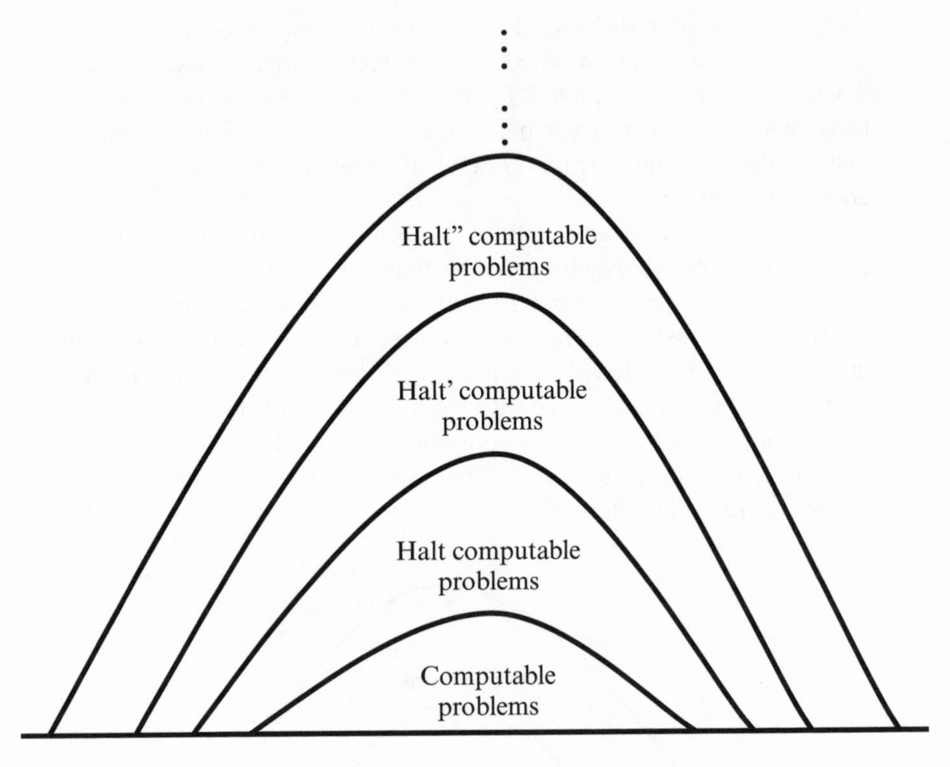

Figure 6.11
A hierarchy of unsolvable problems

At the end of section 5.3 of the last chapter, we introduced the **P** =? **NP** question and showed why it is important and exciting. It would be interesting to know whether this question is solvable if we take oracles into account.

First some definitions. **P** was defined as the set of problems that can be solved by a regular computer in a polynomial number of operations. Let us generalize. Consider any oracle X. Define $\mathbf{P}^X$ to be the set of X-oracle problems that can be solved in a polynomial number of operations. **NP** was defined as the set of all problems that can be solved by a regular computer in at most an exponential or factorial number of operations. Let $\mathbf{NP}^X$

denote the set of X oracle problems that can be solved in at most exponential or factorial number of operations.

In 1975, three researchers named T. P. Baker, J. Gill, and Robert Solovay published a paper containing two very interesting results. They described two oracles, A and B, such that

$\mathbf{P}^{A} = \mathbf{NP}^{A}$

and

$\mathbf{P}^{B} \neq \mathbf{NP}^{B}$.

The first result shows that there is an oracle, A, where hard problems can be done in few operations. The second result states that there is an oracle, B, where there is a hard problem that can be shown to require a large number of operations. So the long-standing **P =? NP** question is solved if you assume different oracles.

Results along these lines continued. In 1976, Juris Hartmanis and John E. Hopcroft showed that an oracle C exists such that the question of whether

$\mathbf{P}^{C} = \mathbf{NP}^{C}$ or $\mathbf{P}^{C} \neq \mathbf{NP}^{C}$

cannot be resolved by the usual axioms of mathematics.[7] It is not very clear what relevance these theorems have to the original **P =? NP** question, but it remains an interesting topic.

6.5 Minds, Brains, and Computers

This chapter is concerned with what is beyond the ability of computers. We might ask whether the human mind can perform tasks that computers cannot perform. Is the human mind more than just a machine? Is it limited like a computer?

We showed that a computer cannot solve the Halting Problem. Can a human being solve it? After all, isn't the human brain a type of computer? For small programs human beings usually can solve the Halting Problem. That is, we can look the program over and see if it will halt. But what about large programs? The Halting Problem is about *any* program. There are thousands of very smart people who work at Microsoft and many of them look over their large programs but fail to see that there are times when the programs will go into infinite loops. Does that mean that human beings cannot find all such infinite loops? What about other computing problems?

These questions are related to the question of the relationship between the human brain and the human mind. The human brain is a highly complex physical machine. In fact, it is probably the most complex physical machine in the entire universe. There is no doubt that mind is somehow related to brain. Whatever happens to the brain definitely affects what happens to the mind. If you doubt that, try to read this chapter after taking a few shots of tequila! Nevertheless, the relationship between the two is not clear. Our mind and our thoughts seem to be something more. We feel like we are more than just a bunch of firing brain synapses. We imagine ourselves to be far more than physical machines following the laws of physics. Human beings feel like they are conscious beings with free will and independence of thought. But are we really free? Do we really have control over ourselves? Ambrose Bierce defines the brain as "an apparatus with which we think we think."[8] Do we really think freely, or are our minds trained to think that they are free of the brain?

If a human mind is simply a physical brain following physical processes, then a human mind will also not be able to solve any of the problems presented here. In contrast, if the mind is more than physical brain, then perhaps the mind can perform more. Which is it?

Kurt Gödel felt that the human mind is more than just a machine. He felt that there are certain statements that cannot be proved by any mechanical system but that these results are nevertheless known/understood by human beings. Gödel said that this shows that human minds are more than just finite machines. If our brains are not finite machines, what are they?

Sir Roger Penrose, a famous professor of mathematics, offers similar arguments that the brain is more than just a machine. Penrose goes on to speculate that perhaps the brain uses the mysterious concept of quantum gravity to explain the seeming ability of humans to perform tasks that machines cannot. He claims that a computer that uses quantum gravity might be able to solve the Halting Problem. Penrose also says that this might help explain consciousness.

Douglas R. Hofstadter, an American researcher, speculates that the human mind has consciousness because it has the capability of self-reference. Since we can think about ourselves and think about ourselves thinking about ourselves, etc., we are capable of feeling that we are an "I." Contrast that with what we have learned in this chapter. This chapter tries to show that the computer's ability to perform self-reference is the cause of its limitations. Can we say that self-reference in computers brings limitations while in humans it causes consciousness? Perhaps. Do human beings really have self-reference? Do we really know what is going on inside our minds?[9]

Many great minds have thought about these questions without reaching any clear consensus.

One can turn around the questions posed above. Rather than asking if the human mind is more than a machine, ask if we can ever get a machine to act more like the human mind. An entire field of computer science, namely *artificial intelligence*, is dedicated to this question. The answer, in a way, depends on the answer to the previous questions. If a human mind is something more than a machine, there is no way to get a machine to really have a mind. On the other hand, if the mind is simply a fancy machine and it only seems like it is doing more than a machine, then we can expect that with enough time and ingenuity, we can get a computer to also seem like it is doing more than a machine. Is artificial intelligence possible? Even if we get a computer to act just like a human being, does that mean the computer will have consciousness?

The problem with trying to achieve artificial intelligence is recognizing whether this achievement has been made. One would need a legitimate definition of intelligence. Computers currently do amazing things that they could not do thirty years ago. Back then, it was largely believed that a computer would never win a match with a world chess master. In May 1997, this prediction was shown to be false. Deep Blue, a supercomputer developed by IBM, won a six-game chess match against the world champion Garry Kasparov. So computers can beat human beings at chess. And yet at present, no robot can beat a human at tennis. What if we looked thirty years into the future? There is no doubt that we would be shocked if we were given a chance to see what computers can do then. As time goes on, and computers gain more skills, we are less impressed with them and say that they are "just following a program." We always want more from our machines. "If only it could do this," then they would have "real intelligence." It seems that the boundary between what is "just following a program" and what is "real intelligence" moves as time goes on. Perhaps we have already achieved artificial intelligence.

Along the same lines as Deep Blue is Watson. In 2011, IBM had one of their language-recognition computers named Watson play against humans in the television game show *Jeopardy*. The computer decidedly beat the humans. Rather than asking if computers can reach the level of human beings, perhaps we should ask if computers have already surpassed humans. After all, a typical personal computer can now beat 99.99 percent of humans at chess. As many companies have shown, computers are much more efficient at answering phone calls and handling other tasks once performed only by people. Rather than seeing this as a decline of the status of man, see this as a triumph of man's ingenuity. Man has programmed machines to transcend their limitations.

This section contains more questions than answers. For most of these questions, this humble scribe resists the urge to push any purported answers. The questions are enjoyable enough.

Further Reading

Sections 6.1–6.3

The undecidability of the Halting Problem and other problems mentioned in section 6.3 can be found in detail in many theoretical computer science books—for example, Cutland 1980, Davis, Sigal, and Weyuker 1994, Sipser 2005, and Sudkamp 2006, to name a few. Rice's theorem can be found in his original paper, Rice 1953.

Davis 1980 and Harel 2003 provide popular accounts of undecidable results.

Section 6.4

Oracle computation can be found in the books listed above. Baker-Gill-Solovay 1975 and Hartmanis-Hopcroft 1976 are the original papers with the results described at the end of section 6.4.

Many of the ideas described in sections 6.2 through 6.4 emerged from the mind of Alan Turing. The life of this genius is wonderfully described in Hodges 1992. This is a very interesting scientific biography and is well worth the read.

Section 6.5

The literature on the questions posed in section 6.5 is immense. Some of the many books and articles that discuss these issues are Rucker 1982, Hofstadter 1979 (which won the Pulitzer Prize), and a most intriguing new book, Hofstadter 2007. Chapter 6 of Wang 1996 contains an in-depth philosophical discussion of Gödel's beliefs. Penrose 1991 and 1994 present his arguments and are full of interesting topics. On the Web, there is also an immense amount of literature both supporting and opposing Penrose's arguments.

Shainberg 1989 is a funny novel written by a neurosurgeon about a neurosurgeon performing a neurological operation on his own brain. The self-reference can drive you to . . . neurosis.

7 Scientific Limitations

Reason rules the world.
—Anaxagoras (500–428 BC)

Twas brillig, and the slithy toves
Did gyre and gimble in the wabe
—Lewis Carroll (1832–1898)

But you can travel on for ten thousand
miles, and still stay where you are.
—Harry Chapin (1942–1981), *W*O*L*D*

Science is the exact reasoning that we use to make sense of the physical world we inhabit. We use science to describe, understand, and sometimes predict physical phenomena. The limitations of the scientific endeavor are, in a sense, the most interesting.

I start with a short discussion of chaos theory and science's ability to predict the future. In section 7.2, I describe several different experiments in quantum mechanics that demonstrate the strangeness of our universe. Section 7.3 gives some intuition about relativity theory and what it tells us about space, time, and causality.

7.1 Chaos and Cosmos

Henri Poincaré (1854–1912) tells a morbid tale of a man walking down a street and getting killed because a roofer accidently drops a tile.[1] Had the man been there a few seconds later or earlier, he might have lived for many years. Had the roofer not dropped the tile or done so a fraction of a second earlier or later, the man would have continued his stroll through life.

What is the moral of the story? An obvious moral is that bad things happen in this world of happenstance. But that is not exactly a balanced judgment. Good things also happen. The vast majority of falling tiles do not hit anyone. The walking man could also have become a mass murderer. In that case, the falling tile is a good thing. Rather, the correct moral one should derive from the story is that small changes in events at one time can cause major changes at a later time. Had the man lingered home with his wife and kids for a few more seconds, he might have lived to play with his grandchildren. Had he walked a little quicker, he could have become a philanthropist who helped many people. Had the roofer's fingers been slightly less moist, the potential mass murderer below might have grown into a full-fledged mass murderer. One can imagine many variations of this shopworn tale for which the outcome would be totally different.

This obvious fact, that slight changes can cause major unpredictable changes, is known to everyone. If you had chosen a 42 instead of a 43 on that lottery ticket . . . If that death row pardon had just occurred two minutes earlier . . .[2] If only that bungee cord had been a little stronger . . . The reason why this fact is obvious to us is that we all live in a big, complicated world, and we know that there are so many things affecting every action that it is impossible to predict the future. But what about small systems in which we can perfectly describe how the different parts interact? Such small systems are described and investigated by scientists. One would believe that in such small systems, we would be able to predict the future. In this section, we will see that even in certain small describable systems, small changes can cause major changes.

Since the time of Newton, our vision of the universe has been that of a large, flawlessly functioning clock. We envision gears and springs that interact perfectly and with total predictability. It was the job of science to understand this functioning and to predict how this clock would continue over time. From Newton on, as the laws of physics were being formulated, there was optimism concerning our ability to know the entire universe. This optimism is expressed by one of the pioneers of mathematics and physics, Pierre-Simon Laplace (1749–1827), who wrote:

> We may regard the present state of the universe as the effect of its past and the cause of its future. An intellect which at a certain moment would know all forces that set nature in motion, and all positions of all items of which nature is composed, if this intellect were also vast enough to submit these data to analysis, it would embrace in a single formula the movements of the greatest bodies of the universe and those of the tiniest atom; for such an intellect nothing would be uncertain and the future just like the past would be present before its eyes.[3]

Laplace and others believed that this progress would continue forever and that eventually every scientific problem would be solvable and the future would be exposed before everyone's eyes. One can simply sit down with the right physical laws and calculate anything. However, by the beginning of the twentieth century, this optimism seemed unjustifiable. Poincaré and others had discovered systems for which it is humanly impossible to predict the future. Such unpredictable systems are called *chaotic*.

In 1961, Edward Lorenz, who was both a mathematician and a meteorologist, was studying computer simulations of weather patterns. He found a few simple equations that could describe certain weather patterns. Lorenz plugged these equations into a computer and studied the outcomes, which were very similar to common weather patterns that can be found in the real world. One day he wanted to review a certain simulation that he had previously seen. Rather than starting the entire simulation from the beginning, he attempted to start the simulation somewhere in the middle. The computer was dealing with numbers with six decimal places. However, to save space, it was outputting numbers with only three decimal places. Instead of typing in 0.506127, he entered the number 0.506. Thinking that the difference is less than one part in a thousand, he expected to get the same weather pattern. To his shock and amazement, the weather pattern that emerged was totally different from the one he intended. Lorenz realized that for these simple equations, the way the different parts interacted with each other, and the way outcomes of some of the equations became inputs to other equations, caused major changes in weather patterns depending on starting positions. In other words, an ever-so-slight change in the initial conditions of the equations can radically alter the rest of the simulation. In the real world, this means that a slight change in a weather pattern now can cause a major change later.

After exploring this, Lorenz went on to write a paper on this phenomenon with the colorful title "Predictability: Does the Flap of a Butterfly's Wings in Brazil Set Off a Tornado in Texas?" The title implies that a small change to the weather caused by a butterfly flapping its wings might cause a severe weather pattern far, far away. Not that the butterfly really causes the tornado, but rather, the flapping of its wings means that there will be a totally different weather pattern. The flapping might send an impending tornado off course and away from Texas. People can never track all butterflies and hence will never be able to predict the weather.

This effect has come to be known as the "butterfly effect" and has entered popular consciousness. The more scientific way of stating this is that the system shows "sensitive dependence on initial conditions"—that is, small changes in the initial setup of the system can cause major changes in the outcome. Systems that have this property are said to be "chaotic." The word *chaos* is from the Greek word for "gap," "lacking order," or "disorder." In contrast, the word *cosmos* comes from the Greek word for "order."

The opposite of chaotic systems—that is, systems that are not so sensitive to initial conditions—are called *stable systems* or *integrable systems*. Figure 7.1 provides a nice way of looking at the difference between such systems. The left-hand diagram shows a stable system with four different points that start and end near each other. In contrast, the chaotic system on the right has four points that start near each other but end in wildly different places.

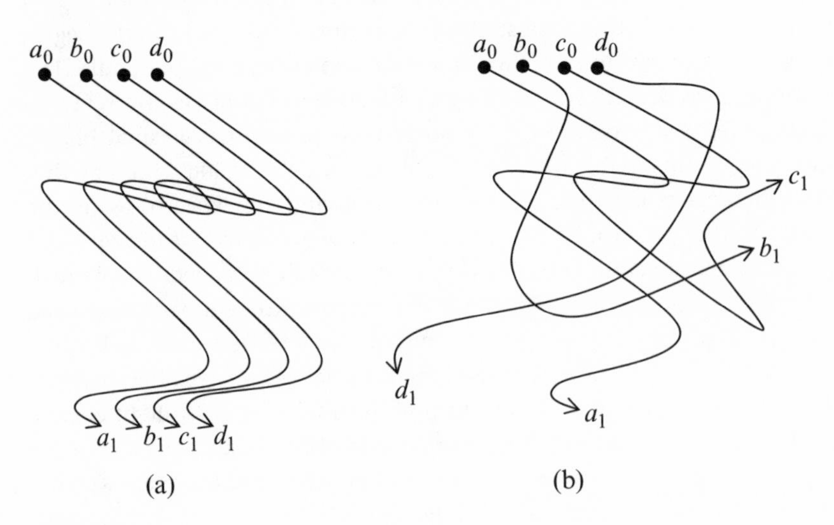

Figure 7.1
(a) Stable and (b) chaotic systems

Once a system is known to be chaotic, we lose the ability to make any long-term predictions about it. There is no way in the world that anyone can keep track of all the flapping butterflies in Brazil. We cannot retain information about a system if it demands infinite precision. Yes, the system is deterministic and we can write equations and formulas that describe its actions, but we cannot use these equations and formulas to predict any long-term outcomes. Chaotic systems force us to make a distinction between determinism and predictability. Determinism is a fact about the existence of laws of nature, whereas predictability is about the ability of human beings to know the future.

Researchers have gone on to show that many systems besides the weather are also chaotic. For example,

- Economists have determined that prices and the stock markets are dependent on small fluctuations.
- Biologists who study population dynamics have found that the rise and fall of certain populations of species are very sensitive to minor effects.
- Epidemiologists have found that the spread of certain illnesses can be affected by small factors such as a particular individual contracting the disease.
- Physicists studying simple systems of fluids have identified chaotic processes.

It can be shown that in all of these different systems there are physical processes that are deterministic but are not predictable.

It's not hard to make one of the simplest chaotic systems in your own toolshed. A pendulum is a rigid rod with a weight on one end. If you permit the rod to swing back and forth, you get a stable system that follows the simple laws of physics. First-year undergraduates spend much time calculating everything there is to know about such a pendulum. However, if you couple one pendulum to the bottom of another and let both rods swing, you get a double pendulum. This very simple system is totally chaotic. Both pendulums will swing in strange and unpredictable ways. The swinging is totally deterministic—that is, a physicist can write equations describing the motion of a double pendulum. The equations will take into account the lengths of both rods, the two weights, and the starting angles. Nevertheless, the system will be chaotic and unpredictable. To appreciate the butterfly effect, start this pendulum at some position and watch the way it swings. Then, attempt to replicate this motion from almost the exact same starting position—you'll see it swing in a very different way. Infinite precision is necessary to place such a pendulum in the same exact position twice. If you're not good with tools or too lazy to make your own double pendulum, you can find lots of cool videos of such contraptions online.

How can a deterministic process produce an unpredictable or seemingly random event? After all, if it is deterministic and we can write a formula that describes its short-term future, then why is it unpredictable in the long-term? To understand this we must remind ourselves that whether a process is deterministic represents an objective fact about the universe. Does this process follow fixed deterministic law? In contrast, whether a process is predictable is a subjective question about the mind. Do you have enough information about this system and its initial conditions to predict its long-term future or will it seem random? What is random to one person might be predicable to another person. My desk may look like a chaotic mess to you, but I know where everything is.

Flipping a coin is the classic example of an unpredictable process. However, if you had a laboratory with an exact machine that flipped a perfectly fair coin with no air interference, then you could flip a coin and know what the result would be beforehand.[4] The lack of precise information about the speed at which you are flipping a coin, lack of information about the exact weight of the coin, and lack of exact information about air interference while the coin is flying cause unpredictability. In a chaotic system, the imprecision of the initial conditions is so strong that the system becomes objectively unpredictable. No human being in the world possessing all the desired computing power can ever determine the the system's future. In conclusion, a deterministic process can cause an event that cannot be predicted.

Chaotic systems have shown that the Newtonian dream of calculating the future of every system is over. Laplace's optimism was for naught: the universe is far more complicated than he thought.

The truth is that science was never really about predicting. Geologists do not really have to predict earthquakes; they have to understand the process of earthquakes. Meteorologists don't have to predict when lightning will strike. Biologists do not have to predict future species. What is important in science and what makes science significant is explanation and understanding.

I once humorously pointed out to my thesis advisor, Alex Heller, that there are subatomic particles in nature that follow equations of motion that human beings cannot solve. And even though humans do not know where the particles will go, the particles seem to know exactly where to go. Professor Heller responded by saying that this shows that science has nothing to do with calculating or predicting. Calculations can be done by computers. Predictions can be performed by subatomic particles. Science is about *understanding*—an ability only human beings possess.

One might try to disregard the butterfly effect by saying that if two initial conditions were close enough, the results would be the same. We saw in the case of the weather that being the same to the third decimal number is not close enough. Perhaps initial conditions have to be the same to the fifth or the tenth decimal number? Although this sounds reasonable, it is, in fact, wrong. The best way to demonstrate that this is erroneous is by looking at something called the *Mandelbrot set*, which was formulated by mathematicians in the late 1970s and can be characterized as one of the prettiest parts of mathematics. The Mandelbrot set is an easily describable set of complex numbers. Start with a complex number c, square it, and add c to the result. This gives you another complex number. Square this number and add c to it again. Iteratively continue this procedure over and over again—that is, take the complex number z, calculate $z^2 + c$, and repeat.[5] Either of two things can happen to the numbers in this iteration:

- They can get bigger and bigger until they go off to infinity.
- The complex numbers can remain small.

The Mandelbrot set consists of complex numbers that remain small after this iteration process. The set is the central black part shown in figure 7.2.

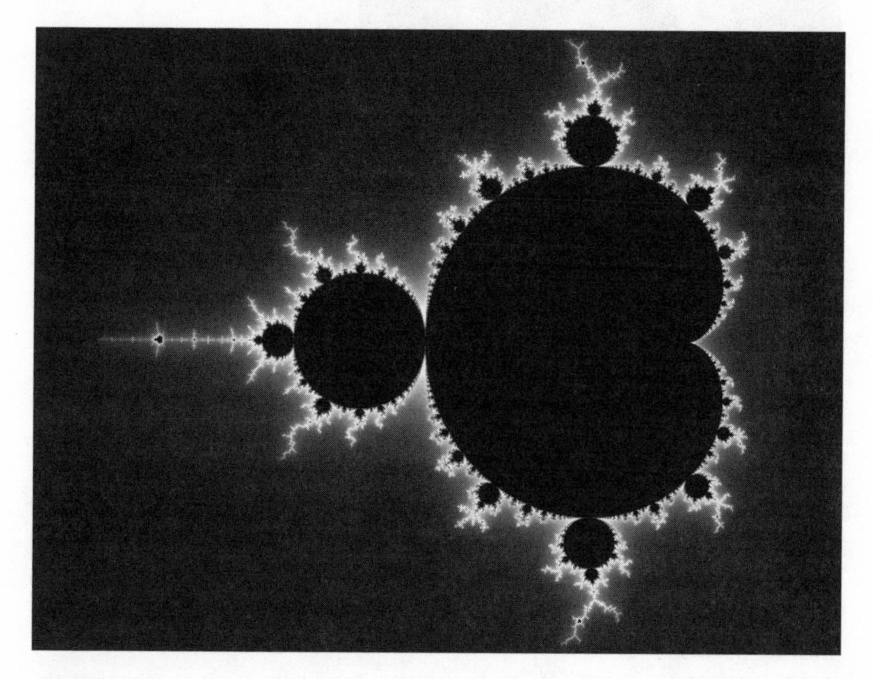

Figure 7.2
The Mandelbrot set

The fascinating part of the Mandelbrot set is the boundary between those numbers whose iterations go off to infinity and those that stay small. The boundary does not have any straight lines. It twists and turns and makes more and more shapes. The shapes are self-similar in the sense that within the boundary of the Mandelbrot set you will find shapes that are similar to the Mandelbrot set. This twisting and turning goes on forever and ever. Such shapes are called *fractals*. Since this is a mathematical shape and not a physical object, one can continuously magnify the image, as in figure 7.3.

Figure 7.3
A close-up of the top part of the neck of the Mandelbrot set

Two 2-dimensional pictures simply cannot do justice to the true splendor of this shape. With modern computers one can magnify the boundary with ease and in real time. It is worth looking online for some videos of the Mandelbrot set.

What is the area of a two-dimensional figure? Since the shape of the Mandelbrot set can be shown in a picture, its area is fixed and finite. It might not be possible to know exactly what the area is, but we can approximate it pretty well. In contrast, since the boundary is forever twisting and turning and getting more and more complicated as you look deeper and deeper, the length of the boundary can be shown to be infinite. This is a seemingly paradoxical situation where there is a finite area bounded by an infinite border.

The complexity of the boundary of the Mandelbrot set shows that the question of whether an iteration of a complex number goes off to infinity is not so simple. The answer depends on many digits of the decimal expansion of the complex number. In fact, it depends on all of its infinite digits. Similarly, when we have any chaotic system, we would need to know the initial conditions with infinite precision in order to make any reasonable predictions. Human beings cannot deal with infinite precision.

This sensitive dependence on initial conditions is related to other interesting phenomena that researchers are currently investigating. Scientists deal with complex chaotic systems that have characteristics that are called self-organizing, emergent, feedback, and so on. These properties make the world a very interesting place.

As an interesting sideline that brings to light some of these properties, let us consider a complex process called *morphogenesis* (from the Greek for "form" and "creation," i.e., the "creation of form"). This is a field of developmental biology that deals with how organisms take shape. One of the first people to study this subject in a serious manner was the computer scientist Alan Turing, who we met in the last chapter. Consider a zygote, a single-cell fertilized egg. It has the DNA of an organism. Through the process of mitosis, the zygote divides into two cells. Each of the cells has an exact copy of the same DNA. This process continues to four, eight, sixteen, etc., cells. As this process continues, something amazing happens. Certain cells become skin cells and some become bone cells. Some cells go on to form the nervous system, while others become the stomach muscles. The question arises, how does each cell know what to become? Why should one cell become part of a nail, while another becomes part of the brain? After all, they all have the exact same DNA. This is similar to handing complete blueprints of a building to construction workers as they enter a construction site without telling any worker where to go or what part of the building they should work on. And yet the building gets built! The zygote becomes a complete organism with many different parts. How do the cells self-organize? If all the cells are the same, why does the heart come out on the left? Turing was able to make some progress on these questions by coming up with certain equations that show how parts of the process work. (He did this before Watson and Crick actually described DNA. He was truly a genius!) The cells differentiate themselves by their relative position within the multicellular organism. They also differentiate themselves by where they are in relation to the outside of the organism. As they continue to differentiate themselves, they affect other cells. This is a type of feedback mechanism. There is extreme sensitivity to the position of each cell within the multicellular mass. The position determines the type of cell it will become. From a single-cell zygote, multicellular life emerges.

Certain systems are deterministic and unpredictable in a special way. Their determinism is a little less clear than the systems discussed above. An example of such a system is the three-body problem.

First some history. Newton taught us how two physical bodies interact. He provided a small, fantastic formula that determines most of the movement that we see in the world. This formula,

$$F = G\frac{m_1 m_2}{r^2},$$

applies to rotations of the planets around the sun, as well as to an apple falling to the ground. In detail, the force (F) between any two bodies depends on the mass of one body (m_1), the mass of the second body (m_2), and the distance (r) between the two masses squared. The G is simply a constant number that is needed to put all the measurements in order.

Obviously, there is a desire to generalize this formula to three bodies. In other words, we would like a formula for three different bodies and the forces between them, as shown in figure 7.4. One would imagine that such a formula would have variables m_1, m_2, and m_3 representing the three masses. There would also be variables for the distance between the first and second bodies (r_{12}), as well as the other distances (r_{13} and r_{23}). This problem of determining how three bodies interact is called the *three-body problem.*

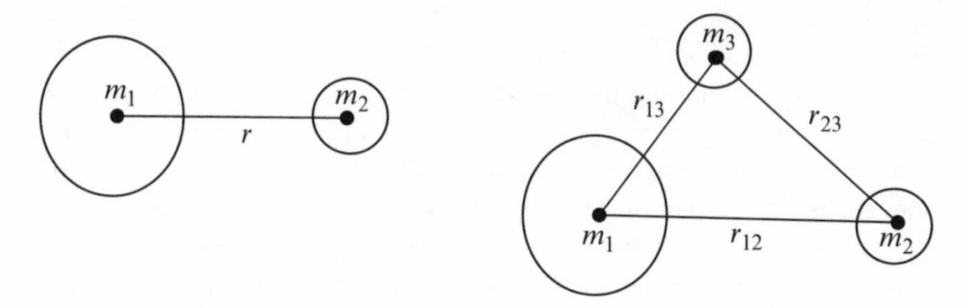

Figure 7.4
The two-body and three-body problems

The three-body problem and its generalization, the n-body problem, is not an abstract problem for absentminded physicists. Rather it occurs all the time. You are held to the Earth because you and the Earth are two bodies that follow Newton's beautiful formula. What happens if there is a pen near you? The pen is being pulled toward the Earth (release it from your grasp and watch it fall to the Earth), and there is a very subtle pull between you and the pen. Is there a simple formula that tells us how the three bodies will interact?

In the late nineteenth and early twentieth centuries, Ernst Heinrich Bruns (1848–1919) and Henri Poincaré showed that there is no simple formula that solves the three-body problem. By "simple" I mean a formula consisting of the usual operations as opposed to infinite sums and integrals. In short, they showed that the three-body problem is essentially unsolvable. That means that the complex relationship between you, the pen, and the Earth is beyond the limits of science.

It can also be shown that the three-body problem is chaotic—that is, it is extremely sensitive to initial conditions. However, such a chaotic system has a somewhat different character than the systems discussed at the beginning of this section. Yes, the three-body problem is deterministic in the sense that the components of the system follow set rules of how to act. Unlike Lorenz's weather formation, for which we can jot down some nice short formulas that describe the motion, no such nice formulas exist for the three-body problem. The system does follow set rules but we cannot easily describe these rules. This is yet a further step away from Newton's dream. Such systems are even further beyond the boundaries of predictability and reason.

Let us take a look at an example to show that the inability to solve the three-body problem is relevant to our world. Consider the Earth and its two largest, most influential neighbors: the sun and the moon. Using observations and Newton's formulas for two bodies, physicists have calculated that the Earth makes a full rotation around the sun in 365.2421897 days. One might argue that it is inappropriate to use Newton's formula for two bodies because the moon is also a player in this drama. In fact, the moon does have an effect on the Earth: the tides are influenced by the moon. However, since the sun is so much larger than the moon, and its influence on the Earth is so great in comparison to the moon's, we might as well ignore the moon when calculating the length of the year.

Contrast that with the calculation of a lunar month. If you investigate how long it takes the moon to go around the Earth you will find statements like "the approximate average length of the lunar month is 29.53 days." What does "approximate average length" mean? Can't someone tell us exactly how long it takes for the moon to go around the Earth? The answer is no! There are two bodies pulling at the moon: the Earth and the sun. Although the Earth is smaller than the sun, since the Earth is closer to the moon than the sun is, its influence cannot be ignored. This makes the problem of determining the moon's path a three-body problem. Such problems are unsolvable and there is nothing we can do about it. In fact, the lunar month can be up to 15 hours longer or shorter than 29.53 days. Since it cannot be calculated exactly, the average is approximated by keeping track of how many days passed in many months and then calculating the average length of the month. Hindu priests have kept records of this for over three thousand years and have a very exact average. The point is that the simple question of the length of the lunar month is unsolvable because the three-body problem is unsolvable.

It should be noted that modern mathematicians have partially solved the three-body problem and even its generalization, the *n*-body problem. Donald G. Saari, Quidong (Don) Wang, and Zhihong (Jeff) Xia, building on the earlier work of Finnish mathematician Karl Sundman (1873–1949), have produced formulas describing these systems. However, these formulas are immensely complex and not made of a finite number of simple operations. They demand an unreasonably large amount of computation to achieve even partial solutions. So, there are equations, but they are essentially useless for long-term predictions.

Statistical mechanics is a related area of physics that describes systems that are deterministic but not predictable. This branch of physics deals with phenomena such as heat, energy, water flows, and other systems where there are a huge number of components. Each component of these systems follows deterministic laws, but the systems are not predictable as with chaotic systems. Within statistical mechanics the nonpredictability occurs not only from sensitive dependence on initial conditions but from the huge numbers of components of the system. There is no way we can keep track of all the water molecules in a glass of tea. Whether we are dealing with hot air atoms in a combustion engine or water molecules hitting some piece of pollen in a flask, the number of items that we would have to track in order to make exact predictions is simply beyond our capability. In order to deal with such large ensembles, physicists must state their laws for such systems in a probabilistic language. With such laws they have become amazingly adept at predicting the large-scale phenomena of these ensembles of elements. The statistical laws coincide with experimental observations. It must be stressed that each component of the system obeys deterministic laws. Every atom bounces in ways that it is supposed to, like billiard balls on a pool table. Every water molecule hits the pollen in a deterministic manner. However, since there are so many such atoms and molecules, and since we can never know where each is, the laws must be given as probabilities. As previously stated, unpredictability or seeming randomness is simply a subjective result of lack of information. It is an expression of the limitations on our knowledge.

In this section I have concentrated on *practical* barriers to predicting the future of certain systems: there is simply no way we can know (with enough precision) the initial conditions of all the components of certain systems to make a valid calculation of the future. Too much information exists to be knowable. There is, however, a curious little puzzle that shows the inherent *logical* impossibility of perfectly predicting the future. This puzzle is a version of our old familiar friend: it is a self-referential paradox.

Imagine for a moment that we are capable of perfectly predicting the future. The simplest formulation of this puzzle is that we program two computers with this ability. Call one computer Mimic and the other Contrary. Both will simply print out the word *true* or *false*. The Mimic computer will predict what Contrary will print at some specific time in the future and print the same thing. In contrast, the Contrary computer will predict what Mimic will do at that specific time and print the opposite.[6] If Mimic will print *true*, then Contrary will print *false*. If Mimic will print *false*, then Contrary will print *true*. This is a paradox. The astute reader will notice that this is nothing more than a simple formulation of the liar paradox that we met in chapter 2:

L_2: L_3 is false.

L_3: L_2 is true.

In a similar vein, we can formulate the same puzzle with one computer. Program a single computer to predict what it will do at some specific time in the future and have it do the opposite. In other words, the computer will negate its own prediction. Such a computer would cause a contradiction and hence cannot exist.

This is nothing more than the *crocodile's dilemma* paradox of classical Greek philosophy.[7] A crocodile steals a child and the mother of the child begs for the return of her beloved baby. The crocodile responds, "I will return the child if and only if you correctly guess whether or not I will return your child." The mother cleverly responds that he will keep the child. What is an honest crocodile to do?

In section 3.2 we saw that reason does not permit us to change the past. Here we see that reason also restricts us from knowing the future.

Let us summarize the types of physical systems that we have discussed. To us, the most interesting aspect of a system is the amount of human knowledge we can acquire about it. Figure 7.5 presents a crude hierarchy of such systems and examples that we have mentioned.

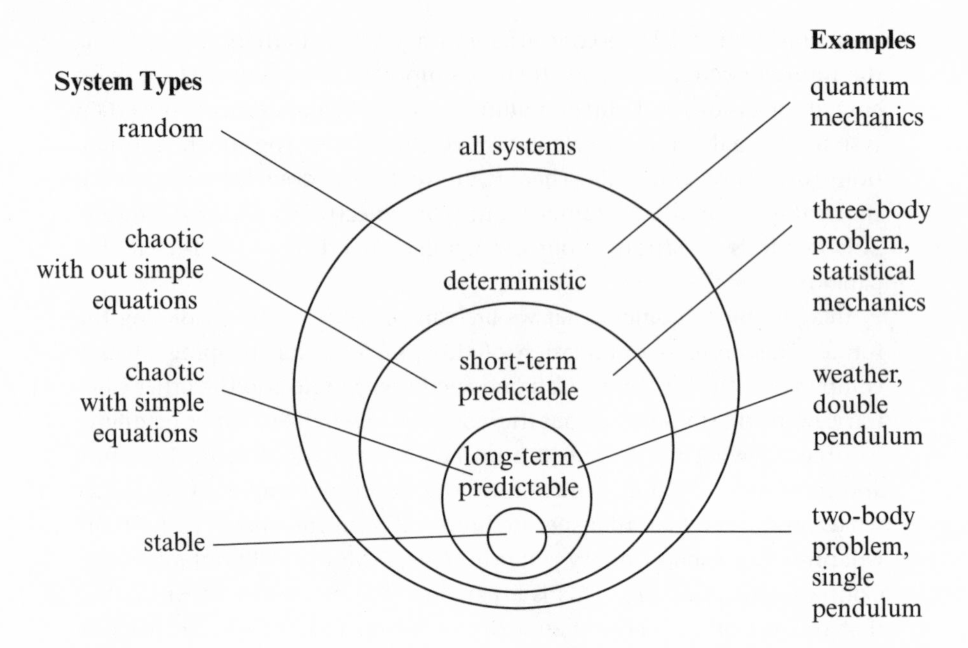

Figure 7.5
Physical systems

Begin with the innermost circle. These are deterministic systems that we know the most about and can predict with ease. They are stable systems. As we have discussed, regular single pendulums and the two-body problem that can be solved with Newton's small formula are examples of such systems. The next level is the central focus of this section, namely chaotic systems. These are also deterministic and we can write simple formulas to describe their short-term behavior, but because they are extremely sensitive to initial conditions, long-term predictions cannot be made. The weather and the double pendulum are just a few of the many examples that we have discussed. Further away from predictability are chaotic systems that do not possess even simple formulas describing the short-term behavior of the system. These systems are still deterministic but because of their complexity and/or the fact that they have a huge number of components, they cannot be described by simple formulas. Quintessential examples of such systems are three-body problems and systems described by statistical mechanics. Finally, stepping outside of deterministic systems, we find random systems. Here, no formula exists that determines the future of a system nor can there be a formula to predict the short-term behavior of the system. The only[8] known example of such random behavior is quantum mechanics, covered in the next section.

I conclude with a little meditation on the number of physical phenomena that can and cannot be explained/predicated by science.[9] In a sense, language, be it spoken or written, be it natural language or exact formulas, is countably infinite. There is no longest word or longest novel, because there is no limit to the longest formula, and so on. This makes language infinite. However, it can be alphabetized or counted, which makes language countably infinite. In contrast to language, which can be used to describe or predict phenomena, let us examine what is really "out there." It is plausible to say that there is an uncountably infinite number of phenomena that can occur.[10] This is stated without proof because I cannot quantify all phenomena. To quantify them, I would have to describe them and I cannot do that without language. So there might be an uncountably infinite number of phenomena and only a small, countably infinite subset describable by science. This is the ultimate, nonscientific (science must stay within the bounds of language) limitation on science's ability. At this point we must take Wittgenstein's dictum to heart: "What we cannot speak about we must pass over in silence."[11]

7.2 Quantum Mechanics

Probably the greatest development in all of physics is quantum mechanics. With the exception of gravity, all physical phenomena are described by this theory. Phenomena ranging from the interactions in an atom to the workings of the sun follow the laws of quantum mechanics. However, quantum mechanics has also taught us that we have a severe limitation when it comes to understanding how the particles of our universe behave. They are extremely mysterious and defy our attempts to make sense of them.

In this section, I discuss some highlights of quantum mechanics and show that our universe is a very strange place indeed. There are ideas and concepts here that are counterintuitive and will blow your mind! Nevertheless, they are all true. It is important to realize that quantum mechanics is not an approximation to a theory. It is the most exact science we know. The weirdness I describe cannot be brushed away. As strange as the results sound, they must be accepted as science and not as science fiction.

Although there are many different, strange, and counterintuitive parts of quantum mechanics, I will show that most of its bizarre features can be understood as consequences of the following intuitive idea:

The Wholeness Postulate: The outcome of an experiment depends on the ***whole*** setup of the experiment.

This makes sense. After all, you would expect that different experiments would yield different outcomes. What is unexpected is the dependency on the entire experiment as opposed to just part of it. I emphasized the word *whole* because, as we will see, most of the strange aspects of quantum mechanics can be understood as simple consequences of what we mean by that word. I will return to this postulate over and over throughout this section.

Rather than getting into the nitty-gritty details behind quantum theory, I will go through several experiments and explain what they tell us about our world. The physical experiments are stressed because we want to emphasize that this is not just some strange theorizing. Rather, we are talking about the real world.

The End of Well Defined: Superposition

The first experiment is called the double-slit experiment. Richard Feynman (1918–1988), in discussing this experiment, waxed lyrical: "We choose to examine a phenomenon which is impossible, absolutely impossible, to explain in any classical way, and which has in it the heart of quantum mechanics. In reality, it contains the only mystery. We cannot make the mystery go away by 'explaining' how it works. We will just tell you how it works."[12] The experiment was first performed by Thomas Young (1773–1829) in the early nineteenth century. Imagine a barrier with two slits in it that we can view from above, as in figures 7.6 and 7.7. In the first figure we close one of the slits and shine a light at the barrier. As expected, the light will pass through the slit and radiate out to the screen on the right. The light will be intense directly across from the open slit and will be less intense farther away from the slit. This is depicted by the curve on the right of figure 7.6.

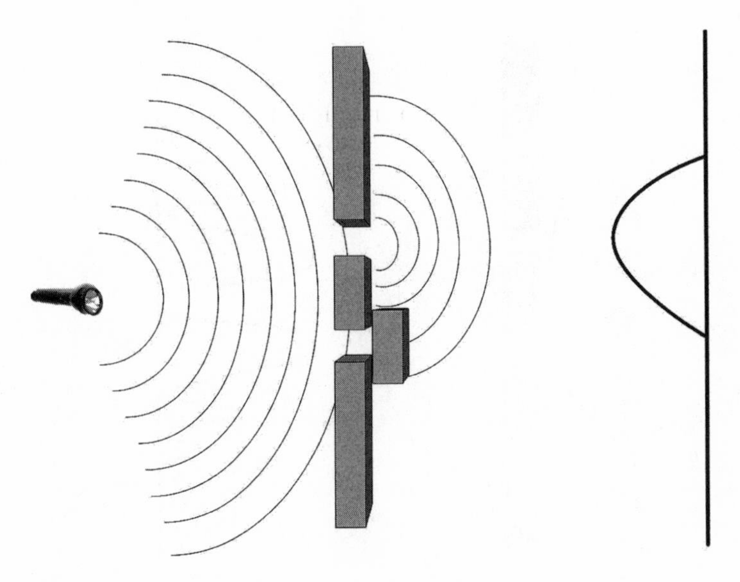

Figure 7.6
Light going through a single slit without interference

If the second slit is opened, something very interesting happens. The light passes through both slits, but rather than having the expected pattern, there will be an alternating pattern where some regions have intense light and some have no light, as in figure 7.7. The reason for this strange light pattern is that light is acting like a wave with crests and troughs. When the crests of the light wave from one slit meet the crests of the other light wave, they add up and the light is intense. In contrast, when the crests meet the troughs, the waves cancel each other out and there is no light at all. When such canceling occurs, we say the light has "an interference effect." This is similar to waves in a pond after dropping in pebbles. So far so good.

Now, for the amazing aspect of this experiment and probably the most mind-blowing result in all of science. Physicists have a way of performing this experiment by releasing one piece of light at a time. A piece of light, or a light atom is called a *photon*, and physicists have become adept enough to be able to fire one photon at a time through the slits. After releasing a photon, it passes the barrier, hits the right-hand wall, and makes a little light. They can perform this experiment millions of times and see the pattern that the photons make on the right-hand side. The remarkable aspect is that an interference pattern is still found. That is, many individual photons will land in the area where there was high intensity, few will land

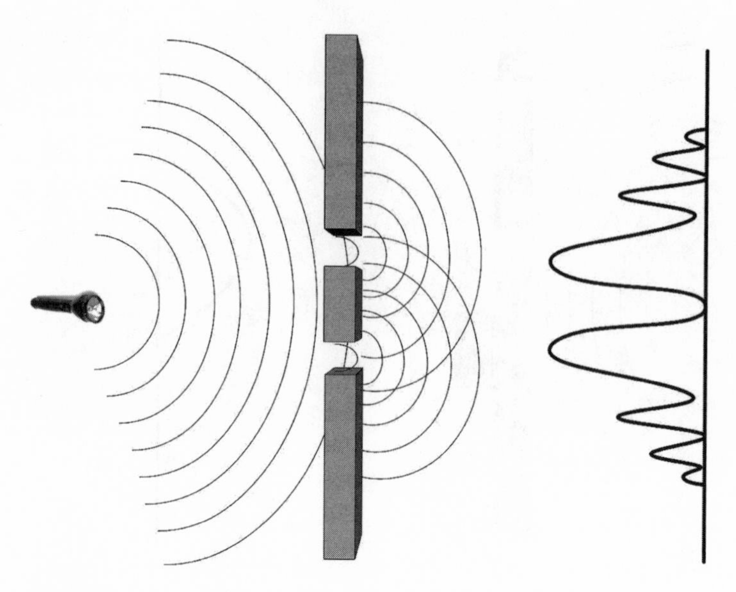

Figure 7.7
The double-slit experiment with interference

in areas where there was low intensity, and none will land where there was total interference. How can this be? When we have many photons, we can say that the photons are interfering with each other like waves in a pond. But when each photon is released one at a time, what can a single photon interfere with to create such a pattern? The answer is that the single photon interferes with *itself*. The individual photon does not pass through the top or bottom slit. Rather, the photon passes through both slits simultaneously and when it (singular) emerges through both slits, it interferes with itself.

How can one object pass through both slits simultaneously? That is the major mystery of quantum mechanics. Usually an object has a *position*—that is, a single place where the object is found. But here, an object can be found in more than one position. The phenomenon of being in more than one place at one time is called *superposition*.

Whenever I open my eyes, I see objects in exactly one place, not in many places. It seems as though we live in a world with position, not superposition. The computer screen I am looking at is only in one place. And yet there is superposition. We might not see it, but we see the consequences of superposition. After all, we do not see wind, but we see the trees bend.

Researchers are not in total agreement as to why we do not see things in superposition. All that is known is that when we examine the results of a quantum experiment, or to use the right lingo, when the system is *measured*, we no longer see a superposition. We say the system *collapses* from a superposition of many positions to one particular position. The *measurement problem* asks why this collapse occurs and is one of the major discussion points in the philosophy of quantum mechanics.

We need a simple example of such a superposition and a collapse. In high school we learned that an electron orbits the nucleus of an atom in one of its shells. This is a little false. Actually the electron orbits the nucleus in a superposition of *all* of its shells. Such a superposition is called a "probability cloud." Like a cloud, it is a bit amorphous. It is only when we measure the position of the electron that it collapses to one particular level, as depicted in figure 7.8.

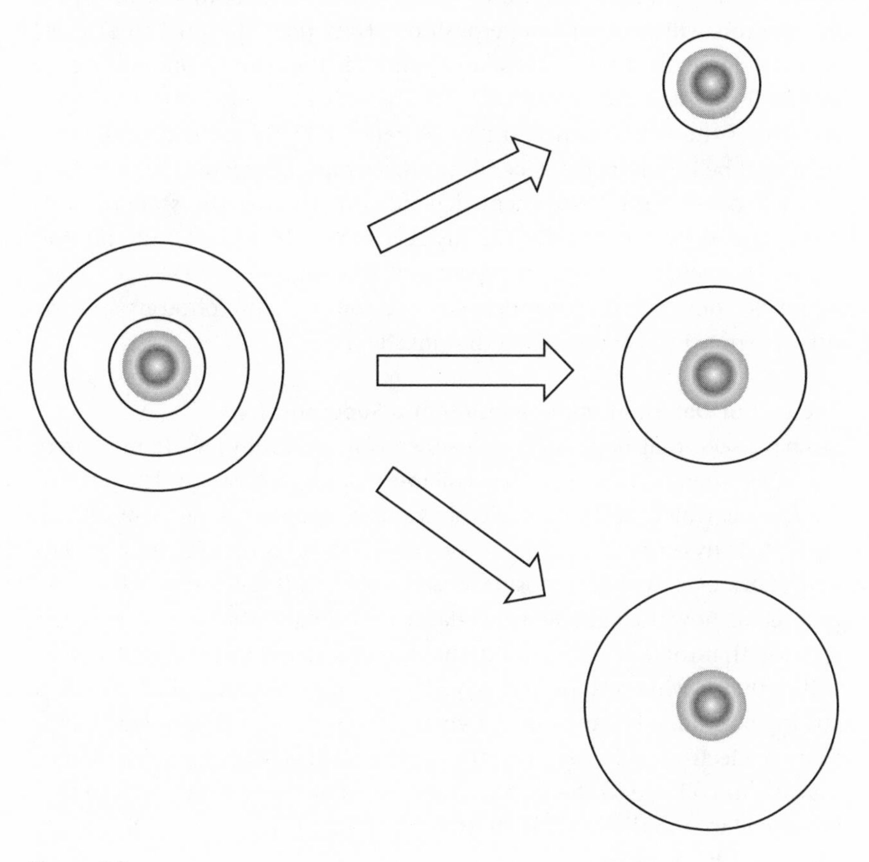

Figure 7.8
An electron in a superposition of orbit that collapses to one orbit

This concept of superposition is the main idea in quantum mechanics. It will be our central concern throughout the rest of this section. The position of an object is not the only property that is subject to such craziness. Many other properties in the quantum world like energy, momentum, and velocity will also have many values simultaneously and then collapse to one value when we measure them. For all these different properties of quantum systems, superposition will be the norm until the system is measured.

Before we leave the double-slit experiment, let us rephrase the experiment in a slightly different way. The photon leaves the light source, and then depending on whether the barrier has one or two slits open, the photon will have a position or a superposition. If only one slit is open, it will remain as a single photon. If, on the other hand, both slits are open, the photon will go into a superposition. How does the photon "know" what to do when it leaves its source? Should it remain as one photon or should it go into a superposition? The answer can be seen from the point of view of the Wholeness Postulate, namely, that the outcome (whether there will be interference) depends on the setup of the whole experiment. The outcome of the experiment depends on whether the second slit is open. This does detract from the mystery. After all, how can the photon "know" the setup of the entire experiment when it leaves the source? There might be some real distance between the source of the photons and the slits. There is no real answer to that mystery.

The End of Determinism: Collapsing of a Superposition

We have seen that objects are in a superposition until they are measured, and when they are measured they collapse to a single position. The obvious question is which of the possible positions a measured superposition collapses to. Physicists tell us that it is random. There is no deterministic law that states exactly which position each object will collapse to. The laws that tell us how the particle will collapse are probabilistic laws. That is, the laws say that there is a probability that it will collapse this way and a probability that it will collapse that way. For example, in terms of figure 7.8, a law in quantum mechanics might say that there is an 11.83 percent chance that the electron will collapse to the outside shell, a 47.929 percent chance that it will collapse to the middle shell, and a 40.241 percent chance that it will collapse to the inner shell. However, what the electron will actually do cannot be determined.

In slightly more detail, a superposition is described as many possible positions of the system. The different positions are indexed by complex numbers—that is, every position has a complex number associated with it. When a superposition collapses, the chances that it collapses to a particular position are determined by that complex number.

The fact that the laws are given by probabilities should not lead one to think that quantum mechanics is somehow an approximation of a real theory. On the contrary, quantum mechanics is the most exact physical theory that we have. Experimental evidence shows that our predictions are correct to many, many decimal places. One must realize that the predictions of quantum mechanics are made about subatomic particles. Experiments are done on a very large number of such subatomic objects. The outcomes show that the many particles follow the probabilities given. So we do not know what each particle will do, but we do know what a large ensemble of them will do.

A distinction must be made between the laws of quantum mechanics and the laws of chaotic systems that we met in the last section. With chaos theory we learned about some processes that are deterministic but not predictable. Here we have processes that are not even deterministic and of course not predictable. If there are no exact laws that describe the actions of all the parts of the system, then we definitely cannot predict where the system will end. It is one thing not to be able to predict the long-term future of a system, but it is far worse not even to be able to tell what will happen in the short term. We cannot determine what a single object in a quantum system will do in the short term. This takes us one more step outside the bounds of reason.

At this point, you might be skeptical about this lack of determinism. After all, all the other laws of physics are deterministic. There must be something that physicists are missing that would explain the seeming randomness of it all. You would not be alone with such skepticism. Albert Einstein, one of the forefathers of quantum mechanics, also did not believe it. He expressed his skepticism with the rather colorful phrase "God does not play dice with the universe." Einstein did not believe that the fundamental laws of physics are random. Supposedly, Niels Bohr responded to Einstein by saying, "Don't tell God what to do." The universe works the way it does and it does not have to satisfy our wishes. Although we might want Einstein to be correct, most contemporary physicists assure us that Einstein was wrong and the universe at its very core is not deterministic and hence random.

There are those who have taken up Einstein's challenge and are looking for laws of quantum mechanics that are somewhat deterministic. They believe these laws are governed by *hidden variables*. That is, there are extra variables in the system that cannot be seen but when they are taken into account, the laws of quantum mechanics are deterministic. This is similar to a chaotic system in the classical world. Consider the lottery machines that work by mixing up balls in a giant jar. Such machines are used because there is no way to predict which balls they will choose. Nevertheless, despite the machine being unpredictable, the laws describing what goes on in the machine are totally deterministic. Every ball bounces around following fixed deterministic laws, but there are too many individual parts of the system for there to be predictability. The exact positions of every ball and every air molecule are the hidden variables in this system. Some physicists posit that quantum mechanics also has variables that cannot be seen. Such hidden variables are a possibility, and if they are true then all of the laws of the universe are deterministic. I return to the possibility of hidden variables at the end of this section.

Let us come back to the collapse from a superposition to a position. There is an easy experiment that can be seen as a demonstration of this collapse. First a little background about light and polarization filters. Light can be thought of as a wave, which can come in many different forms. Consider three typical waves in figure 7.9. The first wave moves up and down and is called a *vertical wave*. The second wave moves left and right and is called a *horizontal wave*. The final wave moves in a *diagonal* direction. I have highlighted three directions, but obviously light can come in any direction. Only the light from a laser has all of its waves uniformly lined up. The usual light that we see has many different waves coming in many different directions.

Sheets of flexible plastic called *polarization filters*, which block light coming in certain directions, are used in fancy sunglasses. They can be oriented in different directions and block light in those directions. One can think of the filter as a measuring instrument that collapses a superposition of different directional light waves into one direction. We will draw the filters as discs with a slit to indicate its direction, as in figures 7.10 through 7.12. A filter in the horizontal direction will permit all horizontal waves to pass through it but it will block all vertical waves. What about the intermediate waves? The closer a light wave is to being horizontal, the more of a chance there is for the wave to pass through the horizontal filter. The diagonal waves, which are halfway horizontal and halfway vertical, will allow have about 50 percent of their light go through and about 50 percent of their light will be blocked.

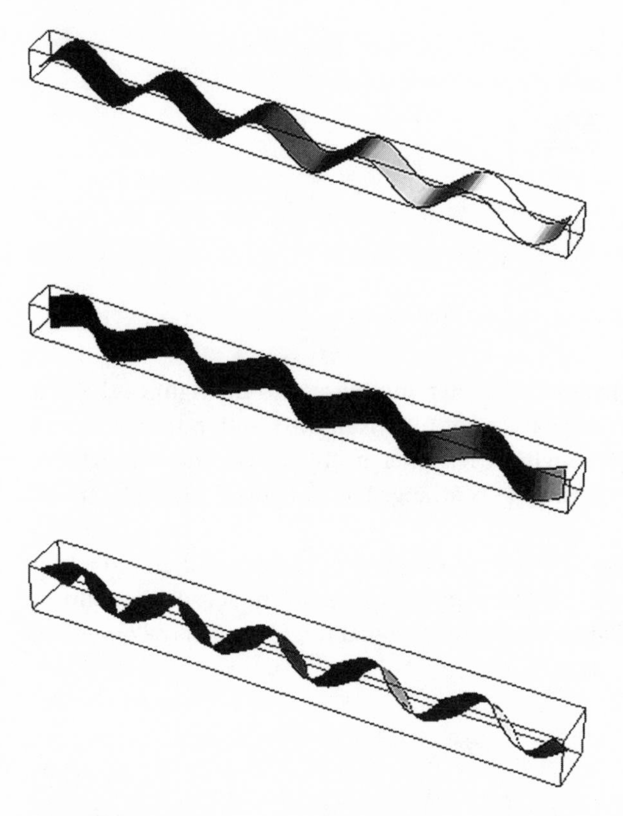

Figure 7.9
Vertical, horizontal, and diagonal light waves

Let us combine these filters for a fascinating result. Take a horizontal polarization filter and place all different types of light through it as in figure 7.10. About half the light will pass through it and that light will be a horizontal wave.

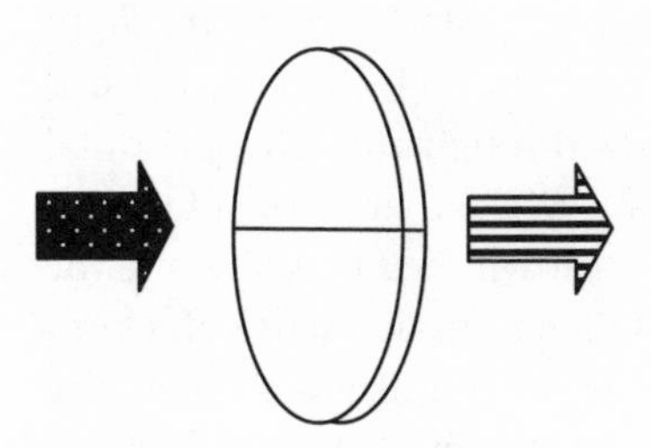

Figure 7.10
Light going through a single polarization sheet

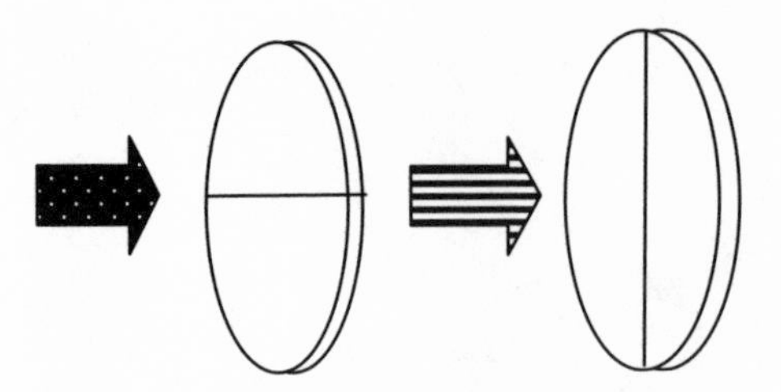

Figure 7.11
Light not passing through two polarization sheets

Now take a vertical polarization filter and place it to the right of the first one as in figure 7.11. Since the right filter is in the position that blocks all the light that passes through the left filter, nothing will come through the right filter. So with the two filters arranged as in figure 7.11, no light will permeate both.

Now for the magic. Take a third diagonal polarization filter and add it to the filters already there. Don't put the filter to the left or the right of the previous filters. Rather, place the diagonal filter between the other two filters as in figure 7.12.

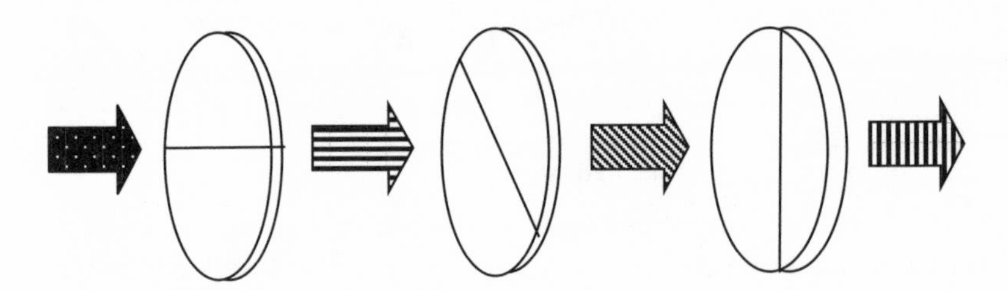

Figure 7.12
Light passing through three polarization sheets

Something amazing happens: whereas two filters can block all the light from going through, three filters permit light through. In fact, there is really no magic here. Whatever light passes through the left horizontal filter comes out in the horizontal direction. When that horizontal light hits the diagonal filter, approximately half of it will be blocked while the other half will pass through in the diagonal direction. In a sense, the horizontal light is in a superposition with respect to the diagonal filter. The middle filter measures this diagonal light and collapses it. Now that diagonal wave will meet the vertical filter on the right. On average, half of the diagonal wave will be blocked while the other half will pass through the vertical filter and come out as a vertical wave. Since the light in figure 7.12 must pass through three filters and on average half the light is blocked by each filter, only one-eighth of the light will pass all three of the filters. Nevertheless, some light will pass. So even though two filters can prevent all the light from passing through, three filters permit some of the light to pass.

The End of Certainty: Heisenberg's Uncertainty Principle

Watch a car speeding down a highway. It is easy to determine both the color and speed of the car as it moves. One can effortlessly figure out a person's weight and height simultaneously. Similarly, with ease one can determine the exact position and momentum of a flying baseball. The point is that it is not hard to determine two different properties of an object. This obvious fact is true for the world we live in but simply fails in the quantum world. There are situations in the subatomic world where one cannot determine two properties at the same time.

Heisenberg's uncertainty principle is one of the central features of quantum mechanics. It says that there are certain pairs of properties of subatomic systems such that it is impossible to know both of these properties at one time. For example, it is impossible to know both the position and the momentum of a moving subatomic particle. The doctrine that states this limit on human knowledge is called *complementarity*.

In detail, given two such properties, X and Y, we will get one pair of answers if we measure X first and then measure Y, and other answers if we measure Y first and then measure X. For example, first measuring the momentum and then the position of a subatomic particle will yield different answers than first measuring the position and then the momentum of that particle. This leads to the obvious question: What exactly are the momentum and the position of the object? Why are we getting two different answers here? Aren't there objective values of these properties that are independent of our observations?

We must stress that our inability to know both values simultaneously is not some problem with our present-day technology. It is not the case that as our microscopes and measuring instruments improve, the uncertainty principle will be less bothersome. These are not technological limitations. Rather, complementarity is an inherent limitation of our ability to know about our universe.

Notice that a radical new element comes into play here. The outcome of measurement Y depends on whether or not the person doing the experiment decided to perform measurement X first. The experimenter is not separate from the experiment. Rather, the experimenter has become part of the experiment and influences the outcomes of the experiment. The person who does the experiment influences the world that he or she is investigating. This is a revolutionary idea. No longer is there a closed system and an experimenter examining that closed system. Now the human experimenter is also part of the system. This can be seen in terms of the Wholeness Postulate: the experimenter is part of the *whole* experiment.

Researchers going back to Bohr take this one step further. They proclaim that it is wrong to say that humans learn the properties when they measure them. Rather, they say that the very act of measurement causes the properties to become well defined.[13] Before the measurement, it is not that we do not know what the property is, rather there is no property to know. Before any measurements, the properties are in a superposition. When X is measured, the X property collapses to a single value while the Y value remains in a superposition. If the Y property is then measured, then it too collapses. The point is that if the measurements were done in a different order, then the values could collapse into different values.

Philosophers discuss a philosophical position called *naive realism*. This is the belief that physical objects have a real existence outside of our minds and these objects have well-defined properties that can be determined when they are observed. This is obvious and every child knows this to be true. However, Heisenberg's uncertainty principle and the concept of complementarity destroy naive realism. Properties of an object do not exist before they are measured, and even when they are measured, the property depends on how it is measured. The realism that you feel is true in the regular world of cars and baseballs. It is not true in the subatomic world. Such realism is naive.

Do you believe all of this? A sane person would be justifiably skeptical. Read on!

The End of Ontology: The Kochen-Specker Theorem

At this point, you are legitimately allowed to feel disbelief and yell in a loud voice, "Balderdash!" You might say that all this talk of superposition is foolish and that when a subatomic object is measured, a property is determined that was there before we measured it. The measurement did not cause the property to come into existence; it was always in existence. Alas, this seemingly sane and bold stance that you are taking is wrong and I will prove it.

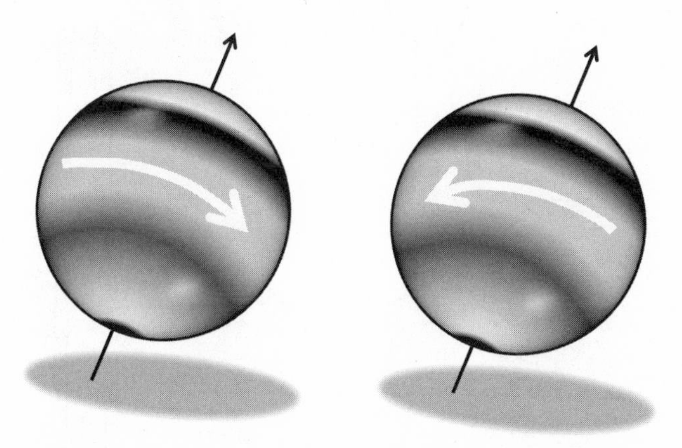

Figure 7.13
Two possible spins for a given direction

We need some preliminary notions. One of the central ideas of quantum mechanics is the notion of *spin*. Certain subatomic particles have spin. This is not the same as the usual notion of a basketball spinning on a finger. It is a little more complicated than that. Nevertheless, given a direction, a particle can spin one way, the other way, or not at all. Figure 7.13 shows the direction with the arrow, and we find it spinning one way or the other. We might call the directions *positive spin* and *negative spin*, or *spinning up* and *spinning down*. As with most properties of quantum mechanics, before a particle is measured or observed, the particle will be in a superposition of both positive spin and negative spin.

We can "see" the particle spinning by performing a *Stern-Gerlach experiment*, as depicted in figure 7.14. A ray of particles is shot out of a source. North and south magnets are set up to test for spin in a certain direction. The ray will then split up with the particles spinning in one direction going to one part of the screen and particles spinning in the other direction going to the other part of the screen. A particle can also not have spin and will simply go forward. One can think of the particles as spinning magnets (or charges) that are attracted by the magnets in the experiment. Measuring spins in different directions can be accomplished by moving the north and south magnets around the ray. It is important to reiterate that the particles in the ray are in a superposition of states where they are spinning positively and negatively in all directions before they are measured by the magnets.

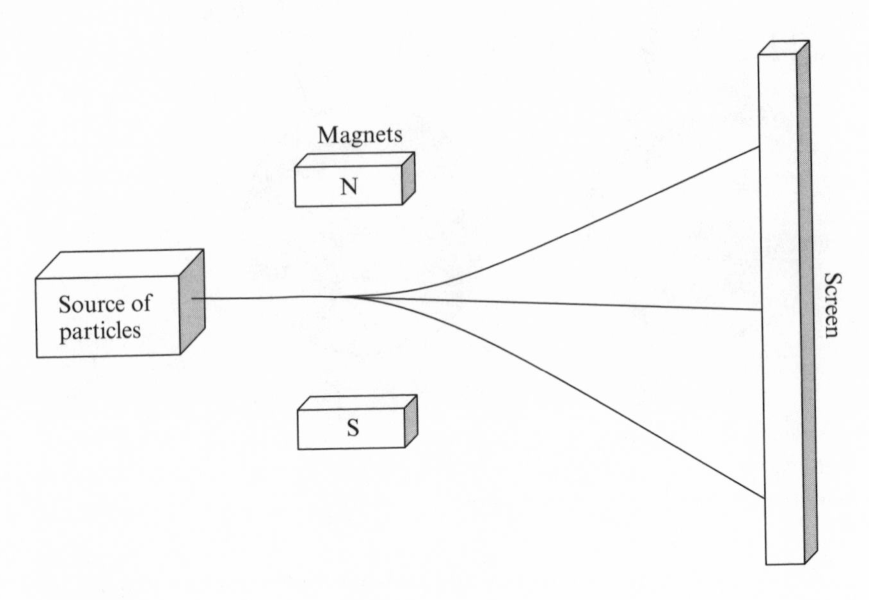

Figure 7.14
A stream of subatomic particles split by their spin

It is the magnets that are performing the observation and causing this superposition to collapse to one particular spin direction. Once a ray is split up, we can measure the spin in other directions and get different spins, as shown in figure 7.15.

There is a form of the Heisenberg uncertainty principle for spin. It says that there are certain directions such that if you measure the spin in one direction and then measure the spin in another direction, you are going to get different answers than if you measure them in the other order. In general, if the two directions are orthogonal to each other, then we can measure both of them and get both values. As long as they are orthogonal, Heisenberg's uncertainty principle will not play a role. If, however, the angles are not orthogonal to each other, then we will not be able to measure those two directions simultaneously.

Now that we have the ideas and language of spin in hand, let us move on to the *Kochen-Specker experiment*. In 1967, Simon Kochen and Ernst Specker described an experiment to show that objects do not have properties until they are measured. They worked with a certain subatomic particle called a "spin-1 particle" that has the following property: if you choose any three orthogonal directions for measuring spin, two of the directions will have spin and the third will not have spin. Since these three directions

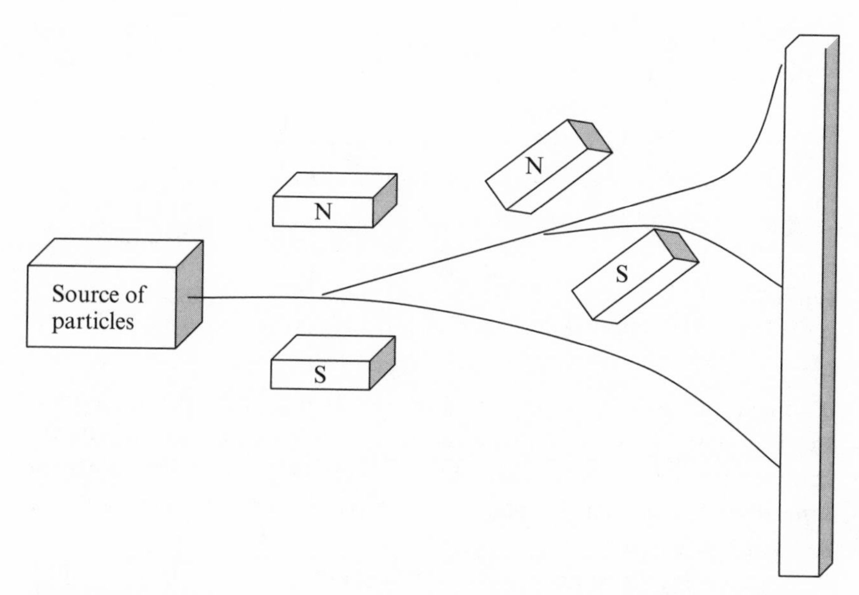

Figure 7.15
Particles split by their spin in one direction and then another

are going to be orthogonal, Heisenberg's uncertainty will not play a role. However, there are many different triplets of orthogonal directions (see figure 7.16). For any triplet you choose, two directions will have spin and a third will not.

Now suppose you did not believe Mr. Bohr and you felt that objects have properties even before they are measured. You think this whole business with measurement causing a property to come into effect is nonsense. Then you would believe that whether or not the particle had spin *in any direction* was a fact that was true even before measurement. In other words, you feel that before any measurement in any direction takes place, there is spin or there is no spin. And when we measure it, we determine what already existed before.

Unfortunately you would be wrong! It is simply impossible to attribute spin or lack of spin before measurements to *all* the possible directions. If you think of a subatomic particle as a sphere, then each point on the sphere corresponds to a direction from the center of the sphere to that point. Saying that the directions have or do not have spin is like assigning 1s or 0s to the points of the sphere. We have the following conditions on assigning the 1s and 0s:

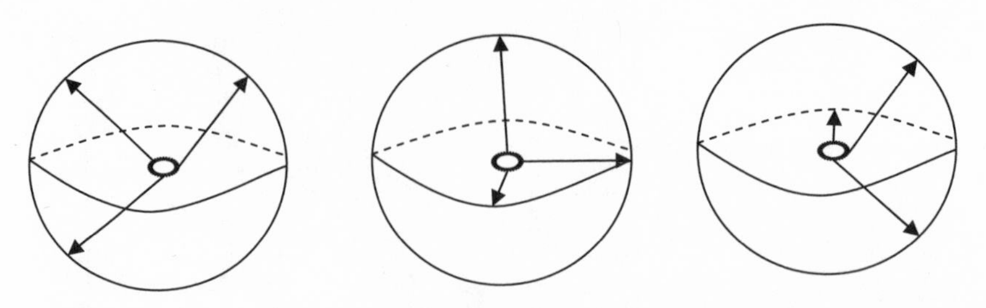

Figure 7.16
A spin-1 particle with three different triplets of orthogonal directions

1. If a particle is spinning in one direction then it must also be spinning in the opposite (antipodal) direction. So if a 1 is assigned to a point on the sphere, then a 1 must be assigned to the opposite point because it is the same direction. Similarly, if a 0 is assigned to one point, it must be assigned to its opposite point.
2. Also, for any three orthogonal directions chosen, two of them will be spinning and one will not. That is, two points will get a 1 and one point will get a 0.

There is simply not enough room for the particle to be assigned such properties. This is a mathematical fact!

It would take us too far afield to provide a rigorous proof of this fact. It suffices to provide an intuition of why it is true. Consider for a moment that the North Pole direction does not have any spin. We can depict this as a 0 at the North Pole of the spheres in figure 7.17(a). From the first proviso the South Pole direction also lacks spin. Now look at the directions orthogonal to the north-south direction. These directions are along the equator of the sphere. By proviso 2, all of those directions must have spin. We depict the spins as a thick line around the equator. In part (b), we further imagine that the direction slightly to the east of the North Pole also does not have spin. This is depicted by another 0. By proviso 1, the direction slightly to the west of the South Pole also does not have spin. The directions orthogonal to this are slightly off the equator and must have spin. Those directions are also depicted as a thick black line. Yet a third direction off the North Pole is depicted in (c). We can further say what does and does not have spin as in (d). In (d), half of the sphere has thick dark lines, and there are two thin lines of directions from the poles to the equators that do not have any spin. We are not done. If you believe that every direction has or does not have spin, you should be able to continue this process and assign to every point either a 0 or a thick black

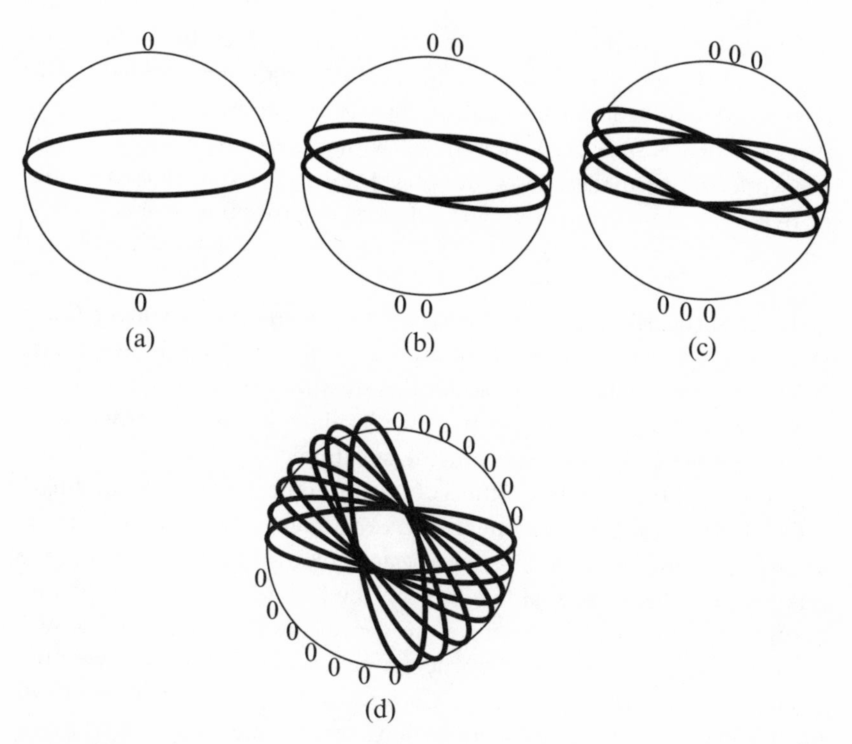

Figure 7.17
The intuition of the Kochen-Specker theorem

point. It should be obvious that such an assignment cannot be made. There simply is not enough room! The black line is too extensive for every 0 point. There is no way we can give every point of the sphere a determination of whether there is spin.

What we have shown is that one cannot take it as fact that every direction has or does not have a spin before measurement. Only after choosing three orthogonal directions and performing an experiment can we determine if there is spin. There was no spin beforehand. The measurement does not tell you what was there before. Rather, the measurement *produces* the outcome.

This is crazy! We have just proved that objects cannot have certain properties until we measure them. We showed geometrically that there is not enough room for there to be such properties. An object only acquires properties after we measure it. Einstein (who died before Kochen and Specker described their experiment, but was nevertheless told of Bohr's ideas that objects don't have properties until they are measured) ridiculed this by asking whether one was really to believe "that the moon exists only when I look at it."[14] But again, the vast majority of contemporary physicists would tell Einstein that, as crazy as it sounds, the moon is only there when it is measured. Heisenberg wrote: "The idea of an objective real world whose smallest parts exist objectively in the same sense as stones or trees exist, independently of whether or not we observe them . . . is impossible."[15]

The End of the Microscopic-Macroscopic Distinction: Schrödinger's Cat

One might try to be flippant about all these problems. After all, what does the "real" world have to do with all this quantum stuff? You have never seen a subatomic particle in one position, let alone in a superposition. How does this idea of superposition in the subatomic world affect the larger world? One of the founding fathers of quantum theory, Erwin Schrödinger (1887–1961), described an interesting experiment that has come to be known as *Schrödinger's cat*. Imagine a sealed box with a piece of radioactive material in it. This material is subject to the laws of quantum mechanics and is in a superposition of "ready to decay" and "not ready to decay." Place a Geiger counter that can detect any decay in the box with the radioactive material. Connect the Geiger counter to a hammer that will break a vial of poisonous gas when the Geiger counter beeps, as in figure 7.18. Now place a living cat inside the box and close the box.

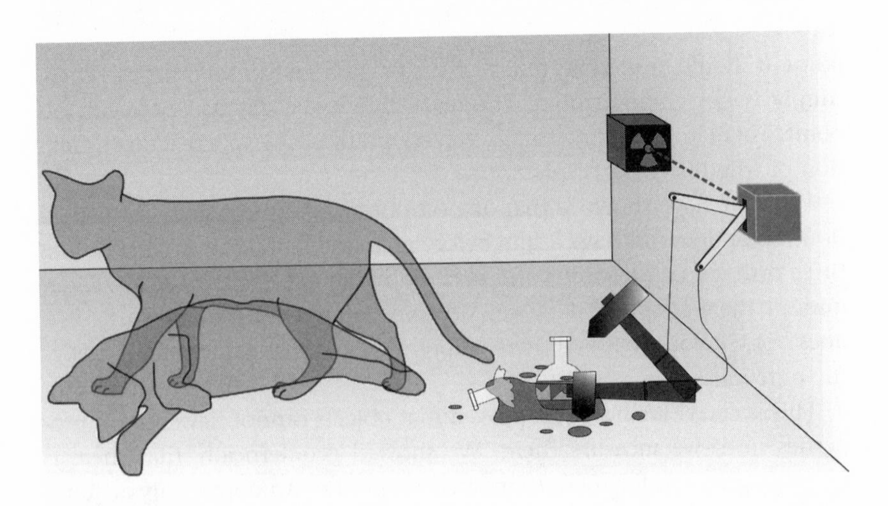

Figure 7.18
Schrödinger's cat
Source: Image by Doug Hatfield, used under the Creative Commons Attribution-Share Alike 3.0 Unported license.

As with all quantum mechanical processes, we cannot determine whether the radioactive material will actually decay. Hence, there is no method of determining whether the Geiger counter will beep. If the radioactive material decays, the Geiger counter will beep, the poison will be released, and the cat will die.[16] On the other hand, if the radioactive material does not decay, then the cat will be alive. Since there is a 50-50 chance for the decay to occur in the time given, there is a 50-50 chance that the cat is dead. That is, before we open the box, the cat will be in a superposition of both being alive and dead. It is only after the box is opened and a measurement is made that one of these possibilities really happens. The experiment has successfully transformed the weirdness of the subatomic world into the everyday world of cats and human beings.

Eugene Wigner (1902–1995) took the Schrödinger cat experiment one step further to get to the heart of quantum mechanics. This experiment has become known as *Wigner's friend.* Imagine Wigner setting up the experiment and placing a live cat in the box. He then closes the box and walks out of the room. Rather than opening the box himself, Wigner has a friend open the box. Before it is opened, we have that the radioactive material is in a superposition, the poison is in a superposition, and the cat is in a superposition. Question: When the friend opens the box, is he also in a superposition of seeing the cat alive and seeing him dead? No human being has ever reported being in a superposition. Does the superposition only collapse when Wigner learns the result, or earlier when the friend learns the result? The obvious answer is that the friend is not in a superposition. Rather, the whole system collapses when the friend looks at it. The one thing that the friend has that no other physical object has is consciousness. Wigner uses this to show that a superposition collapses when any conscious being observes it. Wigner takes this as proof that the only thing in the world that can collapse a superposition to a position is human consciousness. Human beings only observe positions, not superpositions, so it must be something about consciousness. What is it about consciousness that brings about this collapse of a superposition to a position?

This role of consciousness brings to light a criticism of a school of philosophy called *materialism.* A materialist basically believes that this world contains physical objects and spaces between physical objects. And that is it! Most of the laws of physics can be seen from this perspective. A materialist believes that even human beings are simple creatures made out of atoms and molecules that simply follow the laws of physics. Quantum mechanics places simple materialism in jeopardy by highlighting a new entity in the universe called consciousness. This consciousness is not made of physical objects and yet it affects how the universe works. Consciousness causes a superposition to collapse to a position. No longer are there only physical objects and spaces between them. Scientists and materialists must incorporate consciousness into their worldview.

The End of Locality: Entanglement

Another counterintuitive aspect of quantum mechanics is *entanglement.* This concept shows that the whole universe is more interconnected than previously believed.

We first need to learn more about spin. There are important physical laws called *conservation laws* that state that certain measures in a system stay the same. Conservation of energy means that the amount of energy in a system does not change. That is, energy cannot disappear or come out of nowhere. There are also conservation of momentum and conservation of mass/energy. Quantum mechanics shows that there is a conservation- of-spin law. This means that throughout an experiment the amount of spin of all the subatomic particles must remain the same.

What happens when a particle does not have any spin and decays into two particles that do have spin? These two particles will each be spinning both positively and negatively in a superposition. Since there is a conservation- of-spin law, if one particle was measured to have positive spin (or right-handed spin) then in order to maintain the no-spin status of the whole system, the other particle must have negative spin (or left-handed spin). The two possibilities are depicted in figure 7.19.

Which of the two scenarios in figure 7.19 actually occurs? Does the left one spin positively and the right one negatively, or the other way? The answer is that each of the two particles is in a superposition of spinning both ways. Only when one of the particles is measured does it randomly collapse into a particular spin direction. And here is the amazing part: the instant one of the particles collapses one way, the other must collapse the exact opposite way. This is true even if the two particles are light-years apart. That is, in order for the universe to maintain conservation of spin, measuring one particle's spin will collapse the other particle's spin across the universe. Although these two particles are far away, they are entangled with each other. How can this happen?

With the help of two younger colleagues—Boris Podolsky (1896–1966) and Nathan Rosen (1909–1995)—Einstein wrote one of the first papers about entanglement in 1935. It was titled "Can Quantum-Mechanical Description of Physical Reality Be Considered Complete?" and came to be known as "EPR." The goal of the paper was to show that there is

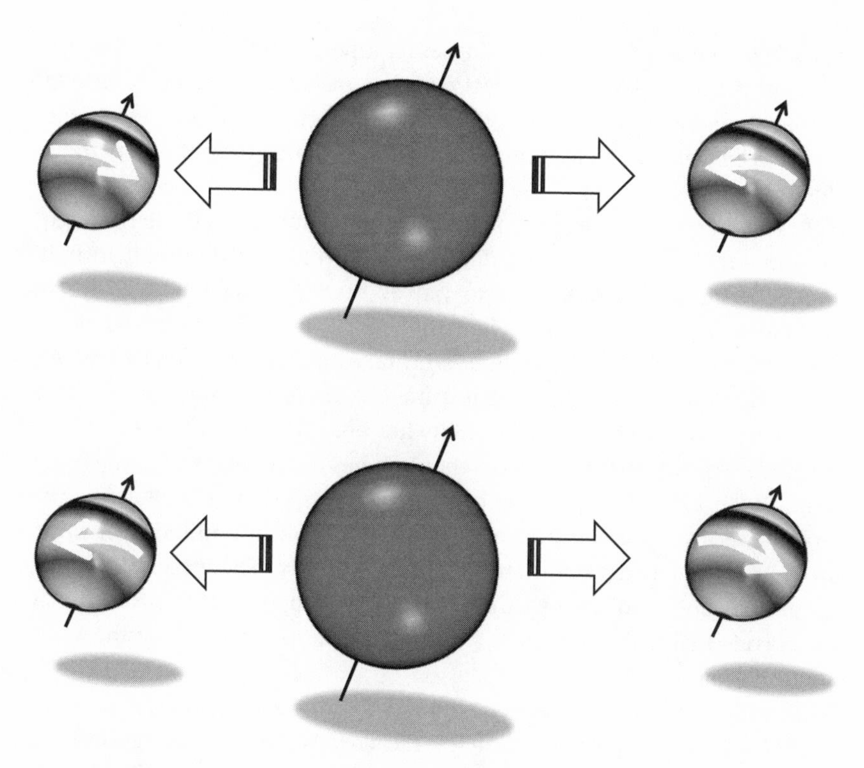

Figure 7.19
Two possibilities for the decay of a particle without spin

something missing in the world of quantum mechanics. Einstein imagined two particles with spin flying apart from a no-spin particle source.[17] Let us envision that the particles are sent across the universe to Ann and Bob, who are going to measure different properties of the particles. Ann measures the spin of her particle at a particular direction. If she finds that her particle is spinning up, she automatically knows that Bob's particle must be spinning down in that direction. On the other hand, if Ann finds her particle is spinning down, she knows that Bob's particle is spinning up.

There is something seriously wrong here. In all of the previously known physics, objects affect other objects that are close by. One object pushes another object or one object affects another object through gravity or some other force. The main idea is that one object has to be near or local to another object in order to affect it. This fact about physics is called *locality*. However, entanglement shows that by Ann measuring her particle, Bob's particle far, far away instantly collapses its superposition of spins. How can Ann measuring her particle affect another particle across the universe? Rather than being local, entanglement shows that quantum mechanics seems to be *nonlocal*. Measurement of one particle instantly affects particles that are not nearby.[18]

Researchers like to compare quantum entanglement to a similar thought experiment. Imagine someone taking a dollar bill and ripping it in half. The experimenter places the two halves in two different sealed boxes without revealing which half went into which box. One box is given to Ann and the other is sent to Bob, who takes his box to Alpha Centauri, the closest star to our galaxy. It is a mere 2.565×10^{13} miles away. Once Bob is on Alpha Centauri, Ann opens her box. If she sees the left side of the dollar she immediately knows that Bob has the right side. On the other hand, if Ann has the right side of the dollar she knows that Bob has the left side. So Ann gained information about something millions of miles away and gained this information instantaneously. There seems nothing mysterious about this. One can say that the properties of the torn dollar bill traveled with it from Earth to Alpha Centauri. Can we say the same about the particles?

Einstein, Podolsky, and Rosen concluded that there are two possibilities in the case of the spinning particles. Either (a) there is some mysterious, nonlocal interaction that is different from any other branch of physics that explains how Bob's particle is affected by Ann's measuring her particle. If this was true, our naive notion of space where distant objects and measurements are independent of each other is wrong. Or (b) something similar to what is going on with the dollar bill is happening with the particles. In other words, the particles are not in a superposition. When they split up at the source they have fixed spin values. That is, the particles have their spin values when they leave their source, and when Ann measures her particle she finds out what her particle's spin values are and instantly knows Bob's values as well.

The EPR paper discounted possibility (a) since Einstein and his colleagues could not imagine that physics could work in such a strange manner.[19] Rather, they preferred to accept possibility (b) as the correct view. In that case, we must ask what is missing in quantum mechanics. Why could quantum mechanics not tell what spin a particle is in prior to measuring it? Einstein and his associates postulated that there must be *hidden variables* that stay with the particles from the time they leave their sources until the time they hit the measuring devices. These hidden variables are like the split dollar bill. They ensure that properties of the particles have a fixed value. Until physicists learn more about such hidden variables, Einstein and his coauthors insisted that quantum mechanics is incomplete and waiting to be finished.

That's the way the physics world remained for almost thirty years, until the brilliant Irish physicist John Stewart Bell (1928–1990) showed that, in fact, option (b) is wrong and only option (a) is possible. In 1964, Bell published a paper, "On the Einstein-Podolsky-Rosen Paradox," which famously showed that no regular hidden variables can explain away the mysteries of quantum entanglement. This result—which came to be known as *Bell's theorem* or *Bell's inequality*—demonstrated that superposition is a fact of the universe[20] and that our notion of space needs to be adjusted.

The intuition behind Bell's theorem[21] is that if we assume that there are hidden variables and that these hidden variables describe the properties of particles, they must satisfy some regular logical truths. In particular, if we allow Ann and Bob to each measure the spin of their particles in three different specified directions, then these spin properties must satisfy certain logical truths. Bell describes what these logical properties are and then shows that they are not satisfied by the quantum mechanics of spin. He concludes that the particles did not have these properties while traveling from the source to the observers. They are in a superposition before measurement.

To understand Bell's theorem[22] we are going to need to step away from the quantum world for a minute and discuss a little classical logic. Consider an object with three different properties that the object can have, call them A, B, and C. For example, look at a person and ask whether they are

- A, male or $\sim A$, female,
- B, Democrat or $\sim B$, Republican, and
- C, young or $\sim C$, old.

Consider a person who is both male and old ($A \wedge \sim C$). He is either a Democrat or a Republican ($B \vee \sim B$). If he is Republican then he is a male and a Republican ($A \wedge \sim B$). Otherwise, if he is a Democrat he is old and a Democrat ($B \wedge \sim C$). We have just proved the following simple property:

If you are male and old, then you are either a male Republican or an old Democrat.

In symbols, this is represented as

$$(A \wedge \sim C) \rightarrow [(A \wedge \sim B) \vee (B \wedge \sim C)].$$

This logical rule is true for any three properties. We can prove this either by examining a truth table for this logical formula and seeing that it is a tautology, or simply by considering A, B, and C. If A and $\sim C$ are true, then either B is true or $\sim B$ is true. If $\sim B$ is true, then we have A and $\sim B$. In contrast, if B is true, then we have that B and $\sim C$ are true.

An implication ($\rightarrow$) about two properties tells us something about the probability of such properties happening. If $Q \rightarrow R$, then the probability that Q is true is less than the probability that R is true. In symbols we write this as $p(Q) \leq p(R)$. For example, it is a logical fact that if it is raining, then there are clouds in the sky. From this we conclude that the probability of a rainy day is less than or equal to the probability of a cloudy day.

Returning to our logical law about A, B, and C, we have that

$$p(A \wedge \sim C) \leq p(A \wedge \sim B) + p(B \wedge \sim C).$$

That is, the probability that $A \wedge \sim C$ is true is less than or equal to the probability that $A \wedge \sim B$ is true plus the probability that $B \wedge \sim C$ is true.

That is enough classical logic. Let us return now to our particles. Consider two particles that are entangled and sent to Ann and Bob. Both experimenters can measure the spin of the particles in one of three different angles. These three directions are going to correspond to the three properties of particles A, B, and C. In particular:

- A corresponds to Ann's particle spinning up at 0°.
- B corresponds to Ann's particle spinning up at 45°.
- C corresponds to Ann's particle spinning up at 90°.

Combining these properties, we determine that

- $A \wedge \sim C$ corresponds to Ann's particle spinning up at 0° and spinning down at 90°.
- $A \wedge \sim B$ corresponds to Ann's particle spinning up at 0° and spinning down at 45°.
- $B \wedge \sim C$ corresponds to Ann's particle spinning up at 45° and spinning down at 90°.

By Heisenberg's uncertainty principle, it is impossible for Ann to measure the spin of her particles in these two different directions, so we are going to have to take into account that Bob's particles are spinning oppositely from Ann's particles. With this our propositions become

- $A \wedge \sim C$ corresponds to Ann's particle spinning up at 0° and Bob's particle spinning up at 90°.

• $A \wedge \sim B$ corresponds to Ann's particle spinning up at 0° and Bob's particle spinning up at 45°.
• $B \wedge \sim C$ corresponds to Ann's particle spinning up at 45° and Bob's particle spinning up at 90°.

Quantum mechanics makes probabilistic predictions concerning such measurements. It says that when the two angles are close to each other, the particles will probably spin opposite each other. That is, if Ann and Bob both measure spin at about 0°, then it is very probable that Ann will measure up and Bob will measure down or vice versa. Another way of saying this is that if they both measure close to 0°, it is very unlikely to find both spinning up. In contrast, when the two measurements are 90° apart, it is more likely that Ann and Bob will both measure up. Quantum mechanics tells us that the probability of having the same outcome of a measurement depends on the angle between the two measurements. If the angle is φ, then the probability of having them both up is

$$\frac{1}{2}\left(\sin\frac{\varphi}{2}\right)^2.$$

In our cases,

• $p(A \wedge \sim C)$ is $\frac{1}{2}\left(\sin\frac{90}{2}\right)^2 = 0.25$,

• $p(A \wedge \sim B)$ is $\frac{1}{2}\left(\sin\frac{45}{2}\right)^2 = 0.0732$, and

• $p(B \wedge \sim C)$ is $\frac{1}{2}\left(\sin\frac{45}{2}\right)^2 = 0.0732$.

If this satisfies the logical and probabilistic laws that we derived, then we have

$$0.25 \leq 0.0732 + 0.0732$$

and that is simply false!

What went wrong here? We showed that there is a basic conflict between the classical logic of three properties and quantum mechanics. How can this possibly happen? The answer is that we really cannot attach propositions about spin while they are in flight. They do not have fixed values then. Rather, the particles are spinning in a superposition of spinning both up and down. They only have fixed values after they are measured. Classical logic, which works so wonderfully with regular objects (like dollar bills in a box), does not apply here. More important than just showing that particles are in a superposition, Bell showed that the particles collapse from a superposition even though the particles are far away. When Ann does a measurement of her particle, Bob's particle collapses from its superposition to have the opposite spin of Ann's measured spin. This means that the usual notion of space that we have is wrong: measurements do affect distant objects.

When Bell formulated this inequality and proved his theorem he did not discuss experiments. He simply stated that this is what quantum mechanics predicted, and it is different from classical logic. A few years later, experimentalists like Alain Aspect and John Clauser confirmed the fact that the subatomic particles followed the laws of quantum mechanics as opposed to the laws of classical logic. Since then, many other experiments have been performed and have shown that Bell's results are not abstract mathematics but actually say something very important about the universe in which we live.

In essence, Bell's theorem is the ultimate expression of the Wholeness Postulate. It says that the outcomes of experiments depend on the whole experiment, including Ann's and Bob's measurements. In other words, we cannot just look at what Bob will measure or what Ann will measure. Rather, we have to consider what each will measure and where their particles came from. If the particles came from a single system with no spin, then the outcomes will take that into account.

There is still a way to believe in hidden variables and the fact that particles have spin properties even before they are measured. Rather than saying that the hidden variables keep track of three different spin values (for the three possible measurements that Ann can perform), we could say that the hidden variables keep track of the nine different measurements that Ann and Bob can possibly take. That is, Ann can perform three measurements and Bob can perform three measurements, which means that a total of nine different measurements can be performed on the two particles. If you assume that there are such hidden variables, then in fact the logical problems above go away. However, we are left with one very perplexing problem: How does Ann's particle know what measurements Bob will perform? After all, Bob could be across the universe. Such a theory is called a *nonlocal hidden-variable theory*. The very fact that such hidden variables need to take into account information that is very far away causes most physicists to disregard this possibility.

Regardless of the existence or nonexistence of nonlocal hidden variables, one thing remains certain: the notion of space where measurements do not affect distant objects is wrong. As we saw above, the EPR paper set up a dilemma. Either (a) the universe we live in is nonlocal, or (b) quantum mechanics is incomplete and contains nonlocal hidden variables. Either way, there are nonlocal effects.

One consequence of entanglement is to end the philosophical position of reductionism. This position says that if you want to understand some type of closed system, look at all the parts of the system. To understand how a radio works, one must take it apart and look at all of its components, because "the whole is the sum of its parts." Reductionism is a fundamental supposition in all of science. Entanglement shows that there are no closed systems. Every part of a system can be entangled with other parts outside of the system. All different systems are interconnected and the whole universe is one system. One cannot understand a system without looking at the whole universe. That is, "the whole is *more than* just the sum of its parts."

Once again, our defender of the sane, rational view of the world, Einstein, found it difficult to accept that distant points of our universe were so intimately connected. He derided entanglement as "spooky action at a distance."[23] But again, we must point out that many contemporary experiments show that Einstein was mistaken. The universe is a lot weirder than even he imagined.

The End of Time and Free Will: Quantum Eraser Experiments

We now know enough to describe some cutting-edge research called "quantum eraser experiments." These experiments take the famous double-slit experiment and go much farther. Remember that in the double-slit experiment there will be a superposition of the photon and an interference pattern will be made. This will happen as long as we permit the photon to go through both slits. If we were somehow to measure which slit the photon went through, then, since we do not see things in a superposition, the photon would have to go through exactly one of the slits and there would be no interference. This is because we do not see superpositions; we can only see their effects.

What if there were a way of "seeing" which slit the photons passed through? Perhaps we can "tag" the photons when they pass through the slit so that we can later tell which tag they have and hence which slit they went through. In that case, there will not be a superposition and there will not be an interference pattern of the photons. In fact, we can do this: photons can be tagged by placing polarization filters next to each of the two slits, as in figure 7.20.

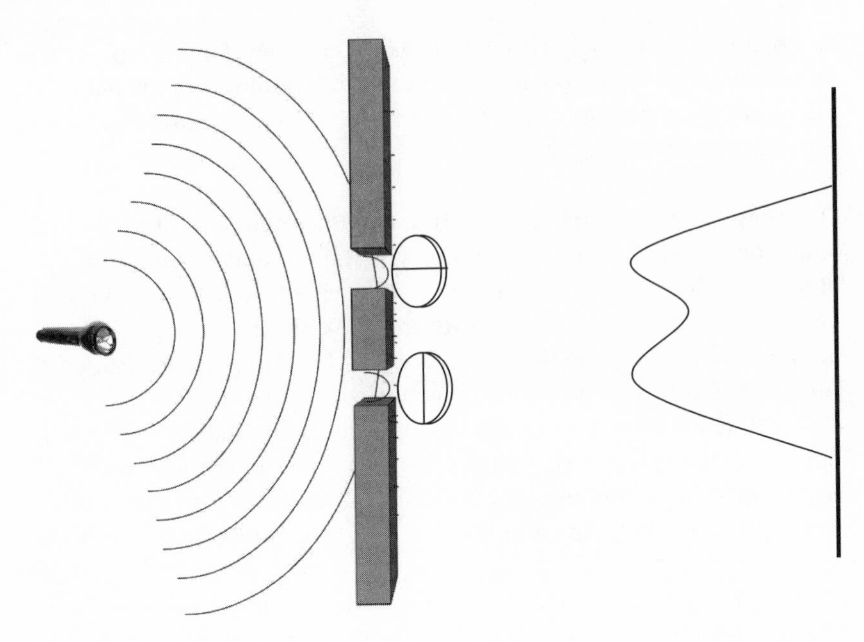

Figure 7.20
The double-slit experiment with polarization markers

Notice that one filter is set horizontally and the other vertically. This will ensure that the photons that pass through different slits are tagged differently, so we can tell which slits they went through. Sure enough, when such an experiment is carried out, since there is information available that would tell us which slit the photon passed through, there will *not* be an interference pattern. The screen on the right will show light without interference.

There is an obvious question: When the photon leaves its source, does it go into a superposition or a position? We saw that if both slits are open and there is no way to tag the photons, then they go into a superstition. If, however, there is a way of tagging the photons, they do not go into a superposition and there is no interference. When the photons leave their source, how do they "know" if there is going to be a tagging device on the other side of the slits? After all, the filters can be far away from the source of the photons. And yet, somehow the photons "know" what to do. In terms of the Wholeness Postulate, this makes sense: the outcome of the experiment depends on whether there are tagging devices in the experiment.

We are not done yet. A large polarization filter can be placed between the other polarization filters and the screen, as in figure 7.21. This polarization filter is set in the diagonal direction.

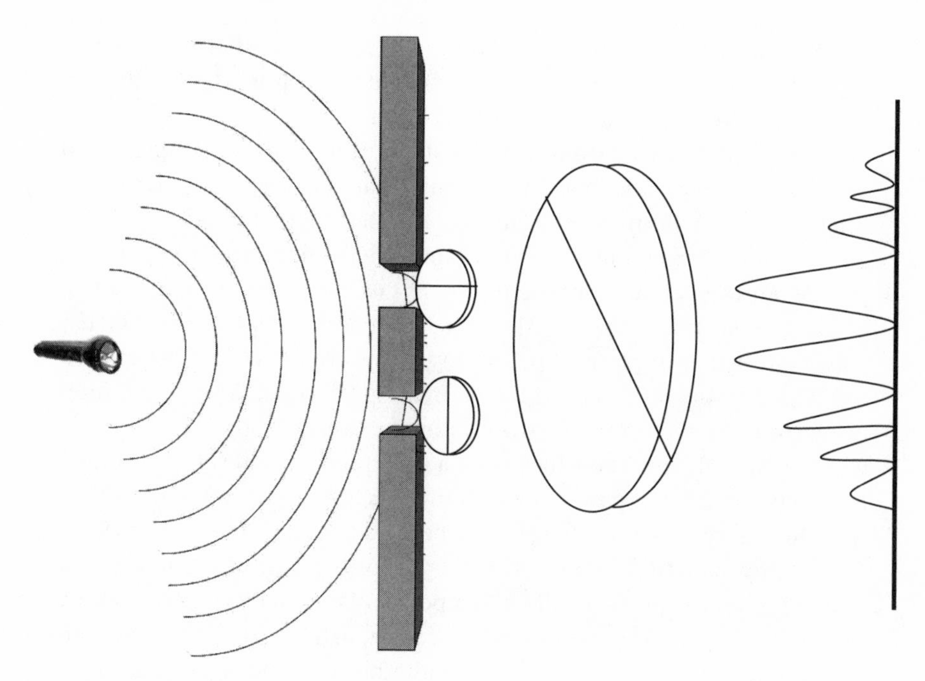

Figure 7.21
The double-slit experiment with a quantum eraser

Let us see what happens as the photon goes on its journey. If it goes through the top slit, it will pass through the horizontal filter and come out horizontal. It will then have to go through the diagonal filter, and if it comes out, it will come out diagonal. Similarly, if the photon goes through the bottom slit, it will also go through the vertical filter, and if it passes through the diagonal filter, it will again come out diagonal. Either way, the diagonal filter "erases" the tagging information of which slit the photon passed through. Without that information or our ability to get that information, the photon reverts to its superposition status and will interfere with itself. Amazingly enough, this is exactly what happens when the experiments are carried out: there is an interference pattern.

Already we have an extremely interesting scenario. When the photon approaches the slits, it has to determine if it should go through one or both slits. This depends on whether there will be a diagonal polarization filter on the right side of the slits. Somehow the photon "knows" what it will find on the other side of the slits. If the filter is not there, it will go through one of the slits and if it is there, it will go through both slits. How does the photon know what will be on the other side of the barrier? This again conforms to the Wholeness Postulate. Here we see that the outcome depends on the setup of the *entire* experiment, including whether there is an eraser filter on the right side of the barrier.

Physicists take this experiment one step further with something called a *delayed-choice quantum erasure*. Imagine that the diagonal polarization filter is far away from the slits and is on rollers so that it can be moved away from the screen quickly.[24] To recap: if we leave the eraser in place we will get an interference pattern and if we take the eraser away there will be no interference. However, the experiment can be set up so that the eraser can be left in place or pulled away after we know that the photon has passed through the slits in the barrier. The diagonal polarization filter can be in place and we can let the photon go. Because the filter is in place, we know that the photon will go into a superposition and will go through both slits. Once it has passed through the slits, we can then pull the diagonal filter away and the photon will be in a position and not form an interference pattern. In contrast, if the diagonal polarization filter is not in place, then the photon will be in a position and go through one of the slits. Once the single photon has passed through the single slit, we can push the filter back in place. Then the photon somehow goes back into a superposition and creates interference.

There are two crazy ways of looking at this: (a) After the photons pass through the slits, by moving the diagonal polarization filter the experimenter is changing what the photons did in the past before they got to the slits. Or (b) somehow before the photons come to the slits, they "know" whether the observer will pull away the filter. In short, either (a) the experimenter changes the past or (b) the photon "knows" the future. Both options are mindboggling.

It is hard to understand what it means to change the past. Such a concept violates our notions of cause and effect and all of science. In contrast, option (b), where the photon acts as if it "knows" the future, fits in well with our Wholeness Postulate. The outcome depends on the whole experiment, including what the experimenter will do while the experiment is in progress. Here *whole* emphasizes that the experiment takes place in time and the outcome depends on the experiment from start to finish. The outcome takes into account whether the experimenter is going to pull away the diagonal filter. As Yogi Berra says, "It ain't over till it's over."

How can the photon "know" what the experimenter will do? What about the experimenter's free will? [25,26] Doesn't the experimenter have free will to decide whether to pull the eraser away?[27] Let us be careful with our language. A photon does not have consciousness or "know" anything.

What we mean is that whatever physical law is controlling the actions of the photon must take into account all the actions of the experimenter. The reason this is amazing is that the laws that govern the actions of the photon must take into account the *future* actions of the experimenter even though such actions do not exist yet. That is, the laws controlling the actions of the photon must take into account the laws controlling the actions of the experimenter. If we are going to assert that the experimenter has free will and there is nothing controlling the actions of the experimenter, then there is nothing controlling the actions of the particle either. In other words,

human beings have free-will $\Rightarrow$ particles have free-will.

Do particles really have free will? Can we believe such a thing? They do not seem to show any freewill action. There is another way of looking at this: What if the particle does not have free will and its actions are totally determined by laws? Well, then a human being also has no free will. Hence the contrapositive:

particles do not have free will $\Rightarrow$ human beings do not have free will.

From a scientific perspective this is not strange at all. After all, human beings are made out of particles. Abiding by the usual dictum of reductionism, scientists would have to say that the tendency of particles to follow the habitual laws of physics implies that humans must follow the habitual laws of physics. If they believe that particles have no free will, then they are forced to believe that humans, made out of particles, also have no free will.[28]

Other Strange Aspects of Quantum Mechanics

There are at least three other aspects of quantum mechanics that are strange and worth mentioning. First, quantum mechanics uses the mathematics of complex numbers. Such numbers are usually written as $a + bi$, where a and b are real numbers and i is the imaginary square root of negative one. This seems shocking because most other physical laws use simple real numbers. After all, the measurements that we make are real numbers. A rod is 18.63 inches long. The temperature of an object is measured at 46.168 degrees Celsius. The projectile is going 265.643 miles per hour. Since we measure properties with real numbers, we expect the laws of physics to be stated with real numbers. They are not! Rather, complex numbers are used in a most fundamental way. It would be very hard to perform any calculations in quantum mechanics without complex numbers.

A second oddity that needs to be reiterated is the total nondeterminism of quantum mechanics. As mentioned in the previous section, every other part of the physical sciences is deterministic. There are formulas that describe the actions of the system. These formulas might not be computable. They might not even be known to us. Nevertheless, the systems follow rigid laws of physics. This is in stark contrast to quantum mechanics, which seems, at its heart, totally random. Why should that be?

And finally, the last strange aspect of quantum mechanics that I will discuss was actually the first one discovered by researchers. In the early twentieth century, Max Planck (1858–1947) found that certain types of energy had only discrete values. Whereas you can turn your thermostat to any value between 72.4 and 72.5, quantum mechanical systems had energy being released in certain units and could not have energy between these units. As time went on, the founders of quantum theory realized that not only did energy have this discreteness, but many other properties of quantum mechanics had this characteristic as well. They found that particles had discrete spinning states, space was discrete, and time was also discrete. Electrons jump from shell to shell but do not cover the intermediate distance. Such jumps are called "quantum leaps."

For me, it is hard to see how these three aspects of quantum mechanics fall under the purview of the Wholeness Postulate. Perhaps they do not.

We have finished our little tour of quantum mechanics. What a strange trip! What have we learned?

- Properties of objects have more than one value at a time.
- There is no way to determine which value will be observed when the property is measured.
- There are pairs of properties for which there is an inherent limitation of our ability to know their values.
- Reality depends on how it is measured.
- Distant parts of our universe are strangely interconnected.
- Experimenters and their free will cannot be separated from their experiments.

What are we to make of this psychedelic world? Quantum mechanics has been around for more than a century and researchers have been busy making it more palatable. This theory has done such violence to our usual intuitions of the physical world that there is no way we can ignore it. I will highlight four major schools of interpretation.

Interpretation of Quantum Mechanics

The Copenhagen School

The most popular school of thought was developed by the founders of quantum mechanics Niels Bohr, and Werner Heisenberg. This school represents the orthodox viewpoint of most physicists and has influenced our presentation. Basically, it says that there is really no underlying physical universe. Values do not exist until a conscious observer measures the property. The value is not there beforehand, rather the measurement causes the value to come into being.

To Bohr & Co. there is no quantum world. There is only an abstract quantum mechanical description. It is wrong to think that the task of physics is to find out what nature is. Rather, physics concerns itself with what we can say about nature. The Copenhagen interpretation does not explain why quantum mechanics is nondeterministic or how the superposition collapses. In fact, they go further and say that such questions are not scientific and must be considered meaningless.

It is important to realize that the Copenhagen interpretation is not some opinion held by a few crazy people on the outskirts of scientific discussion. Rather, this interpretation is considered the mainstream view among people who study quantum mechanics. In fact, it is the other opinions that are considered unorthodox and far-fetched.

There are both positive and negative aspects to the Copenhagen interpretation. The positive side is that if all questions about the meaning of quantum mechanics are deemed illegitimate, then there are no questions. One can go on and simply work with the equations. While this is unsatisfying to me, most physicists who do not want to think about foundational issues like this freedom. They believe that all the problems of quantum mechanics are simply pseudoproblems. The physicist Murray Gell-Mann said in his Nobel Prize acceptance speech that "Niels Bohr brainwashed a whole generation of physicists into believing that the problem [of the interpretation of quantum mechanics] had been solved fifty years ago." Good for them.

The negative aspects of the Copenhagen interpretation are obvious. Someone is, literally, a lunatic if they think that the moon is only there when people are looking at it. How can it be that a subatomic particle does not exist until it is observed, and yet we each seem to have a coherent notion of existence? Also, this interpretation does not really explain how or why consciousness brings about a collapse of a superposition. The worst part of the Copenhagen interpretation is that there is a feeling of unquestioned dogma: "This is the way it is and one should not ask about it." There is something unsatisfying when scientists tell you there is no real explanation for phenomena and the question does not make sense.

Multiverse

The main reason superposition is so strange to us is that we have never seen a superposition. When we look at the universe we see every object in one place with only one value for every property. The measurement problem asks: Why should an object collapse simply by measuring it? Furthermore, why should it seem to randomly collapse to one position and not another? In 1955, Hugh Everett III (1930–1982), a brilliant young physicist, proposed a radical solution called the *many-worlds hypothesis* or *multiverse*. He said that when a measurement takes place, rather than a superposition collapsing to one position, the entire universe splits into many different universes where each universe has one of the possible outcomes of the measurement. For example, when Schrödinger's cat experiment is carried out, the universe splits into two. In one universe you open the box and happily find the cat alive, while in another universe you open the box and are sad and regretful about using animals in your scientific experiments. These two universes do not have any connection and are totally separate. They each contain a version of you that does not know about any other universe.

There are a multitude of advantages to believing in a multiverse. For one thing, there is no more measurement problem. There is no collapse from a superposition to one particular position; rather, there is a collapse of a superposition to every position. Furthermore, although when doing an experiment you cannot determine which outcome a superposition will collapse to within the universe you are in, if you look at the entire multiverse, determinism is restored. The physical law says that the superposition will collapse to every position in some universe. This is a deterministic law.[29] Locality is also restored in the many-worlds interpretation. When Ann makes a measurement on her particle the universe splits into two universes: one where she measures spin up and one where she measures spin down. In each of those two universes, Bob's particle also has the correct spin. So one measurement does not affect the other particle; rather, an entire new universe is created. The mathematics of a multiverse is much simpler[30] and hence satisfies Occam's razor, which tells us to choose the view that is simpler. The multiverse also has other advantages that we will meet when we discuss the anthropic principle in section 8.3. The many-worlds interpretation is not held by the majority of physicists. In fact, it is severely derided by most. Nevertheless, there are several major physicists, like Max Tegmark and David Deutsch, who are firm advocates of this view.

The disadvantages of the multiverse are obvious. Where are these other universes? What is the mechanism by which a universe splits? This is the ultimate in nonlocality: one universe here and one universe a great distance away simply because a little measurement was performed. The idea that there are so many universes is simply staggering. How can one posit the existence of something for which there is no physical evidence?

Hidden Variables

The hidden-variable school of thought can be traced to Louis de Broglie (1892–1987) and David Bohm (1917–1992). These two leaders were unhappy about the nondeterminism in quantum mechanics. Perhaps there are some hidden variables that explain which position a superposition collapses to. They worked on some sophisticated mathematics to actually come up with formulas for deterministic laws on how quantum mechanics works. However, the hidden-variable theory, which arose before Bell's theorem, was mostly ignored by the majority of physicists because the equations they supplied took into account features that were far away from the action. That is, their hidden variables were nonlocal. This was totally unpalatable for physicists before Bell's theorem was formulated. They felt that all of physics must be local and only take into account what is near. It was only after Bell's theorem was stated that people realized that quantum mechanics is nonlocal regardless of whether there are hidden variables.

Even if it all works out, hidden variables do not help naive realism. In other words, that simple idea that an object has a certain property before we measure it and then we discover the property after we measure it, has to be discarded. Even hidden variables will not help us get naive realism back. Nor do they help us with nonlocality. The quantum world is inherently nonlocal.

One of the major advantages of the hidden-variable program is that its laws are deterministic. For the past three millennia a major goal of science has been to give deterministic rules for all phenomena: "If this happens, then that will happen." Why should we believe that after success in other areas of science, there is a domain of physics where determinism fails? Why should the subatomic world be different from the rest of the universe? The hidden-variable researchers have restored this important feature to our physical world.

There are, however, disadvantages to hidden variables that have kept most physicists away. For example, the equations and the mechanism of the hidden variables (pilot waves) are ugly. They are not simple equations that give simple answers. Rather, they are equations that take into account many nonlocal phenomena. Even with this extradistant information, it is not easy to calculate what the outcomes will be. They are like some of the systems we met in section 7.1. Even though the laws are deterministic, they will not be much help in making anything predictable. That is, just because the particles know where to go does not mean we will be able to predict where they go.

Quantum Logic

The final school of thought that I will examine is quantum logic. These ideas were first formulated in a 1936 paper by Garrett Birkhoff (1911–1996) and John von Neumann (1903–1957).[31] Their aim was to modify the laws of logic so that they describe the quantum world. We all know the normal logical rules of living in the real world, but what are the logical rules of the quantum world?

Let us look at an example. Consider the following three propositions:

A = Bob is a Democrat.

B = Bob is young.

C = Bob is rich.

We can combine these propositions to form

A AND (B OR C).

This means that Bob is a Democrat and is either young or rich. We can also form the proposition

(A AND B) OR (A AND C).

This means that Bob is a young Democrat or Bob is a rich Democrat. If you read this carefully you will see that these are just two ways of saying the same thing. In symbols, we have that

A AND (B OR C) = (A AND B) OR (A AND C).

This is an instance of the *distributive law* in logic. It says that AND distributes over the OR. Such instances are a basic fact of the universe, and we implicitly use this law every day of our life.

Now let us turn to the quantum world. Consider the following three propositions about the double-slit experiment:

X = There was interference.

Y = The photon went through the top slit.

Z = The photon went through the bottom slit.

Combine these propositions to form

X AND (Y OR Z)

which means that there was interference and the photon went through the top or bottom slit. In fact, it went through both slits. This is a true statement about the double-slit experiment. In contrast,

(X AND Y) OR (X AND Z)

means that there was interference and the photon went through the top slit (false), or there was interference and the photon went through the bottom slit (also false). This statement is false. In symbols, we conclude that

X AND (Y OR Z) $\neq$ (X AND Y) OR (X AND Z)

So while the distributive law is true in the regular world of people, marbles, and Democrats, it fails in the quantum world of particles.

Researchers have gone on to look at many different aspects of the quantum world and examined it from the point of view of quantum logic. The positive side of quantum logic is the ability to make formal sense of the crazy subatomic world. After all, logic helps us navigate in the larger world; it would be nice to see logic help us in the subatomic world. The negative side is that quantum logic does not really make the weirdness go away. It simply begs the question: Why should we accept the strange rules of quantum logic? Why should the world follow rules that are different from the usual rules that we experience every day? If the quantum world follows quantum logic, then why should the real world follow classical logic? More questions remain.

Summing Up

All four of these interpretations of quantum mechanics are unsatisfactory. They each have negative aspects that make us uncomfortable and doubtful. It could very well be that none of the four interpretations (or any other available interpretation) is correct. It might be that in the future there will be another interpretation that will, in fact, give us the correct view. Or, we simply might never get the correct interpretation. Remember, there is nothing "out there" that ensures that the universe we live in is comprehensible.

At present, no scientific experiments can differentiate which of these schools of thought is correct. Every interpretation has its own way of looking at reality and predicting the results of quantum mechanics. A preference for a favorite interpretation basically depends on what type of properties you believe our universe has. If you believe that the universe is deterministic, then you will not follow the Copenhagen interpretation but rather the hidden-variable interpretation. If you cannot wrap your mind around the existence of billions and billions of different universes, then you must stay away from the multiverse interpretation. In contrast, if you blindly follow Occam's razor, then you might accept the multiverse interpretation with its simpler mathematics. All these different choices are essentially ideas that make you feel good about the universe you live in. Unless someone comes up with some experiment that can show one view to be correct and the others to be false, there is no scientific reason to choose any of them. For now, the correct interpretation of quantum mechanics is beyond science.

Unless you are doctrinaire and accept on faith one of the schools of thought, you have to join the rest of us wavering mortals and realize that the fundamental nature of our universe is simply beyond the limits of reason.

There are some researchers who would like to brush away this entire subject of interpreting quantum mechanics. They have a very pragmatic outlook and only care that the results on the measuring instruments are correct. These *instrumentalists* are only concerned that the equations work and make the correct predictions. They see no reason to pay any attention to *why* the equations work or what the underlying reality of the physical universe is. Their motto is "Shut up and calculate!" They believe that one should not waste time pondering what is going on "under the hood" and question whether a deeper reality even exists. To them, the underlying nature of the real world is either beyond us or is not worthy of thought. They think that the study of the interpretation of quantum mechanics is "only" metaphysics and hence should be thrown into the garbage heap of bad ideas.

David Deutsch has severely criticized such instrumentalists.[32] He imagines some extraterrestrial race of supergeniuses giving the Earth an "oracle" or a "magic box" that tells what the outcomes would be for any possible experiment. That is, we humans input the experiment we want to perform into the oracle and it miraculously tells us what the outcome will be. It will make all the predictions we will ever want or need. This will satisfy the instrumentalists that tell us to shut up and calculate. They would no longer need to calculate. They can simply play with their oracle toy. However, this would be extremely unsatisfying for most human beings. We do not want to know the outcome of an experiment in advance. We want to understand why the universe works the way it does. We want an explanation as to why particles do what particles do. Most people are not satisfied with just knowing the results of an experiment. In fact, Deutsch says that we already have such a magic oracle: the universe we live in. It tells us the outcomes of all the experiments we can perform. But that is not enough. We invent mental models of how and why the universe works the way it works.

While I sympathize with the instrumentalist view that the proper interpretation of quantum mechanics is, at present, essentially unanswerable, and I would defend their constitutional right to ignore these questions, I nevertheless do not like their moral stance denying the basic human tendency to speculate about the world around us. Human beings should continue to try to understand our universe.

While quantum mechanics is a real science and we use its predictions every time we turn on a radio or a microwave oven, we nevertheless cannot truly understand why it works the way it works. Even though the laws of quantum mechanics describe and control most of the physical universe, we still have a hard time making it understandable. This forces us to ask several questions: Why do we find it so hard to understand quantum mechanics? Will we ever get used to the strangeness of quantum reality? What does it say about the relationship between our mind and the world we live in if we cannot make a mental model of this weirdness? As a partial answer to some of these questions, we must come to accept that the weirdness of quantum mechanics does not make it false. After all, we grew up in a world without superposition and so we should expect that our mind will consider superposition strange. We also came to experience the world with locality and so it stands to reason that the nonlocality of quantum mechanics should not make sense. Similarly, other strange aspects of quantum mechanics are beyond our comprehension. We must accept that our mind is not the be-all and end-all of the way the universe should work. Many years before the problems of interpreting quantum mechanics became apparent, David Hume asked, "What peculiar privilege has this little agitation of the brain which we call thought, that we must thus make it the model of the whole universe?"[33] The universe works the way it works and we must adjust to it because it will not adjust to us.

I conclude with the words of Niels Bohr, the father of quantum mechanics. He wrote that some of the notions of quantum mechanics demand "a radical revision of our attitude as regards physical reality."[34] Indeed!

7.3 Relativity Theory

Albert Einstein's theory of relativity has some of the most beautiful ideas in all of science. Most of the ideas can be described with simple thought experiments that are easy to comprehend. Even though they are easy, they have revolutionized our understanding of the world around us.

It is not my aim to actually cover the details of relativity theory. I merely want to discuss the ways that relativity theory has reshaped our view of the universe. For that, we can ignore the equations and concentrate on some of the thought experiments with some helpful diagrams.

Relativity theory comes in two flavors. In 1905, Einstein formulated the *special theory of relativity*, which deals with the universe without gravity or acceleration. Later, in 1914, he generalized this work to the *general theory of relativity*, which deals with gravity and acceleration. I start with special relativity and work my way toward general relativity.

The central idea of relativity theory is that properties of the physical universe depend on how they are measured. There are no absolute measurements. This is in sharp contrast to our naive notions about the universe. We usually say a person that we are looking at has an exact height. They will be perceived differently when we are at different distances. When we are far away they look small and when we are near they look large. In this simple case, we can say that there is an absolute height but there are different relative heights. In contrast to the usual notions, relativity theory tells us that properties of an object really are different depending on how they are viewed. There are no absolutes.

Let us begin by asking a simple question: What is the length of the coast of Norway? It is not hard to get an answer. Simply take a map and a ruler and measure it. The problem is that the coast is not a straight line. There are many curves and turns that make it hard to get an exact number. If you get a larger map or you measure it with a smaller ruler and are able to take into account more of the nooks and crannies of the coast, you will find that the length of the coast will be longer. One could take this question very seriously and actually walk along the coast of Norway and measure it while walking around all the magnificent fjords. Measuring the coastline like this would further increase the length of the shoreline. If one were to ask an ant to walk the coastline of Norway, the coast would be even longer.[35] What is the real length of the coast of Norway? The answer is that the length depends on how it is measured. This strangeness is called the *coastline paradox*, but of course the idea has nothing to do with Norway or a coastline. We could have asked that question about many physical objects.

It is important to realize that this thought experiment is arguing against absolute lengths. One might believe that there is really an exact length of the coast of Norway and that by using better and better instruments we will get better and better approximations to this exact length. This is false. Rather, the length depends on how it is measured. We are going to find similar phenomena within relativity theory.

Galilean Relativity

Einstein's relativity theory was based on the work of earlier giants. Galileo described how our perceptions of the laws of physics remain unchanged when there is constant movement. He discussed movement in an important transportation vehicle for his generation: boats. Different experiments were performed on both a stationary and a moving boat. His writing is so beautiful and clear that it is worth quoting at length:

> Shut yourself up with some friend in the main cabin below decks on some large ship, and have with you there some flies, butterflies, and other small flying animals. Have a large bowl of water with some fish in it; hang up a bottle that empties drop by drop into a wide vessel beneath it. With the ship standing still, observe carefully how the little animals fly with equal speed to all sides of the cabin. The fish swim indifferently in all directions; the drops fall into the vessel beneath; and, in throwing something to your friend, you need to throw it no more strongly in one direction than another, the distances being equal; jumping with your feet together, you pass equal spaces in every direction. When you have observed all of these things carefully (though there is no doubt that when the ship is standing still everything must happen this way), have the ship proceed with any speed you like, so long as the motion is uniform and not fluctuating this way and that. You will discover not the least change in all the effects named, nor could you tell from any of them whether the ship was moving or standing still. In jumping, you will pass on the floor the same spaces as before, nor will you make larger jumps toward the stern than towards the prow even though the ship is moving quite rapidly, despite the fact that during the time that you are in the air the floor under you will be going in a direction opposite to your jump. In throwing something to your companion, you will need no more force to get it to him whether he is in the direction of the bow or the stern, with yourself situated opposite. The droplets will fall as before into the vessel beneath without dropping towards the stern, although while the drops are in the

air the ship runs many spans. The fish in the water will swim towards the front of their bowl with no more effort than toward the back, and will go with equal ease to bait placed anywhere around the edges of the bowl. Finally the butterflies and flies will continue their flights indifferently toward every side, nor will it ever happen that they are concentrated toward the stern, as if tired out from keeping up with the course of the ship, from which they will have been separated during long intervals by keeping themselves in the air.[36]

Galileo was describing many different experiments demonstrating that the laws of physics cannot tell the difference between a boat standing still and a moving (unaccelerating) boat.[37]

Einstein discussed many similar experiments on trains, a major transportation vehicle for his generation. However, I will talk about cars since they are a little more contemporary. While you are a passenger in a car going 50 miles per hour, throw a small ball upward. If the car is not accelerating, decelerating, or making a sharp turn, the ball will gently fall right back into your hands. This is actually an amazing fact since your hand has moved forward a couple of yards in the few seconds that the ball was in the air. In fact, the ball would act the same way if the car was not moving at all. This is exactly Galileo's point about the ship: one cannot detect movement by looking at how things move inside the boat/train/car. (You can, of course, look out the window to determine if you are moving.)

Now think of people standing on the curb watching your car and your ball. What do they see? They do not simply see the ball rising and falling. Rather, they see the ball rising and going forward. After all, the car is moving at 50 miles per hour and the ball will go forward with it. The ball will land in your hand in a few seconds. The point is that as a passenger, you can do some calculations, see that you are throwing the ball straight up, and calculate when it will land. At the same time, the people on the curb see the ball go up and forward, perform some calculations, and determine when and where the ball will land. Although the calculations are different, the laws of motion are the same.

Following the ideas of Galileo, Einstein assumed these results were true for any observer looking at the physical universe:

Postulate 1: All observers at a constant speed must observe the same laws of motion.

Since this idea was already known to Galileo, this postulate is known as *Galilean relativity*. However, Einstein went further with these ideas.

Special Relativity

To understand Einstein's theory of special relativity we must start with a discussion of the speed of light. The very fact that light travels at a finite speed and is not instantaneous is counterintuitive. Simply turn on the light and it seems that the room instantly lights up. Nevertheless, scientists in the seventeenth century realized that light is not instantaneous but travels at a finite speed.

The first experiment that attempted to calculate the speed of light was done in 1676 by the Danish astronomer Ole Rømer (1644–1710). The idea is pure genius and worthy of our attention. Rømer was using this newfangled invention called a telescope to observe the planet Jupiter and one of its moons called Io. The moon rotates around Jupiter at a fixed speed. That means that the time that Io goes behind Jupiter (eclipsed) and the time that it emerges from behind Jupiter (moonrise) should be set times. However, the astronomer noticed that when the Earth was far away from Jupiter, Io's appearance was delayed (see figure 7.22).

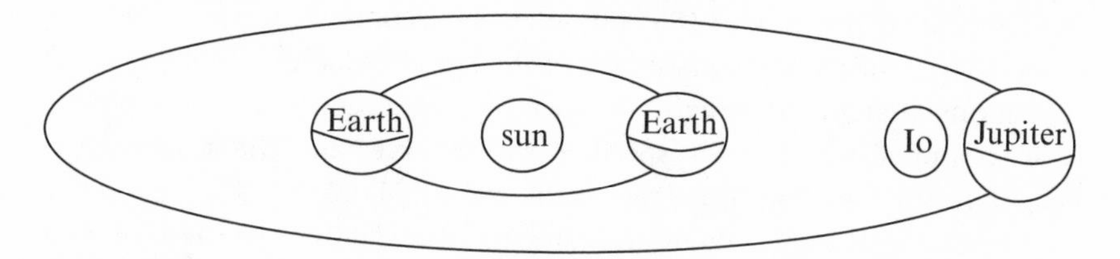

Figure 7.22
Two views of Jupiter and its orbiting moon Io

He reasoned that the cause for the delay was that it took longer for the light reflected off Io to reach the Earth. From his knowledge of the distance of the Earth from Jupiter and his knowledge of the size of the orbit of the Earth, Rømer was able to calculate the speed of light. While his calculations were off, the idea was brilliant and led other scientists to perform more exact experiments.

Eventually it was determined that light (in a vacuum) travels at about 186,000 miles a second. There was, however, something shocking about the speed of light, namely, that this speed is constant regardless of the speed of the source of the light or the speed of the observer. This is contrary to any other phenomenon in the universe. If you are traveling in a car at 50 miles per hour and another car is traveling in the same direction at 30

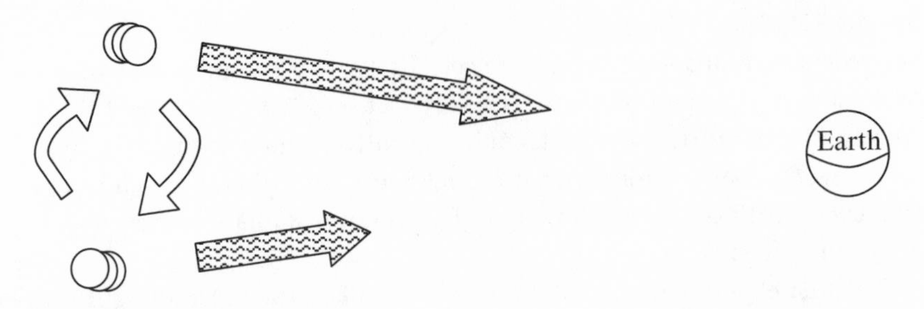

Figure 7.23
Observing the light from a binary star system

miles per hour, you will perceive the other car as going only 20 miles per hour. If you are traveling in a car at 50 miles per hour and a car is coming toward you at 30 miles per hour, you will perceive it as coming toward you at 80 miles per hour. This is true for cars and other objects in the universe. It is *not* true for light. If you are traveling, regardless of whether the source of the light is coming toward you or going away from you, it will appear to be going at 186,000 miles per second.

Einstein realized the consistency of the speed of light by looking at the equations that describe light (Maxwell's equations for electromagnetic waves) and seeing that the speed of the observer and the speed of the source of the light are "not even in the equation(s)."

There were also experimental results that showed that the speed of light will always be perceived at the same rate. The simplest experiment was described in 1913 by the Dutch astronomer Willem de Sitter (1872–1934). He considered a binary star system—that is, two stars close enough that their forces of gravity pull each other into a spin, as in figure 7.23.

If light did not travel at a constant speed, it would move faster from an approaching star and more slowly from a retreating star. Such comings and goings happen all the time with a binary star system. If light did not travel at a constant speed, then the light coming from the star approaching the Earth would reach the Earth before the light coming from the retreating star. In that case, the light would come in a scrambled form. De Sitter reported that no such scrambling occurs. He concluded that the speed of light is constant regardless of the velocity of the source of the light.

As an interesting sidebar, the consistency of light is used in making certain definitions of distances and times. The units of measurements that we use, such as mile, foot, inch, meter, hour, minute, or second, are derived from cultural and historical factors. Researchers would like a more scientific way of describing these units. Since the speed of light is consistent, it is used to determine the official scientific definition of certain lengths. Given that light travels at precisely 299,792,458 meters per second, we can take it as a definition that 1 meter is the distance that light travels in 1/299,792,458 second. What is a second? To answer this, researchers have considered certain vibrations that are done by a cesium-133 atom. A second is defined as the time it takes this atom to make 9,192,631,770 vibrations. This number was chosen because it matches up with what we historically know is a second. These two units of measurement give us exact technical definitions of length and time. But we will see that these definitions are slightly misleading.

Einstein took the consistency of the speed of light as a postulate about the universe:

Postulate 2: All observers will always view the speed of light at the same rate.

These two simple postulates are all we need to derive the results of special relativity.

Length Contraction and Time Dilation

How can we unify these two postulates? How can two observers moving at different speeds agree about the speed of light? First let us meditate a little on measuring speeds. To calculate how fast a car is going, we set a particular distance and measure the time it takes the car to cover that distance. The speed is then the distance divided by the time. So 50 miles per hour means that the car can cover 50 miles in one hour. Imagine that we know that a car is traveling 50 miles per hour but for some odd reason we measure it traveling at only 30 miles per hour. How can that be? We must be making a mistake in our measuring. Since speed is distance divided by time, we must be making a mistake about the distance or the time. Our error must come from the fact that the measuring stick we use to calculate the distance is wrong, or the clock we use to calculate the time is off, or it could be a combination of both. This is the only way we can account for our error.

Now let us return to the speed of light. It is measured by the distance (space) it traveled in a certain interval (time). If there is an "error" in the way the light is observed, then there must be something wrong with the way space and time are measured. Imagine Captain Kirk firing a phaser gun while two space shuttles are observing the action. One space shuttle is stationary and the other is moving in the same direction as the light (as in figure 7.24).

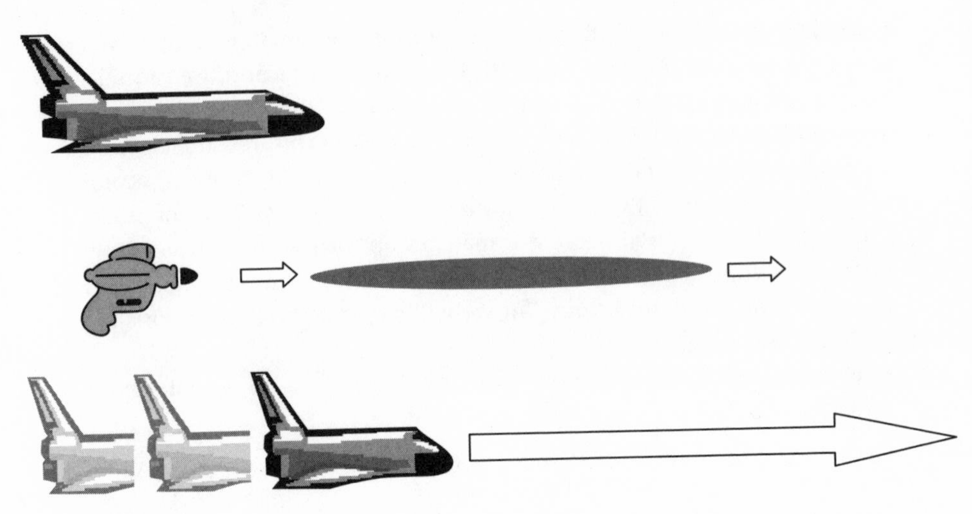

Figure 7.24
Stationary and moving space shuttles measuring the speed of light

Postulate 2 tells us that they both see the light traveling at 186,000 miles per second. Let us say that the stationary space shuttle has measured the "correct" distance and time to calculate the speed. What about the moving space shuttle? Since it is moving, one would expect its passengers to perceive the light going a little slower. But in fact they also see the light traveling at 186,000 miles per second. The only way this "error" can occur is if they measure the distance and time "incorrectly." That is, their measuring rods must have shortened so that the distance they measured is "incorrect," and their time clock must have slowed down so that the duration they measured is "incorrect." In fact, this is exactly what happens! Their measuring rods get shorter in a phenomenon called "length contraction" and their clocks go slower in a phenomenon called "time dilation." Since this is a natural process, it is wrong to call one view correct and the other incorrect. Both views are correct.

The first thing we must realize is that it is not only the measuring rods that shrink: everything in the moving space shuttle shrinks. In fact, the shuttle itself shrinks. Astronauts standing in the moving spaceship will be thinner. When they are lying down in the direction in which it is moving, they get shorter. Since everything goes through this length contraction, it is not noticeable to people in the moving shuttle. In contrast, it is noticeable to the person in the stationary shuttle.

Similarly, not only do the stopwatches used to measure speeds slow down, but all clocks and all processes slow down. These processes include the heart rate and chemical reactions in the bodies of astronauts. Their aging process slows down as well. Again, this will not be noticeable to any observer in the moving shuttle, but it will be perceptible to someone in the stationary shuttle.

It must be stressed that it is not the case that the moving space shuttle *appears* to shrink or *seems* like it is shrinking. Rather, *it shrinks*. This is also true of time dilation. These are not illusions. They are basic facts about the universe we live in: moving objects go through space contraction and time dilation.

These effects are noticeable to a stationary observer only when the rocket moves near the speed of light. We do not see things move even remotely close to the speed of light. Remember, even 1 percent of the speed of light is 1,860 miles per second. There are no vehicles that can even come close to that. In most cases no length contraction or time dilation will be perceptible. However, in a laboratory setting these changes can be measured.

Physicists have formulated equations that indicate exactly how much length contraction objects will experience and how slow time will progress relative to the stationary observer. These equations take into account the velocity of the observer. The faster the movement, the more space contraction and time dilation will occur. What is the limit of this process? What if people could go very, very fast? If they were actually able to go the speed of light, they would shrink to nothingness and time would totally stop for them. That is, they could not exist. This is yet another consequence of special relativity: there is a type of cosmic speed limit. Nothing can move as fast as, or faster than, the speed of light. This is, quite literally, a limitation described by science.

There is an urge to simply wave away all this talk of relativity and insist on absolute space and time. One wants to merely declare that the measurements done from a stationary position on Earth are the absolute measurements and every other measurement is relative. This would be an error. Although it seems like the Earth is not moving, it is, in fact, constantly moving in a wild pattern. Remember that the Earth is spinning on its axis at about 1,000 miles per hour. It is rotating around the sun at about 67,000 miles per hour. Furthermore, our solar system is moving around our galaxy at about a half a million miles per hour. Poke your finger into the air. Wait a second. Now poke your finger "in the same place." Realize that the two places where you poked your finger are hundreds, if not thousands, of miles apart. A stationary observer on Earth is far from stationary. There are no absolute observers, no absolute measurements, and no absolute space and time. All is relative.

How does time slow down? How can clocks possibly slow down? All clocks work with some type of chemical or physical process. Whether it is a battery or a wind-up spring, clocks move forward with such processes. To see an example of a way that moving fast affects time, consider a clock that works using light. Imagine a clock that works with two mirrors and a light pulse bouncing between them, as in the top part of figure 7.25.

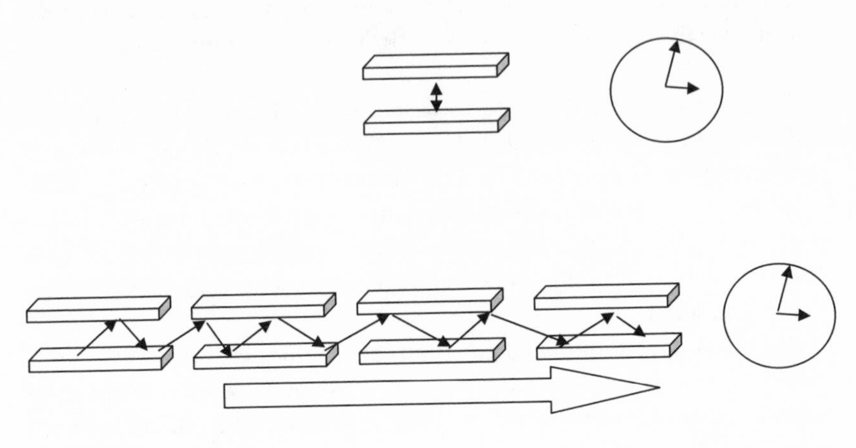

Figure 7.25
A stationary and moving light clock

The clock works by going forward one second every 10,000 times the light bounces back and forth. Since the speed of light is constant in the universe, this would make a good clock. Now imagine the same clock speeding across the universe. The light will still be bouncing back and forth but now the light is bouncing diagonally. The diagonal path is always longer, so the clock will still work but the light will have to travel longer in order to perform its 10,000 laps. Therefore, to the stationary person, the moving clock will appear to be ticking at a slower rate.

There is actually experimental evidence for time dilation. In 1971 four atomic clocks were placed on planes that flew around the Earth. When the planes returned, the time on the clocks was compared to atomic clocks that were stationary on the ground. It was found that the frequent-flier clocks had actually lost time. Science fiction writers take this idea to the extreme with something called the *twins paradox*. Imagine an astronaut who leaves a twin sibling on Earth and then travels through space very close to the speed of light. If the astronaut goes fast enough, they would come back to Earth without aging much while the stationary twin is old and decrepit. In a sense, special relativity permits you to visit the future. Unfortunately, there does not seem to be a way to return.

Consider the implications of time dilation when velocity increases. Examine the path at the left in figure 7.26.

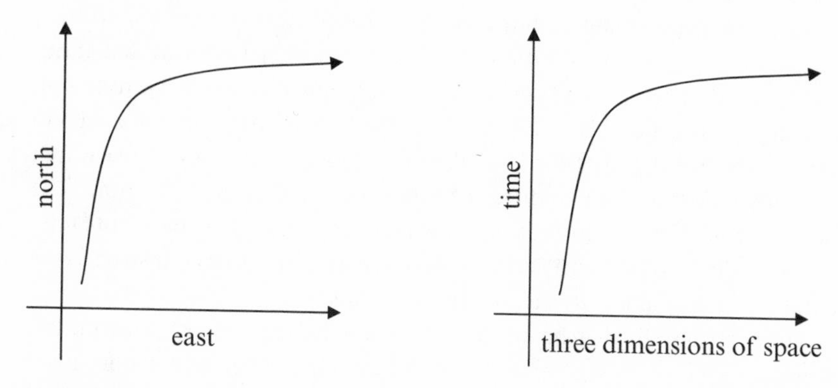

Figure 7.26
An analogy with spacetime

A person is moving northward and only slightly to the east. At some point they start moving eastward. The more eastward they move, the less northward they are moving. That is the way the two dimensions are brought together. Now consider the right-hand diagram in figure 7.26. If a person is moving a little, they do not really change their position in three-dimensional space. At some point they start moving fast and then they are changing their position in space but are moving less in time. The time dilation shows that the faster they move, the slower time flows (as measured by someone who is not moving). This is a way of seeing that the three dimensions of space and one dimension of time are intimately linked. Just as north and east are linked in the left-hand diagram, so too are time and space linked in the right-hand diagram. Einstein showed that space and time are not different entities that can be looked at individually. Rather, they are part of a four-dimensional object called *spacetime*. We usually think of motion as a movement in space while time passes. We now know that the proper way to think about motion is as a path in spacetime. Later, when we study general relativity, this spacetime will be very important.

Simultaneity

Another consequence of special relativity is that the notion of simultaneity, or that two events happen at the same time, is problematic. After all, if two people traveling at different speeds can argue about how much time elapsed, then they can definitely argue about whether two events occurred at the same time. Imagine that two astronauts observe an event and then board spaceships going at different speeds. Since they have traveled at different speeds, their perceptions of how much time has passed are different. Eventually they both land on another planet and are told that a second event happened before they landed. One astronaut may look at her watch and conclude that the two different events occurred at the same time. The other astronaut might look at his watch and conclude that the events happened at different times. Who is correct? There is no correct answer. *Time*, *duration*, and *simultaneity* are all relative terms.

Einstein described a beautiful little thought experiment that underscores this point. Imagine standing in a train station and observing a speeding train pass. When the train is right in front of you, two bolts of lightning appear to you at the exact same instant. The bolts hit the front and back of the train, as in figure 7.27. The distances from the two ends of the train to you are identical and so the light had to travel the same distance. You conclude that the lightning struck the two ends of the train at the same time. But now consider a traveler at the center of the train. They are traveling toward the front and away from the back lightning bolt. The light from the front bolt hits them first. Only later will the light from

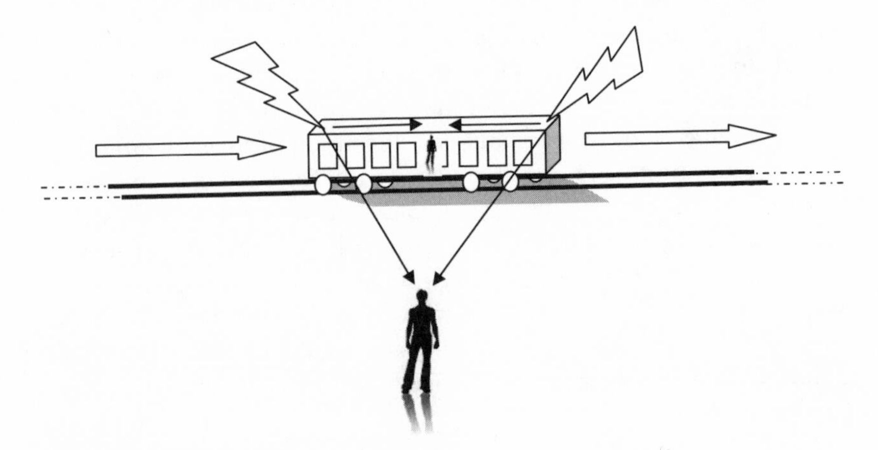

Figure 7.27
Einstein's thought experiment on simultaneity

the back bolt hit them. They therefore come to the conclusion that the front bolt came first.[37] Both you and the passenger agree on the laws of physics but come to two different conclusions about whether the bolts were simultaneous. Who is correct? We can only conclude that both are correct and that the very idea of whether two events happened at the same time depends on who is viewing them.

With the notion of simultaneity destroyed, there is going to be a problem with causality. If you cannot determine what came first, then you will not be able to determine what caused what. How are we to understand the laws of the universe when our very notion of causality is problematic?

Mass-Energy Equivalence

It would be morally reprehensible of us to leave special relativity without mentioning the world's most famous equation:

$$E = mc^2.$$

That is, energy (E) can be exchanged for mass (m) multiplied by the square of the speed of light (c^2). This equation is a direct consequence of special relativity, and it describes how energy and matter can be converted into each other.

Before we begin to understand this, we need to be reminded what mass is. Usually the mass of an object tells us how much matter the object has. Given two balls of the same size where one is made of steel and the other of cork, the ball of steel has more mass. Physicists have two ways of measuring mass: gravitational mass and inertial mass.

Gravitational mass is basically how much the object weighs—that is, how much force gravity exerts on the ball. Obviously, the steel ball weighs more than the cork ball. Such a measurement depends on where you measure it. The same ball weighs slightly more near the Dead Sea than on top of Mount Everest. (I clarify why this is true shortly.) So gravitational mass is relative to how it is measured and is not absolute.

Inertial mass is how much the object resists pressure. In other words, if the ball is pushed, how much will it move? If you push the steel and cork balls with the same amount of force, the cork ball will move at a faster rate. Special relativity shows us that measuring how fast things move is a relative process. That is, it depends on the rate of speed of the observer. We conclude that inertial mass is also not absolute.

These two ways of measuring mass give you the same answer. (In fact, this idea is central in our forthcoming discussion of general relativity.) The fact that both ways of defining mass are subject to relativity shows that the very nature of matter is also relative. Not only are space and time relative terms, but so is mass.

Now to explain the mass-energy equivalence. It would take us too far afield to actually give details of Einstein's famous equation, but we can at least get an intuition as to why mass and energy can be converted into each other. Imagine taking a mass and applying a huge force to the mass while it is in outer space. By Newton's law (force = mass times acceleration or $F = ma$), the mass will accelerate in comparison to the force applied to it. Notice we are talking about acceleration and not about velocity. That means the mass will go faster and faster and there is seemingly nothing to stop it. But remember that special relativity teaches us that nothing can go faster than the speed of light. To ensure that the object never goes faster than the speed of light, the mass of the object increases as it goes along. An increasing mass will guarantee that the object slows down. We have successfully converted force—that is, energy—into mass. In contrast, the process that occurs inside a nuclear reactor is an example of mass turning into energy.

As a result, every object that is moving has more mass than when it is stationary. This extra mass might be imperceptible when we are dealing with speeds that are minuscule compared to the speed of light. Nevertheless a moving object has more mass. Similarly, an object with more energy has more mass. So an electric iron turned on weighs more than one that is turned off.

This equivalence is the basis of nuclear energy and nuclear bombs. The fact that c—and hence c^2—is such a huge number shows that small amounts of mass can be converted into enormous amounts of energy. This is where atom bombs and nuclear reactors get their immense power. The mass-energy equivalence is actually the basis of a lot more: within the sun, there are constant nuclear reactions that are converting mass into energy. This energy comes to Earth and gives us life.

General Relativity

Throughout our discussion of special relativity, we have always restricted ourselves to being an observer moving at a constant speed and not making any turns. Let us drop this restriction. In contrast to people in a car moving at a constant speed, passengers in a car speeding up will feel as if they are going into the seat and when the car makes a sharp right turn, they will feel themselves pushed to the left. This is different from the smoothness of Galileo's ship, discussed above.

To understand acceleration, Einstein posed the following thought experiment. Imagine a child closed up in a box. If she drops a ball, the ball will fall to the floor.

Figure 7.28
The equivalence of acceleration and gravity. Figure by Hadassah Yanofsky.

There are two ways to explain why the ball falls, as in figure 7.28.[38] One possibility is that the box is on the Earth and that gravity is pulling the ball. The other possibility is that the box is in outer space and the spaceship is accelerating. In that case, the ball will fall to the floor for the same reason that an astronaut feels pulled into the seat when blasting off. This is similar to the feeling of a force when you accelerate in your car. Just as the scientist stuck inside Galileo's ship cannot determine if the ship is moving or still, so too will the child in the box not be able to tell if gravity or acceleration is affecting the ball. Einstein concluded that there is really no way to tell the difference between gravity and acceleration. This leads us to his general principle of relativity.

Postulate 3: All observers will observe the same laws of motion.

This means that the laws of physics must be the same regardless of whether the person is feeling gravity or acceleration.

The length contraction and the time dilation of special relativity also occur in general relativity. A person accelerating or decelerating will perceive light traveling at a constant speed, so their measuring rods and clocks must also change. However, the length contraction and the time dilation will no longer be constant. As the traveler speeds up, the stationary observer will observe the moving measuring rods shrinking and the clocks slowing down. In contrast, when the traveler decelerates, the stationary observer will see the measuring rods get larger and the clocks speed up. There is, however, another aspect of general relativity: since acceleration and gravity are the same, measuring rods and clocks will also be perceived to change near any great mass. A spaceship traveling near the sun or near a black hole will shrink and its clocks will go slower. The limit of this is that if a traveler actually entered a black hole with a watch that miraculously survived, it will not move. Of course, the traveler will not notice it either.

As we saw with figure 7.26, space and time are not separate entities. Rather, spacetime is a unified four-dimensional arena where all motion and the laws of physics take place. General relativity makes this spacetime much more interesting. Rather than just being a flat four-dimensional arena, it now has curves and bends. Mass (or equivalently, energy) curve and bend the arena. It is very hard to envision a four-dimensional space and it is even harder to imagine such a space curved. A helpful way to wrap your mind about this concept is to think of a two-dimensional flat rubber sheet. When weights are placed on the sheet, it curves around it as in figure 7.29. The mass curves the very fabric of spacetime and an object near the mass will tend to come toward it. So mass curves spacetime and the curves of spacetime affect the mass. The two spheres have curved spacetime and will attract each other. We call this gravity.

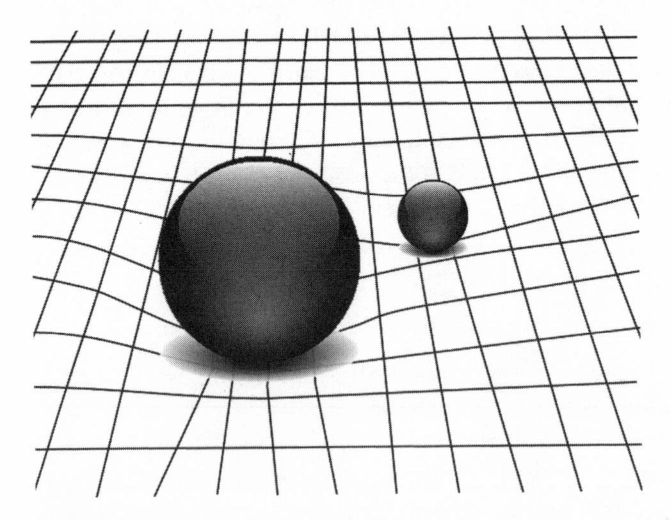

Figure 7.29
Curved spacetime with two bodies. Figure by Hadassah Yanofsky.

This is the general-relativity explanation of gravity. Acceleration can also be understood from this point of view. An object going at a constant pace is equivalent to a straight line in spacetime. When an object accelerates, its path curves and veers away from a straight line.

Five years after Einstein formulated general relativity it was experimentally confirmed in a famous experiment with a solar eclipse. An astronomer named Arthur Stanley Eddington (1882–1944) traveled to the island of Principe off the west coast of Africa to be there on May 29, 1919. There was going to be a total solar eclipse in the southern hemisphere and he wanted to witness this event. A solar eclipse is when the moon passes between the Earth and the sun, blocking the solar rays from hitting the Earth. Eddington calculated the position of the sun when the eclipse would happen and measured the distance between two stars that were near opposite sides of the sun, as in the top part of figure 7.30. He then waited for the sun to come between the stars. At that point, since the sun was so

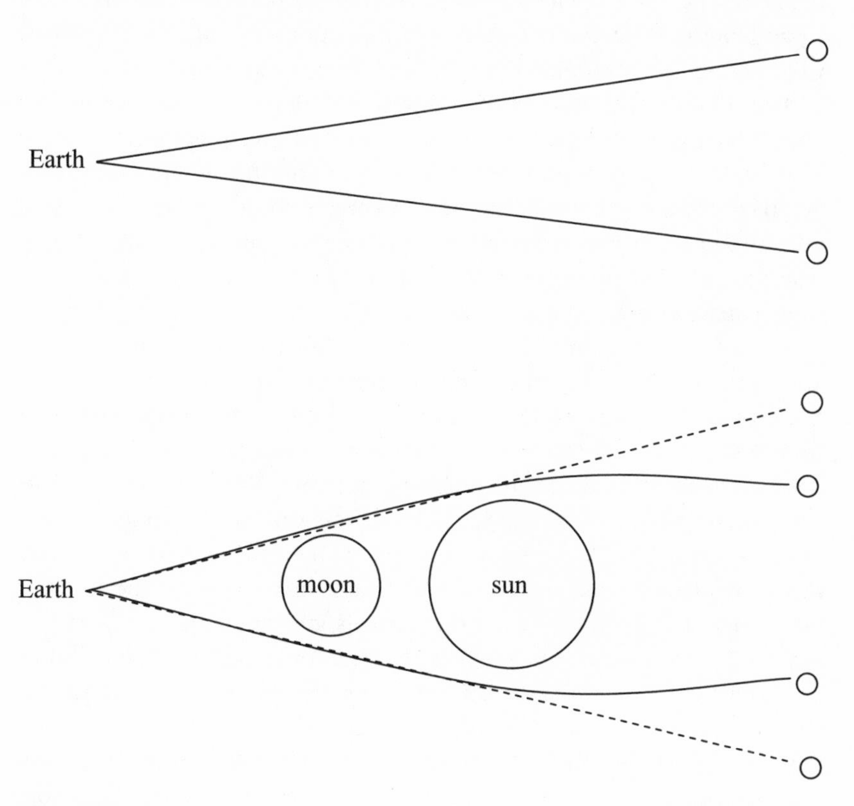

Figure 7.30
Before and during the 1919 eclipse

bright, the stars were not visible. Then the moon came between the Earth and the sun for 410 seconds. During this time, the sunlight was blocked off and the stars became visible again. Eddington was able to measure the distance between the two stars. As Einstein predicted, the stars seemed farther apart. The sun's gravitational pull actually tugged the light rays from the two stars. With those rays being pulled around the sun, the stars actually looked as if they were farther apart than before the sun came between the stars. Light is also affected by gravity because gravity is the curving of space. Eddington's results were broadcast all over the world: general relativity had been confirmed.

Question: Exactly how far apart are the two stars? Answer: It depends on when you are looking at them. The rays of light are curved because spacetime itself is curved. That is why the light is being pulled. You might protest that the positions of the two stars are fixed regardless of what it looks like when the sun passes between them. A similar phenomenon happens when you see a straw bend as water is poured into a glass. The straw seems to bend, but in fact, it is only an optical illusion. Such a protest would be slightly misguided. It's not only that the stars *look* farther apart. Rather, they *are* farther apart. If you wanted to touch the bent straw with your finger, you would not put your finger where you see the straw. You would take the illusion into account and put your finger where it should be. In contrast, if you wanted to travel to one of those two stars, you must take into account the curvature of spacetime. If your spaceship will pass near the sun, you must take its gravitational pull into account. The curvature of spacetime is not an illusion.

There is an interesting way to actually measure the curvature of space. Your weight is different depending on whether you measure it on Mount Everest or at sea level. All other things being equal, an increase in altitude from sea level to the top of Mount Everest (29,035 feet) causes a weight decrease of about 0.28 percent. Someone weighing 190 pounds at sea level will weigh $190 \times 0.0028 = 0.532$ pounds less on top of Mount Everest. Though this might not be enough to notice, it is true nevertheless.[39] This can be explained from the Newtonian point of view: since the distance between you and the center of the Earth is greater at the top of Mount Everest, Newton's equation predicts that the force tugging at you will be less. However, one can also see this from the perspective of relativity theory. If you think of the Earth as making an indentation in the very fabric of spacetime, then the top of Mount Everest is farther away from the indentation and closer to the flat part of spacetime. Therefore, your weight on Mount Everest—that is, the tug on you toward Earth—is less than when you are at sea level.

General relativity has been experimentally proved in many other ways since Einstein formulated it. This idea that space, time, length, duration, mass, energy, gravity and so on are all relative notions is a surprising fact about our ability to know our world.

Unifying Quantum Mechanics and Relativity Theory

In the last two sections, we have discussed the two greatest revolutions in our understanding of the universe: quantum mechanics and relativity theory. These two theories are extremely successful, but they contradict each other in several ways:

- For the most part, the two theories deal with different domains. Quantum mechanics deals with objects that are very small, while relativity theory deals with very large objects. However, there are areas where their domains overlap: places called singularities or black holes. In these overlapping areas these two theories give conflicting predictions.
- These theories also reflect different conceptions of the fundamental nature of space and time. For example, the entanglement phenomenon in quantum mechanics seems to show that space is more intertwined with itself than is the notion of space in relativity theory. Also, general relativity uses the fact that space is continuous, while quantum theory considers space and time discrete.
- The laws of classical physics and general relativity are deterministic, whereas the laws of quantum mechanics are nondeterministic.

In short, it can be shown that

quantum mechanics and relativity theory $\Rightarrow$ contradiction.

As with many paradoxes, this contradiction shows that we are in need of a new paradigm. A new theory is required. This new theory should unify both quantum mechanics and relativity theory. It should make the same predictions as each of the two theories it is replacing. Furthermore, in the domains in which they overlap, this theory should make a single prediction that conforms to observation. It is expected that this theory will provide new conceptions of space, time, matter, and causality.

Although, at present, no such theory is agreed on by everyone, it already has a name: *quantum gravity*. Since this theory will describe both gravity and the quantum mechanical forces, it will be a *Theory of Everything* or a *Grand Unified Theory*. Many different theories are vying for that lofty position. These theories have esoteric names such as *string theory, loop quantum gravity*, and *noncommutative geometry*. Each of these different theories has its own counterintuitive properties. At the moment, string theory seems to be the leading contender. However, it is too soon to tell. Any one of them could be the true Theory of Everything. Or, the true theory may not have been developed yet. Perhaps there will never be a Theory of Everything. One thing seems certain about quantum gravity: it will show us that our naive conception of the universe is wrong and that the universe is a far more interesting place than we believed.

Further Reading

Section 7.1

The story of Lorenz, the basics of chaos theory, the butterfly effect, and many other topics are wonderfully covered in Gleick 1987. Chaotic systems are also discussed in chapter 11 of Tavel 2002. Statistical mechanics is treated in chapter 9 of Tavel 2002. Complex systems that have feedback and self-organization are discussed in Waldrop 1992. Turing's role in morphogenesis can be found in Hodges 1983. The three-body problem and the many attempts to find solutions to it are covered in Diacu and Holmes 1996 and Diacu 1996.

Section 7.2

Contrary to popular belief, it is possible to learn the mysteries of quantum theory . . . you just have to find the right book, because some are simpler or more challenging than others. Here is a list of references from easiest to hardest:

• History: Gamow 1966; Gilder 2009 (an interesting, slightly fictionalized history of the notion of entanglement); Guillemin 1968; and Pickering 1984 (a fascinating book on the relationship between group theory and quantum theory)
• Popular expositions: Greene 2004, 2011; Gribbin 1984, 1995; Pagels 1982; Peat 1991; Penrose 1991, 1994, 2005; and Tavel 2002, chap. 10
• Philosophy: Bohr 1935; Bub 1997; Casti 1989, chap. 7; d'Espagnat 1983; Heisenberg 2007; Herbert 1985; Malin 2001; and Wick 1995
• Simple textbooks: Gillespie 1970; Jordan 1986; Scarani 2006; White 1966; and Yanofsky and Mannucci 2008 (chapter 4 has a short, easy exposition of the basic ideas of quantum mechanics)
• "Real" textbooks: Dirac 1986 (an amazingly readable and deep book, but interestingly, there is nothing about entanglement even in the fourth

edition); Hannabuss 1997 (a more algebraic treatment); Sakurai 1994 (used in graduate-level physics courses); and Sudbery 1986 (for people familiar with some mathematics)

Our version of Bell's theorem was taken from d'Espagnat 1979. It can also be found in a delightfully clear article by Bell himself, Bell 1981. A more exact formulation can be found in Sakurai 1994, section 3.9.

This is, as far as I know, the first popular exposition of the Kochen-Specker theorem. However, a readable, philosophical, and slightly technical exposition is given in Held 2006. It is formally proved in section II.12 of Manin 2010.

Section 7.3

There are some very popular and readable expositions of relativity theory. Here are some resources from easier to more difficult: Gardner 1997; Greene 2004; Rindler 1969; Schwartz 1989; and Tavel 2002, chaps. 5–8.

The lightning-and-train thought experiment comes from chapters 8 and 9 of Albert Einstein's readable 1920 exposition, *Relativity: The Special and General Theory.*

There is a fascinating BBC Horizon documentary titled *How Long Is a Piece of String?* Clips from the documentary are available at http://www.bbc.co.uk/programmes/b00p1fpc.

8 Metascientific Perplexities

All logical arguments can be defeated by the simple refusal to reason logically.
—Steven Weinberg[1]

Not only is the universe stranger than we imagine, it is stranger than we can imagine.
—Arthur Stanley Eddington

It is the harmony of the diverse parts, their symmetry, their happy balance; in a word it is all that introduces order, all that gives unity, that permits us to see clearly and to comprehend at once both the ensemble and the details.
—Henri Poincaré

Scientists are not the only ones who deal with the physical world. Philosophers and other researchers are interested in how the universe works and how we learn about it. They are interested not only in what the structure is, but also in why there is structure and how it is described.

Philosophical questions about the relationship between science, the universe, and our mind are dealt with in section 8.1. Section 8.2 discusses the relationship between science and mathematics. Section 8.3 takes up the question of why the universe seems so perfectly suited for life and rationality.

8.1 Philosophical Limitations of Science

In this section I explore different aspects of the philosophy of science. This large and fascinating branch of philosophy covers the nature of science and the way it progresses. Rather than attempting a detailed survey of the philosophy of science, I simply cherry-pick several topics at the core of the field and examine how they pertain to the limits of science.

The Problem of Induction

One of the major issues in the philosophy of science is the problem of induction. Simply stated, why should one believe that if every swan ever seen is white, then all swans are white? The problem of induction asks what right we have to generalize from our few observations to a universal law. If we observe a phenomenon over and over, why does that mean it is always true? There is no logical reason to come to such a conclusion. It could very well be that pink swans exist and we simply did not see them. There is no logical reason why swans are white and not pink.[2]

We use induction every moment of our lives. We turn the light switch on with the expectation that the light will go on. We turn on the shower with the expectation that water and not mud will come out of the faucet. We plan our schedule with the assumption that the sun will rise tomorrow simply because it has happened every morning until now.

In all these different cases, we are making conclusions from a limited set of observations. We have only seen some swans in our lifetime. We have not seen all of them. The sun has risen every morning until now, but we do not know about the future and yet we make predictions about the future. Why do our past experiences give us any reason to think that the future will be the same?

This is not a new problem. More than two hundred years ago, David Hume showed that there is no logical reason why induction should work. One might counter and say that the reason why the light goes on when the light switch is turned on is because the switch causes the circuit to be completed and the bulb must light up when electricity passes through it. Hume would counter that this long chain of reasoning is simply a series of cause-and-effect operations that work only because we assume they work by induction. Each action has in the past caused a particular effect that we assume is going to happen in the future. Hume says that a person who is using induction is making an assumption that the universe is somehow uniform over time. There is no reason to believe this assumption.

Induction is going from observing many single instances to a general rule. Going in the opposite direction—from a general rule to a conclusion about a particular instance—is called *deduction*. If there is a general rule that says that all swans are white, then we can safely conclude that a particular swan is white. In contrast to induction, deduction is a reasonable process. From the statements "All men are mortal" and "Socrates is a man," it is a logical deduction that "Socrates is mortal." There is no way one can deny this reasoning. The major problem with deduction is that the general rules usually come from induction.

The problem of induction is at the very core of science. Scientific laws are formulated by looking at phenomena and generalizing them to universal rules we call laws of nature. There are, however, no real reasons why we have the right to come up with such generalizations. The law that Newton gave us that describes the motion of pairs of bodies was not formulated because Newton examined all pairs of bodies in the universe. Rather he formulated the law by using understanding and induction on what he saw. In fact, that law is simply false when applied to *all* pairs of objects. Quantum mechanics has shown us that subatomic particles do not follow Newton's simple law. General relativity has also shown that Newton's laws are not the whole story. We conclude that Newton's laws were formulated with induction and they turned out to be false. They did not work for very small or very large objects, as revealed to us by the physics revolutions of the twentieth century.

These abstract epistemological topics are at the core of the contemporary battle concerning global warming. While most scientists look at the data available and come to the conclusion that human beings are causing the Earth to get hotter, some scientists are not convinced. They say that there is not enough data to arrive at that conclusion. They see other times in history where there were ice ages and thaws. They do not see the current global warming as different from those other ages. Such scientists feel that much more data, perhaps even data that we will never be able to acquire, must be examined before we can come to such a conclusion.

Not only science, but our entire worldview is built from induction. We observe phenomena and formulate theories about the true nature of the world. Every time we close the refrigerator door, we are sure the light goes off, even though we do not see it go off. As Wheeler wrote, "What we call 'reality' . . . consists of an elaborate papier-mâché construction of imagination and theory fitted in between a few iron posts of observation."[3]

Philosophers have formulated different responses to the problem of induction. The most popular response is to agree that induction does not always give absolute truths, but it does give probabilistic truths. If all the swans that have been seen so far are white, it is highly probable that all the swans that exist are white. Furthermore, the more white swans that you see, the more sure we will be that all swans are white. As for the sun rising tomorrow, there are no logical proofs to prove it. However, because the sun has risen every morning until now, the probability that it will also rise tomorrow is very good and you can "bet your bottom dollar that tomorrow there'll be sun!"

Another possible answer to the problem of induction is that while making inductive inferences might not be logical, it is definitely a human activity. In other words, humans have learned over time how to go from particulars to general rules. Not all inductive laws that human beings make are perfectly true, but many are true. While it might not be strictly reason, it is nevertheless a justified human activity.

One of the leading philosophers of science in the twentieth century, Karl Popper, believed that there is no real solution to the problem of induction. He claims that science does not work by scientists trying to verify laws made via induction. We don't just look at what happened before and generalize to make scientific laws. Rather, scientists make conjectures that can be shown to be wrong (i.e., they are falsifiable) and try to show that these conjectures are wrong. We will glimpse more of Popper's ideas shortly.

You might try to ignore the problem of induction and just say "it works!" After all, every time humans used induction in the past it worked, so induction must work. This pragmatic solution will not do. We are looking for a reason to believe in induction and you are saying that *it worked in the past so by induction it will always work*. But this is using circular reasoning: you are using induction to justify induction. David Hume summed up why this reasoning is not permitted: "It is impossible, therefore, that any arguments from experience can prove this resemblance of the past to the future, since all these arguments are founded on the supposition of that resemblance."[4] In other words, we are assuming that the universe stays the same in order to prove that the universe stays the same. That is not legal.

Another example that shows the seeming disconnect between reason and induction is called the *ravens paradox* or *Hempel's paradox*. Consider the following statement:

All ravens are black.[5]

Every time you see a raven that is black you are confirming this statement. Assume this statement is true and consider an object that is not black. Because of the truth of the statement, this nonblack object is definitely not a raven. We are led to a statement that is logically equivalent:

All nonblack objects are nonravens.[6]

These two statements say the same thing in different terms. If an observation confirms one of them then it automatically confirms the other one.

On the other hand, if we find a nonblack raven then we have falsified both statements. Now, consider a green sweater. This object is not black and it is also not a raven so the green sweater is a confirmation of the second statement. Every time we see a green sweater—since green is not black and a sweater is not a raven—we are essentially confirming the second statement, which is equivalent to the original statement about black ravens. This is somewhat bothersome. How can it be that when we see a green sweater or a blue ball we are confirming the statement that all ravens are black?

We can even go further. When we see a green sweater we are also confirming the following statement:

All nonblue objects are nonravens.

The green (nonblue) object is a sweater (a nonraven.) This statement is equivalent to

All ravens are blue.

So by just looking at this sweater we are confirming that ravens are black and also blue. These are just two of the infinite number of statements confirmed by the observation of a green sweater. What is worse is that, as far as we know, ravens are not blue and these two statements are, in fact, false. How can green sweaters be so helpful in our ornithology observations?

There are various possible resolutions given for the ravens paradox. One solution is to simply agree with the conclusion of the paradox and say that when you observe a green sweater you are essentially confirming the statement that all ravens are black. However, one must think of it as a probabilistic confirmation. Say for a minute that there are a million ravens in this world. Every time you see a raven and it is black, you are one-millionth closer to confirming the statement. In contrast, there are vastly more nonblack objects in the world. When you see any of those objects you are getting closer to showing that the set of ravens is a subset of the set of black objects, as in figure 8.1. However, since there are so many nonblack objects in the universe, the confirmation of the statement is minuscule.

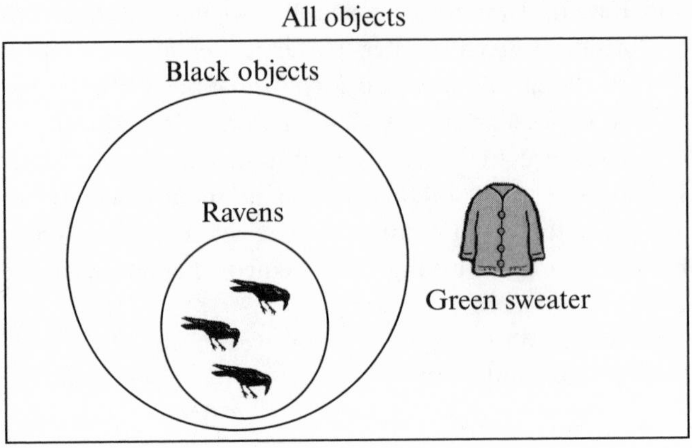

All objects

Black objects

Ravens

Green sweater

Figure 8.1
The ravens paradox

Regardless of which resolution one accepts for the problem of induction, you have to agree that inductive reasoning, the heart of the scientific endeavor, is beyond the bounds of reason. This is not to say that induction is false. Induction clearly works. We nevertheless have to be cognizant of the fact that it is not strictly a reasonable process.

Simplicity, Beauty, and Mathematics

Induction is not the only method that scientists use to find the laws of nature and to explain the inner workings of the universe. They also use other methodologies to select scientific theories. It is important to study these methodologies and their relationship with reason.

One of the oldest and most powerful tools that scientists keep in their toolbox is called *Occam's razor* or the *principle of parsimony*. William of Ockham (1285–1349) was an English philosopher who tells us not to assume more than we need to. That is, if we can explain something with fewer assumptions, then we should not assume more.[7] We should, metaphorically, use a razor to cut away any unneeded assumptions. There might be many different ways of explaining a phenomenon and we should always use the simpler explanation.

When Copernicus promoted the notion of a heliocentric world, it was not because of empirical evidence that implied the Earth moved. It certainly did not feel like the Earth was moving. Copernicus also did not emphasize the heliocentric worldview because it made better predictions about the universe. It did not (since he thought the planets traveled in a circle around the sun rather than in ellipses). Rather, his argument, which turned out to be correct, was that a heliocentric universe was simpler than a geocentric universe. There were no complicated epicycles in the heliocentric world as there were in the geocentric world.

A major problem with using Occam's razor is that it might not be correct. Choosing the simplest theory does not always work. For example, Copernicus thought that the planets follow a circular path around the sun, while Kepler showed that the path is actually an ellipse. Occam would have preferred the simpler circular path. Nevertheless, the laws of nature disregarded Occam's choice and follow the more complicated path. Another example of the failure of Occam's razor is that there are fewer equations in Newton's formulation of gravity than in Einstein's formulation of gravity. Fewer equations means it is simpler and so Occam's razor predicts it to be true. Nevertheless, physicists tell us that Einstein's theory is the right one to choose. Occam might counter that the number of equations is not the right measure of simplicity. Maybe he is right. Maybe not.

There are different types of simplicity. On the one hand, there is *simplicity of hypothesis*: given two theories, choose the theory that uses fewer presuppositions. Another type of simplicity is *simplicity of ontology*: given two theories, choose the one that assumes fewer physical objects exist. For example, given two theories, one that assumes the existence of ether and one that does not, choose the one that does not need the ether. Fewer physical objects are better.

These different types of simplicity sometimes work against each other. It could happen that a theory that has fewer hypotheses demands more ontology and vice versa. A case in point is the theory of a multiverse.[8] This is the belief that the universe that we see around us is one of many universes. The mathematics describing a multiverse is simpler than the mathematics of a universe. However, there are obviously more physical objects in a multiverse than in a universe. Everett's multiverse increases the number of objects, but the amount or type of math needed to describe this universe is much simpler.

Why does Occam's razor work, in general? Why should we always choose the simpler theory? Many people believe that the reason Occam's razor is so effective is that the universe is simple rather than complex. Appropriately, we should choose the simpler explanation. However, there really is no logical or rational reason to believe that the universe is simple. Maybe it is, in fact, complex. It certainly looks complex.

Another methodology that scientists use to find and select different theories is beauty. Scientists insist that a theory must, in some sense, be beautiful. The world-famous physicist Hermann Weyl is quoted as saying, "My work always tried to unite the true with the beautiful, but when I had to choose one or the other, I usually chose the beautiful." Paul Dirac had similar sentiments: "It is more important to have beauty in one's equations than to have them fit experiment . . . It seems that if one is working from the point of view of getting beauty in one's equations, and if one has really a sound insight, one is on a sure line of progress."[9]

What exactly is beauty? The term is just as hard to define in science as in regular life. Some physicists have equated beauty with elegance,[10] which is an equally indefinable concept. Some have said that beauty is related to simplicity, which is basically what Occam's razor is all about. And still others have said that a theory is beautiful if it exhibits a lot of symmetry or harmony. There is much disagreement because no one has a sure-fire explanation for what exactly to look for or why this property works at picking good theories.

One of the problems with beauty is that beauty does not always work.[11] The universe is not as pretty as scientists imagine.[12] Bertrand Russell, in his inimitable humorous way, put it like this: "Academic philosophers, ever since the time of Parmenides, have believed that the world is a unity. . . . The most fundamental of my intellectual beliefs is that this is rubbish. I think the universe is all spots and jumps, without unity, without continuity, without coherence or orderliness or any of the other properties that governesses love."[13] As with simplicity, there is really no reason to believe that the universe is always beautiful and symmetric.

Yet another sieve or tool that scientists use to select physical theories is mathematics. They want their theories to be as mathematical as possible. A theory is not really acceptable to physicists until they see nice equations. Whereas in earlier times math was considered a language or a tool to help with physics, nowadays mathematics is the final arbiter of a theory.[14] Physicists have placed their faith in the symbols and equations of mathematics. If the math works, then the physics must be correct. The culmination of this faith in the role of mathematics in choosing physical theories is the current hot physical theory termed *string theory*. This theory posits the existence of very small strings that wiggle, shake, combine, and separate. It is shown by looking at the mathematics of these strings that all the known forces in the physical universe can be explained with strings. This is one of the only theories that not only describe the forces of quantum mechanics, but also gravity that plays a major role in general relativity. Yet another advantage of string theory is that it deals well with problematic infinities. In other physical theories that try to unite quantum mechanics and relativity, the equations somehow develop uncomfortable infinities. String theory does not have those crazy infinities. For all these reasons, many physicists are excited about the developments in string theory. This seems to be the long-sought Theory of Everything. There is, however, only one problem with string theory: there is not a shred of empirical evidence that it is true. While it makes great mathematics, there is no observation that we can make (at present) that shows that the world is, in fact, made out of little strings. That does not mean the theory is false. Absence of proof does not mean proof of absence. It could very well be that the world is made out of very small strings and string theory is correct. On the other hand, string theory could just be an elaborate fiction. For the time being, we simply do not know. Can we just follow the mathematics without having empirical evidence?[15]

In what way are we justified in using simplicity, beauty, and mathematics as heuristics? One possible pragmatic justification is that these heuristics have worked in the past and we should just continue to use them in the future. After all, for the most part, science has progressed fairly well using these methods, and we should expect that it will progress just as well if we continue using them. Alas, this justification does not hold water. This argument uses induction and, as we have seen, induction is not reasonable. Another possible justification is to say that the reason why these heuristics work is that the universe we live in is, in fact, simple, beautiful, and mathematical. While it definitely seems that this is true, it is far from certain. For all we know, the universe might be complex, ugly, and nonmathematical.

As with induction, we must recognize that these methodologies for advancing science are essentially beyond reason. They work and we will continue to use them, but there is no logical reason to do so.

Karl Popper and Falsifiability

Noticing the problems of induction and the other problematic aspects of the scientific processes, Karl Popper formulated a description of the way science really works. He agreed with Hume and the others that the problem with induction makes it impossible to verify a scientific theory. In contrast, it is relatively easy to show that a scientific theory is *incorrect*. All you have to do is find one instance where the theory is shown to be false. If a theory implies some phenomena and if observation shows that these phenomena do not happen, then we have no recourse but to conclude that the theory is false. In the notation of this book,

Theory $\Rightarrow$ false phenomena

means the theory is false. Popper felt that rather than trying to verify a theory, scientists should try to show it is false.

Popper was influenced by Eddington's famous experiment with a solar eclipse, which we discussed in section 7.3. He was seventeen years old in 1919 when the experiment showed that Newton's notion that space is a flat place where phenomena take place is not true. Eddington showed that large objects like the sun will bend light and that Newton's notion of space needs to be abandoned. Einstein's theory of general relativity was seemingly "more correct." Popper was awed by the fact that after hundreds of years of following Newton, scientists were willing to abandon Newton's ideas because they were shown not to work this one time. He sharply contrasted this with theories in politics, morality, and religion, where a doctrine is held by its believers despite much evidence showing it is false.

This ability of science to be shown to be incorrect—in other words, the fact that science is *falsifiable*—was for Popper the defining property of science. It is the demarcation between science and other disciplines. Science makes predictions that can be shown to be wrong by observation. He described disciplines that do not make falsifiable predictions as *pseudo-sciences* and felt they do not reach the lofty status of science. Popper's primary examples of pseudosciences were Marxism and psychoanalysis. These two influential intellectual movements made many predictions about politics, economics, and human nature. However, whenever their predictions turned out to be false, Marxists and psychoanalysts were always able to show how their theories really explained these abnormalities. In a sense, a pseudoscience is too powerful since every outcome of any experiment and every phenomenon can be explained by it. It is only in real science where some prediction can be wrong and show that an entire theory is wrong and must be discarded.

For Popper, science progresses by making falsifiable conjectures and predictions. Scientists then go out and check these conjectures and predictions by doing experiments. If the experiments show that the conjectures are false, the scientists move on to other conjectures. However, if the experiments do not show the conjectures are false, this does not mean the theory is true. Rather, this only shows that the theory has not been falsified . . . yet.

Not everyone takes Popper's ideas to heart. People have argued whether this is really the way that science progresses. Most scientists are very happy with the probabilistic solution to the problem of induction and freely use other methodologies to select different theories. Once a theory is selected, it is usually felt to be confirmed if there is enough evidence showing it is correct. In real life scientists do not wait around until their theories are falsified.

Another criticism of Popper is that falsifiability is not the be-all and end-all that he makes it out to be. While it is used by many scientists, there are times when it does not work so well. As we will see in the next section, when Urbain Leverrier (1811–1877) looked at the motion of the newly found planet Uranus, he saw certain irregularities that did not fit well with the laws of Newton. An orthodox Popperian would have advised Leverrier to abandon Newton's laws and look for other laws. Luckily Leverrier ignored such hypothetical advice and held onto the laws of Newton. Instead of discarding Newton's laws of planetary motion, he used them to find another planet that caused Uranus to veer off course. Science is a complicated process.

Popper's ideas bring to light another problem in the philosophy of science: How does a researcher tell if a new and radical idea is right or is crazy? Is this new idea visionary or is it simply cuckoo? Obviously one should go out and perform experiments to test the idea. However, what if the idea cannot be tested or if the results of experiments are inconclusive? At what point should a theory no longer be considered radical and start being considered the truth? When should a theory be considered a quack theory? At one time most scientists believed in substances called ether and phlogiston. Planetary epicycles were once common knowledge. And yet these ideas turned out to be false. In contrast, at one time the ideas advanced by Copernicus, Pasteur, and the Wright brothers were considered totally insane, while now they are legitimate science. While it is true that most quacks remain quacks, nevertheless, the history of science has taught us that our ability to differentiate good science from bad science is not perfect.

Let's say we have two different theories that explain a certain phenomenon. How are we to tell which is the correct theory? This is the case when there are two different-shaped pegs that fit into the same hole. Which one belongs there? Obviously if there is any experiment or observation that falsifies one theory then it is to be discarded. Does that mean the other one is true? Perhaps there are other theories besides these two theories. Before any are falsified, what are we to do? This is a limitation to our ability to know the laws of nature. It is a limitation to reason.

Popper's definition of science makes the goal of achieving absolute truth through the scientific method unachievable. More importantly, for our purposes, what it shows is that what we think we know about the universe is not necessarily true. It simply has not been falsified yet. While it could very well be that the scientific theories that we currently do have are, in fact, the ultimate truth, it could also be that our current theories will one day be falsified. We simply do not know. Our theories could be the truth and could just be another tentative stage in the development of science.

We will never know even if we do have the right answer. Eddington's experiment did not show that Einstein's theory of general relativity is the correct theory. It simply showed that Newton's theory did not work well for large objects like the sun. Whether or not Einstein's theory is correct is something we do not know. If it is incorrect, then it might one day be shown to be false. In contrast, if it is correct, Popper offers no method for us to know this. We are constantly in a state of waiting to be falsified. For Popper, all scientific knowledge is provisional and not absolute.

Thomas Kuhn and Paradigms

In 1962, Thomas S. Kuhn (1922–1996) published one of the most influential books in the philosophy of science. *The Structure of Scientific Revolutions* changed the way people think about science and how it progresses. According to Kuhn, science takes place within a paradigm—that is, a group of ideas and language used by all the researchers in that field. The science that is done within such a paradigm is termed *normal science*. The community of scientists who are working within that paradigm will accept this science and agree with it. This is how the vast majority of science is done.

However, normal science is not the only story. As time goes on, certain anomalies will be found within the working paradigm. While these anomalies will not be ignored, scientists will nevertheless stick to the paradigm and make only minor changes to the ideas present in the paradigm. They will try to fix up and slightly modify the paradigm rather than transform it. With more time, these anomalies will build up into a massive crisis. There will be something wrong with the paradigm and a revolutionary change must be made. This is what Kuhn calls *revolutionary science*. The paradigm will change and this change has come to be known as a *paradigm shift*. At first the revolutionary scientists will be shunned by their peers. They will be speaking a different language and will have a different worldview. The new paradigm will seem strange and perhaps even unreasonable.

However, it will explain more and have fewer anomalies than the old one. Eventually the new paradigm will gradually be accepted by the scientists in the field.

After some time, the revolutionary new paradigm will become the normal paradigm and all the new or flexible workers in the field will work in this paradigm. Their science will become the new orthodoxy. Eventually certain anomalies will be found with this paradigm and this process will go on and on. Throughout the development of a certain field, there will be long periods of normal-science equilibrium, punctuated[16] by revolutionary paradigm shifts.

There are many examples of such paradigm shifts. The paradigmatic example is the shift from the Ptolemaic geocentric system to Copernicus's heliocentric system. This revolution took hundreds of years until it became the new paradigm. Another revolution was the change from Newton's worldview to Einstein's general relativity in the early twentieth century. The change from classical mechanics to quantum mechanics was also a major paradigm shift. An example from biology is Louis Pasteur's germ theory. In all these cases, there was a major change from an old point of view to a new one.

Another major idea in Kuhn's work is his notion of *incommensurability* between two paradigms. He felt that since there are different languages and different worldviews, it is virtually impossible for people who accept different paradigms to communicate. Some philosophers go further and say that it is wrong to compare different paradigms. Why should we say that one is more scientific or more rational than another? Each paradigm works for its own time.

Kuhn came to some of these ideas by studying the works of Aristotle. He realized that from the point of view of a Newtonian scholar, Aristotle is totally wrong and a bad physicist. But if you look at Aristotle's physics from the point of view of a scholar in Aristotle's time, he was doing very good physics that was worthy of the two millennia in which he was admired. Kuhn writes: "Might not the fault be mine rather than Aristotle's, I asked myself. Perhaps his words had not always meant to him and his contemporaries quite what they meant to me and mine." Kuhn realized that he had to look at the context that Aristotle was working in: "The central change cannot be experienced piecemeal." One has to look at the larger picture in order to appreciate Aristotle.[17]

Kuhn's book[18] was considered . . . dare we say it? . . . revolutionary. Its most controversial thesis was that science is not a search for some deep-seated notions of truth. Rather, they work within a social structure. During the normal-science phase, they pose and answer questions based on the current paradigm. Their beliefs are formed by an educational process that unquestionably accepts the correctness of a particular paradigm. From this point of view, one can ask if there is a search for truth when one does normal science or if we are simply working within a culturally constructed paradigm. This was taken further by some philosophers to mean that science was on equal footing with some less scientific fields.

Another problem posed by philosophers is the rationality of paradigm changes. Within normal science, it is quite rational to see the next step. But when the entire paradigm has to be discarded for a new one, there is no set rational way to find a new paradigm. Some philosophers believe that a paradigm shift is essentially an irrational process. If one were to believe this, then science is no longer a reasonable quest to understand the universe we live in. Rather, its main changes are not ruled by reason.

What about progress in science? If one takes incommensurability seriously, then the notion that science progresses as time goes on is in doubt. Some people have, in fact, taken this position and do not believe that the science based on current paradigms is in any way better than the science based on previous paradigms. Here I have to take exception and criticize such ideas. There most definitely is scientific progress. Newton's system is better than Aristotle's system and, in turn, Einstein's system is better than Newton's system. Later systems explain more phenomena than earlier systems. Later systems explain phenomena better. No matter what some philosophers say, we will never abandon the heliocentric view and return to geocentrism.

Truth is another topic that is important to Kuhn, his followers, and some of his critics. Most people would say that as time goes on, our scientific theories might not be exact, but we are getting closer and closer to something we might call *the truth*. This truth does not live within some paradigm, but it is in some sense "out there" and does not depend on the paradigm scientists use to understand it. However, there are philosophers who disagree with this standard view. They say that there is no underlying truth and one can only view the physical world through the prism of some type of paradigm. Such philosophers argue that every paradigm until now has been wrong and there is no reason to believe that the current paradigm is somehow the correct view of the world. They would say that science is not progressing toward anything. Rather, as it changes paradigms, it is simply moving away from its past. The idea that there is a fixed set of ideas in the world independent of some paradigm was dubious to Kuhn. He felt that objective truth does not really exist. (While I cannot prove that these ideas about the ultimate nature of reality are wrong, I, and probably most scientists, find such ideas to be simply false.)

The End of Science

There is a feeling among some researchers that the work of scientists over the past few centuries has revealed all the mysteries of our universe and the job of science will be complete very soon. Scientists today have understood and described all the known forces that run the universe. They have unified most of them and shown how they are really the same force. They have explained the wonders of the different chemicals and their interactions. It has been demonstrated how the different materials in the universe are made of the same types of subatomic particles and what happens when they combine. Large parts of human and animal physiology are very clear to us. In short, we seem to know a lot of what makes the universe tick. These thinkers feel that in a little while all scientists will have to do is to "dot the i's and cross the t's." There won't be any major scientific question that will remain open.

Many people scoff at such thoughts. They say that the end of physics has been "just around the corner" many times before. All those previous predictions turned out to be false and so will this one. For some two centuries after Newton's work was done, physicists also believed that only the details remained to be finished. When the twentieth century hit, quantum mechanics and relativity theory showed them that they were wrong and that there were many new phenomena that needed explanations. Perhaps there will be many new phenomena in the years to come.

This argument is not foolproof. Just because people predicted in the past that science will soon end, and these predictions turned out false, does not mean that the current predictions will also fail. For thousands of years mankind was looking for the source of the Nile and failed at finding it. Then, one day, we did find the source of the Nile.[19] Similarly, there was the boy who cried "wolf" many times when there was, in fact, no wolf. Eventually, however, the wolf did come. Similarly, thinkers can predict that the end of science is near many times, but now it could really be true. The reason why it failed in the past is that the solutions were simply not found. There were new phenomena that had to be discovered and new explanations that had to be revealed. Now, perhaps, all the phenomena are known and all the explanations are understood. Or maybe not.

There are many reasons to believe that the end of science is very close. We really are closer than before in describing many forces. Our knowledge of the subatomic world is far superior to our earlier knowledge. As time goes on, we have combined more and more forces and shown that they are really the same. This shows much economy in our theories that Occam would have appreciated. String theory and other theories seem to be legitimate Grand Unified Theories that have combined all the others. Also, our theories seem more mathematical than ever.

Similarly, there are many reasons to believe that science has a long way to go and might never end. If science were ending very soon, one would think that some parts of science would have already ended. And yet, we do not see any major field of science complain of a lack of work or even close up shop. Every field still asks and sometimes answers good questions, so why believe that eventually all of science will end? Immanuel Kant described another problem with the end of science being near: "Every answer given on principles of experience begets a fresh question, which likewise requires its answer and thereby clearly shows the insufficiency of all physical modes of explanation to satisfy reason."[20] In other words, even if we have all the answers to the questions that we have today, in the future we will have many more questions. Science is, in a sense, self-perpetuating.

Whether science will end depends, in part, on some of the answers that one accepts for the questions posed in this section. Do you have enough observational evidence to inductively come up with a theory that is a final theory? If we accept Popper's falsification scheme, then science might have ended and we will never know about it since the theory that we have cannot be absolutely verified. We simply must wait for eternity until we know that our theory has never been falsified. In contrast, if Popper is wrong, then we might come to the end of science and know that we are there. If Kuhn's paradigm view is correct and paradigms must continually change, then we will never get a final theory.

But the answer to whether science will end also depends on the very structure of the universe. Is there some type of final explanation that exists and that scientists are trying to find it? Or, in contrast, is there simply no deepest level of explanation?

There do not seem to be any knockdown answers to any of these questions. No side of the argument is more attractive than the other.

There are many different possibilities for the end-of-science questions, including the following:

- Science can, in fact, end very soon and we will know and understand all the mysteries of the universe.[21]
- Science can end and there will be no new answers, but we simply do not understand all the mysteries. That is, there are no essential new results in science, but we still do not have the answers to all our questions. As I have stressed many times in this book, the ultimate nature of our physical universe might simply be beyond the limits of human reason.
- Science can go on forever and we will still not arrive at the ultimate answer. There is simply an infinite chain of explanations, one after another. Each explanation will be deeper than the previous one.
- Science can go on and we will not know that we already have all the answers to the important questions. That is, science could be working on the unimportant questions and we simply do not realize it. Every scientist thinks that she is working on the most important question in the world, no matter how trivial or unimportant it seems to most other people. That is simply the nature of the profession. Perhaps we are all under an illusion about the contemporary status of science.
- Science can never end, but its progress gets slower and slower.
- Science could have already ended and we are now only dealing with the small problems, but we just do not know it.

There is no doubt that this list of possible scenarios for science can be extended indefinitely. I am sure there are many other plausible developments that we cannot even imagine. It is impossible to even make a complete list of the possibilities for the end of science, let alone to determine which of the possibilities will happen. It is hard to predict the future. Just 100 years ago we did not have anything like computers, the World Wide Web, microwaves, televisions, nuclear submarines, and so on. There is no way we could have predicted what science and technology had in store for us. Similarly, it is impossible to predict what will happen in the century to come. We cannot tell if or how science will end.[22]

This topic clearly has implications for our limits-of-reason theme. If science will never end or if science will end, but we still will not know all the answers to the big questions, then those answers are necessarily beyond the limits of reason. In contrast, if one day the universe does reveal all of its mysteries to its human inquisitors, then the limits of reason are not so severe.

All the different topics in the philosophy of science that we have discussed have one theme in common: they show that science is a human activity. It is created by finite, flawed human beings attempting to search for the ultimate truth. The data sets that we examine are limited, the theories that we come up with are tentative, and the equations that we find are incomplete. We are not promoting some type of silly postmodernist belief that science is not real. Rather, what we are saying is that the ways human beings find and describe these laws of nature are simply human. We do not have access to magic oracles or time machines that will help us peek into the future. Rather we are looking at the evidence we have and trying to make sense of the world we live in.[23]

8.2 Science and Mathematics

There is a rather deep puzzle at the heart of the scientific endeavor. As anyone who has ever studied the sciences knows, a large amount of mathematics is needed to understand the physical world. Science uses mathematics as a language to express itself, and without that language science is impossible. This can be seen by looking at the course prerequisites for college science classes. Physics and engineering majors need several semesters of advanced calculus. Computer scientists must study discrete mathematics, linear algebra, probability, and statistics. Modern chemists have to know a fair amount of topology, graph theory, and group theory. The necessity of studying mathematics was described nicely almost four hundred years ago by one of the greatest scientists of all time, Galileo Galilei:

> Philosophy is written in that great book which continually lies open before us (I mean the Universe). But one cannot understand this book until one has learned to understand the language and to know the letters in which it is written. It is written in the language of mathematics, and the letters are triangles, circles and other geometric figures. Without these means it is impossible for mankind to understand a single word; without these means there is only vain stumbling in a dark labyrinth.[24]

This leads to the obvious question: Why is mathematics so essential to an understanding of the physical world? Why does math work so well? Why does the physical world obey mathematics? These simple questions have been asked by the greatest scientists and thinkers in every generation. Paul Dirac (1902–1984) wrote:

> It seems to be one of the fundamental features of nature that fundamental physical laws are described in terms of mathematical theory of great beauty and power, needing quite a high standard of mathematics for one to understand it. You may

wonder: Why is nature constructed along these lines? One can only answer that our present knowledge seems to show that it is so constructed. We simply have to accept it. One could perhaps describe the situation by saying that God is a mathematician of a very high order, and He used very advanced mathematics in constructing the universe. Our feeble attempts at mathematics enable us to understand a bit of the universe, and as we proceed to develop higher and higher mathematics we can hope to understand the universe better.[25]

In 1960, the physicist Eugene Wigner published "The Unreasonable Effectiveness of Mathematics in the Natural Sciences." This paper posed interesting questions about the relationship between mathematics and the natural sciences. These questions now go under the name of "Wigner's unreasonable effectiveness." Wigner did not come to any definitive answers to his questions. He wrote that "the enormous usefulness of mathematics in the natural sciences is something bordering on the mysterious and . . . there is no rational explanation for it." He concludes the paper with these words:

The miracle of the appropriateness of the language of mathematics for the formulation of the laws of physics is a wonderful gift which we neither understand nor deserve. We should be grateful for it and hope that it will remain valid in future research and that it will extend, for better or for worse, to our pleasure, even though perhaps also to our bafflement, to wide branches of learning.

The mystery was stated perfectly by Albert Einstein, who wrote:

At this point an enigma presents itself, which in all ages has agitated inquiring minds. How can it be that mathematics, being after all a product of human thought which is independent of experience, is so admirably appropriate to the objects of reality? Is human reason, then, without experience, merely by taking thought, able to fathom the properties of real things?[26]

The power of mathematics can be seen by looking at some historical vignettes where mathematics has made amazing predictions about the physical world.

The Discovery of Neptune

On March 13, 1781, the English astronomer William Herschel (1738–1822) pointed his telescope to the heavens and found a new planet, which came to be called Uranus. The motion of this new planet had certain unexplained irregularities. The French mathematician Urbain Leverrier realized that another planet must be influencing the orbit of Uranus. He sat down and used Newton's laws to calculate the exact position of this heretofore unseen planet. Leverrier sent a letter to the German astronomer Johann Galle (1812–1910) telling him about this planet and exactly where to look for it. The letter arrived on September 23, 1846, and on that very night, Galle aimed his telescope at the exact place he was told and found a planet. Galle immediately wrote to Leverrier: "The planet whose place you have [computed] *really exists*." This planet was named Neptune. Nothing but pure mathematics was used to find it.

The Discovery of the Positrons

In 1928, Paul Dirac jotted down an equation to describe some of the properties of an electron. This work was remarkable because it took into account both quantum mechanics and special relativity theory. With the usual way of thinking about the Dirac equation, one arrives at the properties of the electron. One of the properties of this subatomic particle is that it has a negative charge. Dirac wondered what would happen if you played with other solutions to this equation. This is similar to considering solutions to the simple equation $x^2 = 4$. The obvious solution is that $x = 2$. There is, however, another not so trivial solution where $x = -2$. From a simple contemplation of the equation and the possible solutions, Dirac posited that there might be another particle with similar properties to an electron but with a positive charge. In 1932, Carl Anderson (1905–1991) did some experiments that showed that such a particle actually existed. This particle was called the positron. Anderson won the Nobel Prize for this work in 1936.

So by simply following mathematics, one finds out things about the physical world. Dirac wrote:

> A good deal of my research in physics has consisted in not setting out to solve some particular problem, but simply examining mathematical equations of a kind that physicists use and trying to fit them together in an interesting way, regardless of any application that the work may have. It is simply a search for pretty mathematics. It may turn out later to have an application.[27]

Dirac added, "As time goes on, it becomes increasingly evident that the rules which the mathematician finds interesting are the same as those which Nature has chosen."[28]

String Theory

A purported Theory of Everything that will unite quantum mechanics and relativity theory is string theory. This is the theory that takes as the basic building blocks of the universe minute strings that wiggle around. These strings shake, rattle, and roll while combining and separating to make up all the quarks, protons, electrons, and other particles of everyday life. By just looking at the different properties of these strings and the way they interact, theoretical physicists have successfully predicted most of the properties of the physical world. This theory can be used to describe all the forces of quantum mechanics as well as the gravity force of general relativity. There is only one small problem: there is no empirical evidence that shows that string theory is correct. It is a purely mathematical theory derived from looking at the geometry of strings interacting. No one has ever "seen" a string or shown that they exist. Detractors of string theory declare that it is "just mathematics" and has nothing to do with the physical world. Defenders point to other branches of mathematics that have made correct predictions about the physical world. To them, this theory will also be proved true in the future. They say, too, that string theory is one of the few theories that successfully unite quantum mechanics and general relativity. Whether the world is made out of little strings or not, it is still amazing that pure mathematics can describe all the properties of the physical universe.

Another way of seeing the mysterious relationship between science and mathematics is to look at a few instances where *entire fields* of mathematics developed long before physical applications of the fields were found.

Conic Sections and Kepler

The ancient Greeks loved geometry. In contrast to the ancient Egyptians, who used geometry to literally measure (*metron*) parts of the Earth (*ge*) for agricultural or legal reasons, the Greeks studied geometry purely for intellectual purposes. One of the brightest stars of Greek geometry was Apollonius (about 262–190 BC), who was born in Perga in southern Asia Minor. He studied what would happen if you take a cone and intersected it with a flat plane, as in figure 8.2.[29]

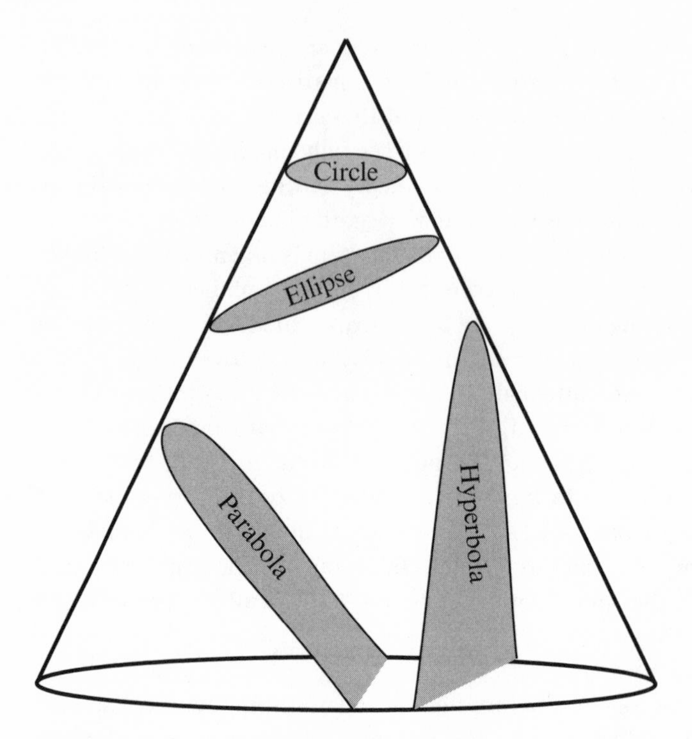

Figure 8.2
Planes intersecting cones and the shapes they make

The curves that such intersections form are called *conic sections*. If the plane is lined up with the cone, the curve formed would be a circle. If the plane was slightly skewed, an ellipse would be formed. By further manipulating the plane, one can create parabolas and hyperbolas. Apollonius wrote a book about conic sections in which he stated about 400 theorems concerning the different properties of such curves.

Eighteen hundred years after Apollonius lived, Johannes Kepler (1571–1630) was trying to figure out how to make sense of Copernicus's radical new idea of having the sun in the center of the universe with the planets flying around the sun in giant circles. There was a terrible problem with Copernicus's new system: his predictions were wrong. The antiquated geocentric system of Ptolemy made better predictions than the new heliocentric system of Copernicus. Kepler realized that Copernicus's mistake was in thinking that the planets traveled around in circles. Rather, the orbits of the planets were ellipses. Because many of the properties of ellipses were worked out almost two millennia earlier, Kepler went back to study the ancient works of Apollonius to determine the properties of planetary motion. Once it was realized that the planets went in the well-understood elliptical motion, the positions of the planets were easy to predict. A historian of science wrote that "if the Greeks had not cultivated conic sections, Kepler could not have superseded Ptolemy."[30]

How is it possible that the abstract writings of a long-dead Greek mathematician could help explain the motion of the planets?

Non-Euclidean Geometry and General Relativity

Classical Greek geometry found its most lasting form in the writings of Euclid (323–283 BC). His *Elements* was one of the most successful textbooks of all time. He began the book with ten obvious axioms. The first four are as follows:

1. A line segment can be drawn between any two points.
2. Any straight line segment can be extended to a straight line.
3. Given any straight line segment, one can draw a circle having the segment as radius and one endpoint as center.
4. All right angles are congruent (the same angle).

These laws are represented in figure 8.3.

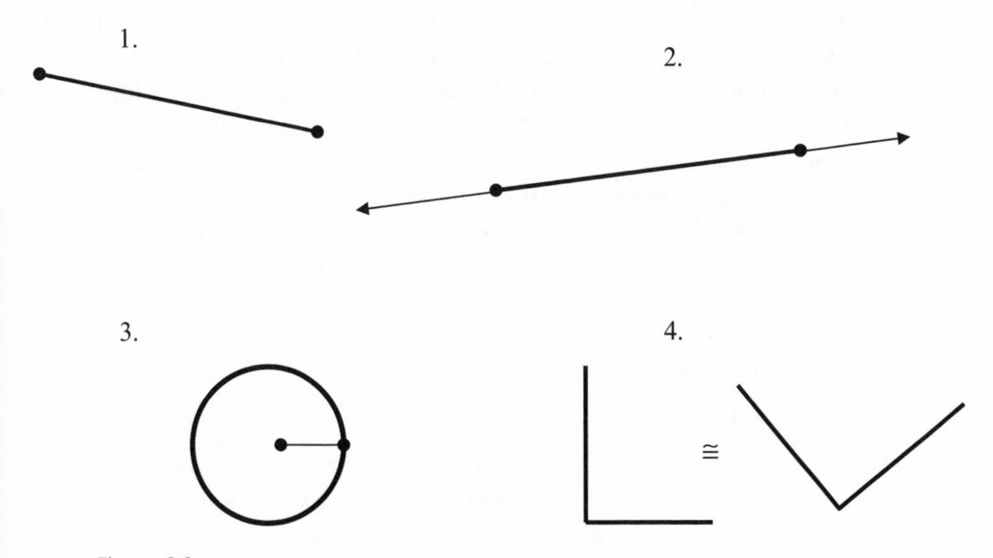

Figure 8.3
Euclid's first four axioms

Euclid's fifth axiom, which has come to be known as the *parallel postulate*, is worth careful study:

5. If two lines are drawn that intersect a third in such a way that the sum of the inner angles on one side is less than two right angles, then the two lines will inevitably intersect each other on that side if extended far enough.

This axiom is depicted in figure 8.4. In can be restated that if two lines are not parallel, they will eventually cross.

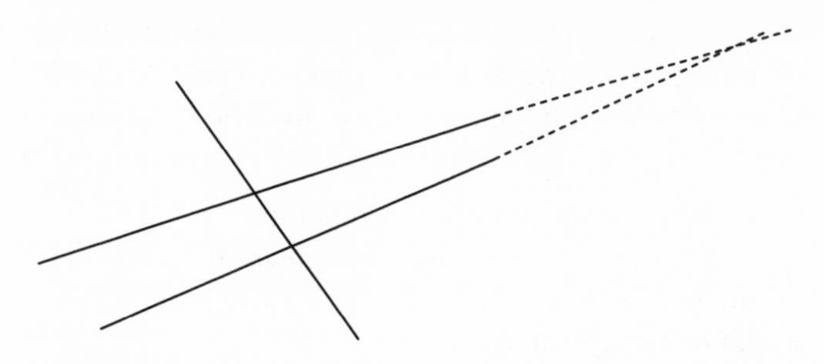

Figure 8.4
Euclid's fifth axiom

The fifth axiom has a different feel than the other axioms. The first four axioms are simple and easy to state. In contrast, although the fifth axiom is obviously true, it is more complicated and a little bothersome. The fifth axiom is the only axiom concerned with something (a crossing) that could occur some distance away from the angles under discussion. Euclid himself was suspicious of this axiom and tried to avoid using it as much as possible. Although mathematicians felt that the axiom was correct, they thought that it was more a consequence of the other axioms than an axiom itself. Throughout the centuries that Euclid's text was used, many people tried to derive the fifth axiom from the other nine axioms. That is, they tried to show the following derivation:

axioms 1–4 and axioms 6–10 $\Rightarrow$ axiom 5.

They were unsuccessful. In 1767, French mathematician Jean-Baptiste d'Alembert (1717–1783) lamented the fact that for 2,000 years mathematicians could not prove the fifth axiom from the other axioms. He called this the scandal of geometry.

In the seventeenth century Girolamo Saccheri (1667–1733), Johann Heinrich Lambert (1728–1777), and Adrien-Marie Legendre (1752–1833) tried using a different method to prove that the fifth axiom was a consequence of the other axioms. Their method of proof is something that readers of this book are already comfortable with: a proof by contradiction. They tried to show that the fifth axiom is a consequence of the other axioms by assuming the other axioms and also assuming that the fifth axiom is *false*. Their goal was to show that this would to lead to a contradiction. In short:

axioms 1–4, 6–10 and the falsehood of axiom 5 $\Rightarrow$ contradiction.

With this derivation in hand, we would have to say that the reason for the contradiction is that the truth of the fifth axiom is a consequence of the other axioms and that our assumption that the fifth axiom is false was wrong. But a funny thing happened on the road to proving this result: no contradiction could be found! These mathematicians went on and on deriving different strange theorems but could not find a clear contradiction. They were so confident that there must be some contradiction there that they "fudged their results" and found dubious artificial contradictions.

At some point, one of the greatest mathematicians of all time, Johann Carl Friedrich Gauss (1777–1855), and others realized that there is a reason why no one could find a clear contradiction: the fifth axiom can be true but it also can be false. That is, this complicated axiom did not depend on the others or was "independent" of the other nine axioms. When the parallel postulate is taken as true, the system studied is classical Euclidean geometry, and when it is taken as false, the system is called non-Euclidean geometry. Gauss never published anything on the topic so credit is usually given to the Hungarian Janos Bolai (1802–1860) and the Russian Nicolai Ivanovitch Lobachevsky (1793–1856) for founding this branch of mathematics. The topic was advanced by the German mathematician Bernhard Riemann (1826–1866).

Years later, Albert Einstein was looking for ways to express the ideas of general relativity. He was stuck. He could not find the proper language to describe the curves that space makes to influence the way matter moves (gravity.) Einstein's friend and teacher, Marcel Grossmann (1878–1936), suggested that he look into the abstract field of non-Euclidean geometry. To his shock, Einstein found exactly what he was looking for. The ideas and theorems of non-Euclidean geometry were precisely what he needed for general relativity. As Einstein wrote, "To this interpretation of geometry, I attach great importance, for should I have not been acquainted with it, I never would have been able to develop the theory of relativity."[31] What are we to think of this? Here we have a bunch of mathematicians playing mind games with the usual axioms of geometry. They worked with an axiom system that used an "obviously" true axiom that was assumed to be false. Decades later that same system would miraculously help describe laws of the physical universe. Why should this be?

Abstract Algebra and Quantum Theory I: Complex Numbers

In high school, many students spend hours in math class learning to solve polynomial equations. Eventually they learn that some equations simply do not have a solution. The simplest such equation is

$$x^2 + 1 = 0$$

For any x, we have that x^2 is more than or equal to zero and adding 1 is definitely a positive number. So there is no x that could possibly satisfy this simple equation. In the sixteenth century, Gerolamo Cardano (1501–1576) posed the following question: What if we imagined that there was a solution to this problem? That is, let's work with a number i (for imaginary) and let this number be the solution to this simple equation. In other words, if

$$i = \sqrt{-1}$$

then plugging it into the equation, we get that

$$(\sqrt{-1})^2 + 1 = 0.$$

Obviously, this number cannot exist. But imagine that it does. We might then multiply this i by real numbers and create numbers like $2i$, $3i$, $-5.7i$, etc. These are called *imaginary* numbers. Pressing on, we can add real numbers to imaginary numbers and get numbers like $7 + 3i$, $6.248 - 8.7i$, or for any real numbers a and b we have $a + bi$. These numbers are a combination of both real and imaginary numbers and are called *complex* numbers. Mathematicians spent many long, lonely years working out many of the properties of these artificially manufactured numbers. All along physicists and others ignored these eccentric mathematicians and the strange curiosities they played with.

Hundreds of years later, when physicists were trying to describe the strange world of the quantum, they found that they needed these very strange complex numbers in an essential way. It turns out that the superposition, which is at the core of quantum theory, is described by these complex numbers. More specifically, the many positions of a quantum state are indexed by complex numbers. Those strange curiosities are needed to describe our world.

Abstract Algebra and Quantum Theory II: Noncommutative Operators

Irish mathematician William Rowan Hamilton (1805–1865) was looking at complex numbers and saw the way they extended the real numbers to two dimensions. Hamilton wondered if there was a way to extend the real numbers to three dimensions. In 1843, he was able to form *quaternions* or *Hamiltonian* numbers. Rather than just looking at i as in complex numbers, Hamilton postulated i, j, and k as special numbers. The quaternions are then numbers

$$a + bi + cj + dk$$

where a, b, c, and d are real numbers. He required that we not only have $i^2 = -1$, as with complex numbers, but also $j^2 = -1$, $k^2 = -1$, and $ijk = -1$. While Hamilton was working out the properties he noticed that such numbers do not satisfy one of the normal properties of numbers: they are not commutative. That is, for all real or complex numbers x and y, it is a fact that

$$xy = yx.$$

We say that such numbers are commutative. However, Hamilton noticed that there are quaternion numbers x and y such that

$$xy \neq yx.$$

This operation of multiplying such numbers is *noncommutative*. This is very strange; after all, nearly every child knows that when you multiply regular numbers, it does not matter what order the numbers are multiplied in. Such noncommutative operations remained mathematical curiosities for many decades. Examples and properties of such operations were worked out by mathematicians such as Hamilton, Hermann Günther Grassmann (1809–1877), and Arthur Cayley (1821–1895) but ignored by physicists and regular people.

In the early twentieth century, when physicists were trying to formulate Heisenberg's famous uncertainty principle (see section 7.2), they found this idea of noncommutative operations very helpful. In particular, imagine having two different properties of a quantum system that you want to measure. Call them X and Y. Measuring X and then measuring Y will give you different answers if you measure Y first and then measure X. That is, essentially the results of XY are different from YX. In symbols $XY \neq YX$.

Abstract Algebra and Quantum Theory III: Group Theory

A final short example of the use of abstract algebra in quantum mechanics is group theory.[32] In the middle of the nineteenth century, mathematicians studying whether or not a polynomial equation has a solution formulated the notion of a group. This is a mathematical object that describes certain symmetries. Many years later, when physicists were working to understand quantum theory, they found group theory was invaluable because it describes the workings of all subatomic particles.

In all three of these cases—complex numbers, noncommutative operations, and group theory—mathematicians were defining different structures that seemed to have nothing to do with the physical world. They were using these structures to deal with their own mathematical problems. And in all three cases, these structures are now used by physicists to make sense of the quantum universe.

In all these historical vignettes, mathematicians were playing little mind games with mathematical curiosities that later turned out to be useful for physicists dealing with the physical world. This is the core of Wigner's unreasonable-effectiveness question. Why should this be? Why does science follow mathematics in such a way? Steven Weinberg put it this way in his wonderful book *Dreams of a Final Theory*:

> It is very strange that mathematicians are led by their sense of mathematical beauty to develop formal structures that physicists only later find useful, even where the mathematician had no such goal in mind. . . . Physicists generally find the ability of mathematicians to anticipate the mathematics needed in the theories of physics quite uncanny. It is as if Neil Armstrong in 1969 when he first set foot on the surface of the moon had found in the lunar dust the footsteps of Jules Verne.[33]

Scientists and philosophers have given many different answers to explain the apparent mystery of the connection between the realms of mathematics and the sciences. Let us look at several of them.

A Deity

One of the oldest answers to this question is that a deity exists who set up the universe in this way. The universe was created with perfect laws and these laws were written in a perfect mathematical language. This mathematical language is made to be understandable to human beings. Johannes Kepler stated this clearly and succinctly: "The chief aim of all investigations of the external world should be to discover the rational order and harmony which has been imposed on it by God and which He revealed to us in the language of mathematics."

Pope Benedict XVI, echoes these ideas:

Was it not the Pisan scientist [Galileo] who maintained that God wrote the book of nature in the language of mathematics? Yet the human mind invented mathematics in order to understand creation; but if nature is really structured with a mathematical language and mathematics invented by man can manage to understand it, this demonstrates something extraordinary. The objective structure of the universe and the intellectual structure of the human being coincide; the subjective reason and the objectified reason in nature are identical. In the end it is "one" reason that links both and invites us to look to a unique creative Intelligence.[34]

In other words, science follows mathematics because they both emerge from the mind of a divinity. Neptune follows the fixed laws of motion that were set up by a divinity. Leverrier was able to calculate the exact position of Neptune because the same divinity that set up the planet's motion also set up the mathematics of calculus discovered (not invented) by Newton. Group theory works perfectly for quantum mechanics because the divinity set up quantum mechanics using group theory. These laws of mathematics and physics are timeless and hence there is no mystery as to why one occurs before the other. They are all part of one divine mind.

While this solution is gratifying for one who already believes in a deity, for those who do not, it is unsatisfying. First, it does not banish the mystery. In fact, a deity or a divine intelligence is more mysterious than Wigner's unreasonable effectiveness. Scientists looking for a scientific explanation of the connection between mathematics and science will find the existence of a deity beyond the scope of their reasoning. They prefer a solution that is less metaphysical and more testable.

A Platonic Realm

A slightly less metaphysical explanation for Wigner's unreasonable effectiveness is a solution that goes back thousands of years. The Pythagoreans comprised an ancient Greek school that held that numbers and relationships between numbers had some mystical control over the physical world. To them, the essence of the universe was mathematics. Parts of this ideology were taken on by Plato and came to be known as Platonism. To Plato and his followers through the centuries, abstract entities such as mathematical objects and physical laws exist in some platonic realm. The physical world was a meager shadow of the real world, which exists and could be called "Plato's attic." All the laws of planetary motion, general relativity, and quantum mechanics are neatly found in Plato's attic waiting for inquisitive human beings to discover them. To a Platonist, mathematics is not a human invention. Rather mathematics exists independent of human beings within this perfect platonic realm. In this attic, all the theorems of Apollonius about ellipses, all the properties of non-Euclidean geometry, and all the features of complex numbers are perfectly laid out. And in this realm, the physical laws are all perfectly stated in the language of mathematics.

One of the founders of electromagnetic theory, Heinrich Hertz (1857–1894), put it like this: "One cannot escape the feeling that these mathematical formulas have an independent existence and an intelligence of their own, that they are wiser than we are, wiser even than their discoverers, that we get more out of them than was originally put into them."[35] The writer Martin Gardner (1914–2010) gave a solid defense of Platonism, saying that ". . . if two dinosaurs met two other dinosaurs in a clearing there would have been four there even if no humans were around to observe them. The equation $2 + 2 = 4$ is a timeless truth."[36] In general, Platonists find non-Platonists frustrating. If mathematics does not conform to anything "out there," then it is just scribbles on a paper. Why should different people scribbling on different papers separated by thousands of miles and in different ages agree? Why should their scribbles miraculously not contradict each other? Platonists answer that the equation $6 \times 7 = 42$ is inherently true. It is not something that a bunch of people just seem to agree on. A perfect circle really exists in Plato's attic and in this—and only in this—perfect circle, the ratio of the circumference to the diameter is pi.

For a Platonist, Wigner's challenge is answered with ease. The reason why the universe follows mathematics is that the natural laws are set in this platonic realm, and mathematicians learn their trade by peeking into this very same realm. The above examples of mathematics being formulated before physical laws are simply times when mathematicians peeked a bit earlier than physicists.

While many people accept Platonism as the correct dogma, there are some problems with this purported solution. The first problem is: How do you know that such a realm exists? As with all metaphysical presuppositions, we must be suspicious of all specious claims. If we can explain our world without this platonic realm, why invoke it? Occam's razor places the existence of Plato's attic in doubt. However, even if we give a Platonist the benefit of the doubt and accept that this mystical attic exists, there are many other mysteries to contend with. Who set up this wonderful magical realm? How do mathematicians peek into this realm? What is the mechanism that physicists learn from this realm? How does this platonic realm have control over our physical world? In short, one can see the problem as in figure 8.5. Wigner is trying to understand the connection between

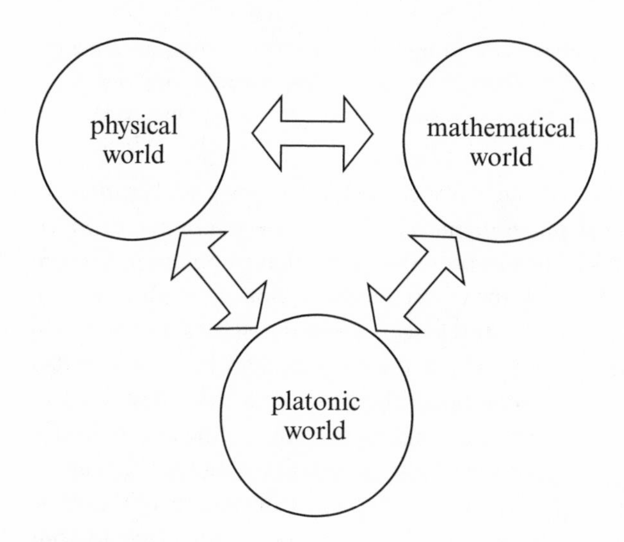

Figure 8.5
The three worlds of a Platonist

the physical world and the mathematical world. Platonists invoke a *third* platonic world to solve the problem. Now we have to deal with the connections between three worlds as opposed to the previous two worlds. This is more of a mystery. Not less.

While science cannot prove or disprove that a platonic realm exists or that a deity set up the universe, it does try to find other, more scientific, explanations.

A Paucity of Mathematics

One of the more intriguing explanations for the mysterious connections between science and mathematics asserts that the connection is, in fact, rather questionable. Most physical phenomena cannot be described by our mathematics. As we saw in section 7.1, the vast majority of physical systems cannot be formulated in mathematics. What will the clouds look like tomorrow? Why were this week's lottery numbers the way they were? Who will win the next presidential election? All of these are legitimate questions about physical phenomena to which no mathematician in the world can give definitive answers.

There are many branches of science that study and aim to predict physical phenomena but find much of mathematics unhelpful. Israel M. Gelfand (1913–2009), a world-famous mathematician who worked in biomathematics and molecular biology, has been quoted as saying:

Eugene Wigner wrote a famous essay on the unreasonable effectiveness of mathematics in natural sciences. He meant physics, of course. There is only one thing which is more unreasonable than the unreasonable effectiveness of mathematics in physics, and this is the unreasonable ineffectiveness of mathematics in biology.

Other branches of science such as sociology, psychology, and anthropology also study physical phenomena but do not use mathematics in an extensive way. A physicist would protest and say that those disciplines are not "real" sciences ("After all, those disciplines do not use mathematics"). However, those studying the "soft sciences" do study physical phenomena. They would legitimately counter that the mathematics that exists today cannot help them with the complicated phenomena that they want to study. Mathematics is only good for helping with the predictable behavior of balls rolling down ramps or subatomic particles passing through one of two slits. In contrast, how a crowd would react to a certain event, or how a human would react to a relationship, is far too complicated for our mathematics. Mathematics does not predict all phenomena. It only helps with predictable phenomena. Or, as it is slightly humorously phrased, "God gave the easy problems to the physicists."[37]

In fact, much of the world around us does not fit ever so snugly into the world of mathematics. The beautiful Brooklyn tree outside my window cannot be described by any figure in any geometry textbook. Although $1 + 1 = 2$, it is not true that if you take one heap and combine it with another heap you will get two heaps. Rather, you will get only one heap. Years ago I had to buy a package of size 4 diapers. The store was out of that size so instead I bought two packages of size 2 diapers. Needless to say, this did not work well. From this we can conclude that $2 \times 2 = 4$ does not work in the important realm of diapers.

Another demonstration of the disconnect between the realm of mathematics and the realm of science is that there are vast tracts of mathematics that never get applied to the physical world. Some parts of number theory and set theory have remained "pure" and have never been applied. In fact, I would imagine that the majority of pure mathematical papers never get applied to the real world. Rather than saying that earlier mathematics mysteriously gets used in later physics, say that a later physicist might *choose some parts* of earlier mathematics to describe some of the phenomena she is trying to describe. For example, Kepler used some of the earlier work of Apollonius. But Kepler ignored many other writings of Apollonius. Kepler chose what he needed. It is not that mathematicians create exactly what will be used by physicists. Rather, mathematicians create an immense amount of mathematics and only some of it is chosen by the physicists. This makes much of the mystery go away.

This solution to the unreasonable effectiveness—that the connection between mathematics and science is tenuous—is somewhat problematic. It is true that there are many branches of science whose dependence on mathematics is not direct. Nevertheless, these branches of science are based on other branches that do depend on mathematics. For the most part, sociology is not based on mathematics[38]. But sociology depends on psychology which in turn depends on neurology and cognitive science. These disciplines are closely related to neurochemistry and computer science, which depend heavily on mathematics. While sociologists might not need to learn a lot of mathematics to practice their trade, if they wanted to understand the foundations of sociology, they would have to study much mathematics. As of now, our mathematics is not complicated enough to deal with all the complexities of sociology. Perhaps in the distant future, such mathematics would be possible.[39] Perhaps not. So while many parts of science do not use mathematics, the parts of science that do use mathematics, in a sense, *generate* the foundations of all of science. And for those generating parts, Wigner's mystery remains.

Mathematics Comes from Physics

Probably the most popular answer that researchers give to explain the mystery of Wigner's unreasonable effectiveness is that mathematics comes from observing the physical world. It is not mysterious that we can describe the physical world with mathematics since it was in that same physical world where we learned mathematics.

For example, a child learns that two apples plus three apples equals five apples by looking at sets of apples. She also observes that when two sticks and three sticks are combined they form five sticks. By seeing this over and over, human beings abstract out that two plus three is five, or in symbols, $2 + 3 = 5$. This gives human beings the beginnings of the addition operation. This operation is going to work in many places in the physical world. Similarly, many mathematical objects and operations come from seeing different phenomena in the physical world. It is no wonder that these same mathematical objects and operations are used in describing the physical world.

Let us look at an example in a little more depth. If we see 7 boxes and each box has 8 red marbles and 3 blue marbles, then we can just multiply the 7 boxes times 11, the number of total marbles in each box. At the same time we can find the sum total of all marbles by looking at the sum of 7 times 8 red marbles and 7 times 3 blue marbles. After seeing many similar counting arguments we can formalize this symbolically as

$$7 \times (8 + 3) = 7 \times 8 + 7 \times 3.$$

This statement is more abstract since it is not about red and blue marbles anymore. It could be about dogs and cats, or boys and girls. We have abstracted out the original content of the statement. Once we see similar rules for many different numbers, mathematicians have further abstracted this statement to

$$a \times (b + c) = a \times b + a \times c.$$

This rule has nothing to do with 7, 8, 3, or any particular number. It is simply a fact that multiplication "distributes" over addition and is true for any numbers. Now, with this rule in hand, we might think of it as a statement of pure mathematics or we might apply this rule to any part of the universe. The fact that this rule can be applied is not mysterious since it was formed by looking at the physical world. Notice that the rule

$$a + (b \times c) = (a + b) \times (a + c)$$

will *not* be applied to the physical world simply because this rule—that addition distributes over multiplication—was not seen to be true in the physical world. Rules and mathematical operations that are commonly experienced will be found in the universe and rules that are not seen or experienced will not be found. This must be true.

Let's analyze what happened with our distributive rule. There was a certain phenomenon in the physical world that was observed about marbles. A human being observes it and makes a model of this true fact with numbers. The human then further generalizes it for all numbers. This is a truth that is perhaps shared with other human beings. Many years later, another phenomenon is observed. This second phenomenon also satisfies some type of distributive law and the same model is used to characterize that second phenomenon. Abstract mathematics becomes true on its own and does not relate to the original way it was discovered. Once it is discovered, it can be applied anywhere. It makes sense: the mathematics was learned in the physical world and is applied in the physical world. Where is Wigner's mystery?

This mechanism can help explain some of our historical vignettes at the beginning of this section:

- Humans saw circles and ellipses all over their physical world. Apollonius saw that he could describe and model many of these shapes with conic sections. It is no wonder that Kepler was able to describe another physical phenomenon, the motion of planets, with these same conic sections.
- One can easily work with little strings and the way they interact. If you think about little strings long enough, then you can describe their geometry. If the universe is made of little strings, then it is obvious that we can describe it with the mathematics we learned of little strings.
- Let us examine Euclidean and non-Euclidean geometry. When we are interested in a flat surface, the geometry of Euclid works perfectly. But what happens if we are interested in a curved surface? Consider the longitude lines on a globe. There we have many lines that seem to be parallel with each other but nevertheless meet at the North and South Poles. This is opposite the spirit of Euclid's fifth axiom. It turns out that non-Euclidean geometry works well with curved surfaces. It is not as shocking as we thought that Einstein used non-Euclidean geometry to describe the curvature and shape of space.

It is not strange that physical phenomena are perfectly described by mathematics, since mathematics is an abstraction and a series of generalizations of what is observed in the physical universe. Once we formulate these ideas as mathematics, its connection to the original physical impetus for the discovery is lost. The mathematics becomes abstract and about nothing in particular. Because these concepts are about nothing, they are about everything. We do not care how the ellipses are created, whether they are in the shape of a walnut, whether they come from the intersection of a plane and a cone, or whether they are the path of a planet around a medium-sized star. With an understanding of an ellipse, we know its properties and it can be applied everywhere.

Some of these ideas about the symbiotic relationship between mathematics and physics are summarized in an extremely clever piece of art (figure 8.6) made by the mathematical physicist Robbert Dijkgraaf.

Figure 8.6
The symbiotic relationship between physics and mathematics

A few minutes of analysis of this wonderful comic are worthy of our time. The top two sections have calendars dated 1968, while the bottom two take place thirty years later when the same researchers have aged. The left side is in a physics professor's office, while the right side is in a mathematician's office. Now look at what is on the chalkboards.[40] What was on the physicist's board in 1968 is being contemplated by later mathematicians. On the other hand, what mathematicians studied in the 1960s is being studied by the physicists in 1998. The path from the upper right to the lower left is what we described at the beginning of this section: the mysterious ability of mathematical ideas to somehow arise in physics. In contrast, the path from the upper left to the lower right is the possible explanation for this mysterious connection: mathematics comes from looking at the physical world.

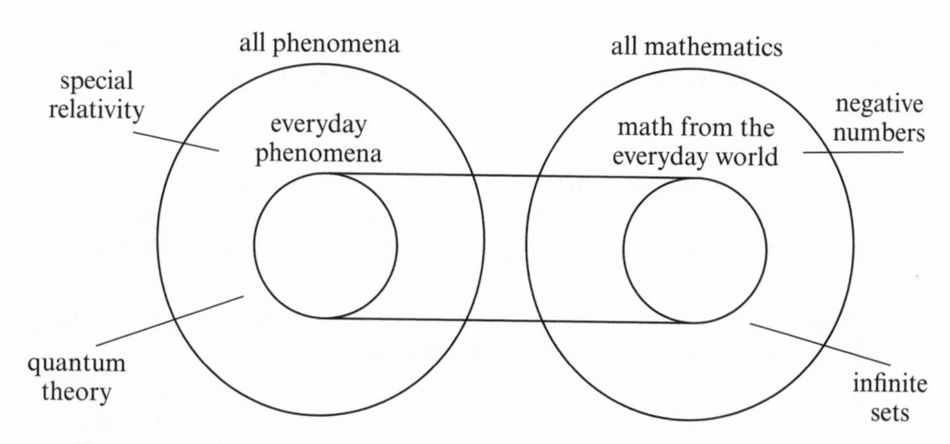

Figure 8.7
The relationship between phenomena and mathematics

While the explanation that mathematics comes from the physical world seems to make sense, it is far from perfect. Mathematics comes from the usual world of our everyday experience and yet the mathematics that we formed in such usual circumstances gets applied to places very far from such experiences. (See figure 8.7.) For example, special relativity tells us what happens when objects move close to the speed of light. We never travel anywhere near those speeds. Why should our everyday-experience mathematics help us with the strange phenomena of special relativity? Another example is quantum theory. As we saw in section 7.2, the quantum world is very different from our world of everyday experience. While walking down the street, we never see objects that are in a superposition or see "spooky action at a distance." Nevertheless, the mathematics that comes from our everyday experience is very helpful at predicting quantum events. We are back to Wigner's mystery.

Another problem with this refutation of Wigner's unreasonable effectiveness is that not all mathematics is a generalization of everyday experience. Some mathematics is a creative leap away from everyday experience (again, see figure 8.7). The simplest example of this is negative numbers. If you have five oranges and you take away eight oranges, how many oranges do you have? There is no such thing as negative three oranges. You cannot take away eight oranges from five oranges. Although they seem obvious to us now, negative numbers were a surprising invention of the Middle Ages. For millennia before medieval times, we counted and traded but did not have concepts like negative numbers. This culturally constructed mathematical set of numbers is nevertheless currently used in every branch of physics. As mentioned, the positron was discovered simply by looking at negative solutions of the Dirac equation. Another example of mathematics that did not spring from everyday experience involves the different notions of infinite sets that we met in chapter 4. There does not seem to be any physical example of an infinite set. Nevertheless, the ideas of infinity are used in every physics and engineering textbook. Every building (that remains standing) and every rocket that is created uses notions of infinity in their construction. So mathematics that does not come from everyday experience is still remarkably effective in the physical sciences. Why?

There are some other reasons given for the unreasonable effectiveness, but these are the main ones. One can combine the final two answers—that is, there is a paucity of mathematics in general science and when there is mathematics in the physical world it comes from intuitions learned in the physical world. To me, the combination of these two reasons gives a satisfactory answer to Wigner's mystery.

The answers to Wigner's unreasonable effectiveness lead to much deeper questions. Rather than asking why the laws of physics follow mathematics, ask why there are any laws at all. Why these laws and not other laws? Why is it that what we learned by looking at different marbles is also true about many other physical phenomena? Why do the planets follow the shape of an ellipse, or any other repeating shape for that matter? Why not a square or a circle? These deep—and perhaps unanswerable—questions are the central topic of the next section.

8.3 The Origin of Reason

Imagine driving along a highway late at night and deciding to stop at a hotel to get some shut-eye. You choose one of the many indistinguishable large hotels on the side of the road and ask for a room. The hotel is mostly empty and the manager gives you a randomly chosen room. You sleepily enter your room and to your surprise you find the closet and all the drawers are full of clothes. There are shoes, slippers, and a robe. Shockingly all the clothes are exactly your size and taste. The slippers are worn out just the way you like them. The shoes are an exact fit, both in style and size. The robe is exactly the material and color you like. How could this be? From all the hotels and all the rooms in the hotel, why does this one have everything perfectly made for you? This sounds like the beginning of a suspenseful movie thriller.

Let's imagine some scenarios that could explain these strange coincidences. It could just be a fluke. It happens to be that the person who was in the room right before you was the exact same size as you and the maid did not properly clean out the room. You simply found that person's leftover stuff. Had you chosen any other room, had any other person been in the room before, or if there had been a slightly more competent chambermaid, you would have found either different clothes or a cleaner room. What are the odds? Nothing like this ever happened to you before. This is a rather an improbable fluke.

It could be that a bunch of crazy marketing executives have decided to make everyone's hotel experience more "homey." After much nonsensical research they decided to leave "typical" clothes and stuff in every room. It just so happens that you are the exact size and your style is exactly the same as the "typical" person. One can easily check this nutty theory by looking into other rooms and seeing whether they also have the same stuff there. Alas, you cannot get into the other rooms to check. Yet another possibility is that the hotel managers have decided to make every room homey with different style and size clothes. Either by luck or on purpose your room fits you. Again, a peek into other rooms will show whether this is true or false.

If you wanted to get all conspiratorial, you can say that there is a vast network of alphabetical intelligence organizations like the CIA, KGB, and NSA that have been following you around with black helicopters and satellites. When they saw that you were going into a particular hotel, they turned on their omnipresent listening devices to find which room is yours. Or perhaps the pleasant person behind the counter of the hotel is "in on it." Once they knew what room you were going to, they put in all the different clothes right before you entered. What these nefarious agencies want from you is still not exactly clear . . . but . . . I would get out of that room quickly.

Of course, this fun little excursion into moviedom is not real. But there is a question in science that is similar to this strange scenario and is very real. For reasons that are not very clear, the universe is shockingly well suited for rational human beings. The laws of physics seem to have been designed so that complicated beings with the ability to reason could exist. Why should this be?

The laws of physics could have been different and yet they are not. What would happen if they were slightly different?

• If gravity were stronger, then stars would collapse into black holes more quickly than they do. Stars are mostly made up of hydrogen and helium.

In a process called *stellar nucleosynthesis,* heavier elements such as carbon and oxygen are formed. If gravity would collapse stars into black holes more quickly, there would be no time for these important elements to be formed. The universe would not have the complex building blocks necessary for life.

- If gravity were weaker, then stars and planets would not be held together and nucleosynthesis would not take place. Again, there would be no building blocks for life.
- If the Earth were a little closer to the sun, all the water needed for life would become a gas.
- In contrast, if our orbit were a little further away from the sun, the water on our planet would freeze and again life would be impossible.
- Physicists talk about a force that works against gravity and its strength is measured by the *cosmological constant.* It is necessary that this constant have a certain value or the universe as we know it could not exist. Astrophysicists have calculated the value that the cosmological constant must have for complex living beings to exist. They have found that if the value were slightly different, we would not be here to worry about it. The value is calculated to the 120th decimal place. It seems our universe is perfectly suited for us.
- If the average IQ of human beings were ten points lower, where would the human race be? Would we have been able to place someone on the moon? Would we be able to make and use the computers and other machines that are so much a part of our lives? If the average IQ were thirty points lower, would we be able to ask these questions about why the universe is the way it is?

This list can go on forever. The universe would be radically different if these constants of nature or the laws of physics were slightly different. Life, and in particular human life, would probably not be possible if there were slight changes in the universe. Scientists call this mystery that the laws of physics are so perfectly formed the *Goldilocks enigma* or the *fine-tuned universe.* Rather than the laws being "too this" or "too that," the laws are "just right." They are like a fine-tuned musical instrument perfectly set to make the cacophony of intelligent human existence possible. Why are the laws just right?

There are many overlapping questions that are usually lumped together. We cannot even begin to start addressing these difficulties until we disentangle the questions. The questions can be classified into three levels, one inside another, like a Russian (matryoshka) doll:

Question 1: Why is there any structure at all in the universe?
Question 2: Why is the structure that exists capable of sustaining life?
Question 3: Why did this life-sustaining structure generate a creature with enough intelligence to understand the structure?

Let's consider these questions in more detail.

Question 1: Why is there any structure at all in the universe?
Why are there laws of nature? Why should there be any regularity in the way physical objects interact? The actions of the physical universe have a certain pattern and these patterns repeat themselves regardless of time and place. There is a certain consistency and regularity to the laws. They are applied in every way and at all times. We can perform an experiment here and on Mars and get the same results. We can perform an experiment on Sunday and Tuesday and see the same outcomes. Why should there be such a consistency and repetitiveness of the laws? Why does the universe seem so normal to us rather than as some type of psychedelic Salvador Dali–type painting? It is hard to even imagine the universe if it lacked such structure.

We can ask even deeper questions. Rather than wondering why there are laws that act on physical objects, ask why there are any physical objects at all. What if the entire universe was a vast vacuum without physical objects? And yet there are physical objects in our universe. Why should that be? We can, of course, go off the deep end and ask why there is a universe at all. There is really no reason why anything should exist. Philosophers state these types of questions with the catchy phrase, "Why is there something rather than nothing?"[41]

Question 2: Why is the structure that exists capable of sustaining life?
Given that there is a universe and there are physical objects in the universe that follow habitual laws of physics, why should there be any life in the universe? The laws of physics are set such that there is a possibility for a process we call life. In order for there to be life, the universe must have enough complicated material so that life can grow and develop. All the life forms that we know of are made out of carbon. In fact, organic chemistry, which is essentially the study of living things, is the chemistry of carbon. The vast majority of material in the universe is hydrogen and helium, which is too simple to create complicated structures. The universe must be able to convert the simple materials to complicated materials. Another requirement that the universe must have for life is time. There needs to

be enough time for the complicated process of life to occur. The laws of the universe need to be very fine-tuned so that such complicated materials exist and long enough time for such a procedure to occur. We happen to live in a universe with exactly such laws. Why are the laws of physics so propitious for the development of life?

Scientists have determined that if certain physical laws were a little different, the complex materials and processes needed for any type of life could not exist. We can conceivably argue about the definitions of life (see below) and how different these laws of physics can be, but nevertheless there seems to be an amazing confluence of strangely exact laws that make life in this universe a possibility. Why should this be? We can ask a deeper question: Just because the constants and the laws of physics are just right for life, why should life still happen? Just because a pair of dice can land on snake-eyes does not mean they, in fact, will land on snake-eyes. Why *does* life exist in a universe where life *can* exist?

In contrast to the first group of questions, where it was hard to imagine the universe (or lack of one) in such a chaotic state, it is easy to imagine the world without any life. As far as we know, there does not seem to be any life outside of our little blue planet. That means, for the time being, the rest of our vast universe is as dead as a doorknob. One can easily imagine some lawyers-turned-politicians declaring global thermonuclear war with the consequence of all life on Earth—not just human life—being destroyed. This will be a lifeless universe.

Question 3: Why did this life-sustaining structure generate a creature with enough intelligence to understand the structure?[42]

Given that there is a well-working universe and that there is life in this universe, why should some of that life have intelligence?[43] The human brain is probably the most complicated object in the universe, and as such, is capable of some of the most amazing feats in the universe. The mind is also the only object in the known universe that wonders at the universe. Not only does it try to control the forces of nature that are around it, but it also tries to understand those forces. It is not just that we are intelligent and that we can play chess and understand some of the jokes on *The Simpsons*. Rather, we can go on and try to figure out why the universe is set up the way it is. Is it merely a coincidence that the universe spawned a creature capable of partially understanding the universe? At the present time we do not have total understanding of the universe around us. We also might never achieve total understanding of the universe, but we nevertheless understand part of it. Homo sapiens are also the only known creatures in the universe aware of their own existence. Why should this be?

When we take certain medicines or have too much tequila, our intelligence goes down. Our ability to understand or follow reason is impeded. This shows that our reasoning ability is very much a physical process, and as a physical process it is subject to the vicissitudes of the laws of nature. Why is it that our mind is so attuned to learning about nature? (Or is it? The vast majority of us would rather watch *The Simpsons* than study the laws of nature.) Slight changes in our environment make us less reasonable creatures. (What about changes that could make us into *more* reasonable creatures and *better* scientists? Would we know more about the universe then?)

Just because the universe is well suited for intelligence does not mean that the universe has to have intelligence. After all, before humans existed there was life on Earth without intelligence. Similarly, if we continue to mistreat each other and our planet, intelligent life will be destroyed and only cockroaches will remain. The universe and its perfectly formed laws will still be around even though there will be no one to study the laws or be in awe of them. There is nothing to ensure the continued existence of this fragile species.

It is not only that the laws of physics are well suited for reasonable creatures, but in a sense we learned reason from the laws of physics. As we saw in the last section, humans learned mathematics from the physical universe. Five apples added to three apples gives us eight apples, which teaches us that $5 + 3 = 8$. Again, in the last section we saw the intimate relationship between the common notion of an ellipse and planetary motions. Why should there be such regularity in the universe so that ellipses that we draw on a piece of paper should also show up in the heavens? It might make sense that we understand the quantum world by looking at the tools of abstract algebra. But why should the quantum world, in fact, work with abstract algebra that we learned in other places? The physical universe also follows logical rules. For example, if you know that A or B is true and then you learn that B is, in fact, false, then you know that A is true.[44] The very fact that we can observe such regularities with the laws of nature makes reason possible. Why do the laws of science have this perfect form? In essence, we are asking why there is any reason at all.

Let me clarify. In the last section I emphasized the idea that humans learned mathematics from the regularity of the physical universe. This explained why there really is no mystery that mathematics can describe the physical universe. In this section I am asking a deeper question: Why does the physical universe have any regularities at all? Why should we be able to learn mathematics and logic from the physical universe?

To answer these questions, researchers have formulated a set of ideas that go under the collective name of the *anthropic principle*. This says that the fact that sentient beings exist can be used to tell us something about our universe. It tells us that the universe must contain enough structure for intelligent human beings to exist. If this structure were absent, we would not be here to ask the questions.

The *weak anthropic principle* says that the observed universe must be of a form that would permit the existence of intelligent human observers. In other words, not all universes are possible. In a sense, the fact that we can make observations and we are somewhat intelligent tells us that the universe we live in has to have the complexity and the time necessary for humans to be born. Our intelligence is a type of sieve that tells us what type of universe we are going to find. A psychologist taking averages of IQs at a university will find that IQs are higher than those of the average population since university students generally have higher IQs. Similarly, when we survey the universe we should expect to find a universe that has beings capable of asking these questions—since we are asking these questions. Most researchers agree with the weak anthropic principle since it says something that is obviously true.

Some physicists have gone on in this line of reasoning and formulated a more powerful idea that is controversial: the *strong anthropic principle*. Rather than just saying that the universe is a certain way because we see it, they say that the universe *must be this way* because the universe must contain intelligent life. Via some mysterious force, the universe must create intelligent life. It should be noted that the vast majority of physicists do not accept the strong anthropic principle. They do not see any reason to believe that this universe must have brought about intelligent life. While the weak anthropic principle is hard to argue with, the strong anthropic principle is usually considered beyond the pale of science.

The anthropic principle says something very interesting about our position in the universe. Since the Enlightenment, science has deprived human beings of their special status in this universe. Copernicus showed us that we are no longer at the physical center of the universe. Rather, the Earth is simply an average planet circling some insignificant star. Darwin took away the status of humans as the most important creatures in the animal world. For him, a human being was simply one of many creatures that sprang from random mutations. And finally, it was Freud[45] who took away reason's status as a uniquely human ability and showed that it is simply a handmaiden to the mysterious animalistic subconscious. By incorporating all these findings, science has come to accept something termed the *principle of mediocrity* or the *Copernican principle* that says that what we observe in the universe is not special in any sense. We are a typical species on an ordinary planet spinning around a commonplace sun in a usual galaxy. There is nothing special about us.

That is, until now. In a sense, the anthropic principle has returned humans to the center of the universe. The Copernican principle has proved to be wrong. The very fact that intelligent human beings exist places restrictions on the type of universe we live in. The universe is the way it is because *we* have certain distinguishing features: we are alive and we think. Since there is human reason in the universe, the universe has the form that it has. One can study the universe by looking at intelligent life and seeing what laws the universe had to follow in order for such intelligence to emerge. The very fact that humans can observe and understand the universe shows that humans with their reason are at the center of the universe. We might not be in the physical center of the universe, but our rationality makes us the center.[46] (If we are not the only ones in the universe, then the intelligence of other beings is also at the center of the universe.)

The weak anthropic principle is a bit unsatisfying. Yes, it explains why we have to see the universe a certain way, but it does not explain *why* the universe is the way it is. It could have been another way and we would not be able to see or report it. The philosopher John Leslie compares this to a man surviving a firing squad. There were ten soldiers firing guns at the man, and he reported surviving the event. In analogy to the anthropic principle, the man would answer that had he not survived the firing squad, he could not report surviving it. Because he survived, he can tell us about it. While what he is saying is certainly true, it is not what we want to hear. We want to know *why* he survived. Were the soldiers bad at shooting? Did someone change the bullets for blanks? Was it just a strange accident? So, too, with the universe, we want to know *why* the universe is set up so that intelligent beings are capable of reporting about it.

Over the years researchers have proposed several possible explanations for the anthropic principle.

A Deity

Deists have no problem explaining why the universe is the way it is: a deity created the universe out of nothing and set it up to have life—more importantly, intelligent life. The universe and its laws were created exactly as they are because an omniscient, omnipotent deity wanted to see the human drama unfold. The deity knew the exact requirements for the human species to arise and established an appropriate world. As it says in the book of Psalms (19:2), "The heavens declare the glory of God; And the firmament shows His handiwork." In fact, this explanation has been used for millennia as a proof of the existence of a deity. It has been called the *argument from design* or the *teleological argument*.[47] All three of the questions about the structure in the universe are easily answered by the existence of a deity.

While this explanation is satisfactory for deists, those who do not believe in a deity will find this solution unsatisfying. Such a deity raises all kinds of other, more mysterious, questions about the nature of a deity. For such nonbelievers, there is a need for a more physical, scientific, and testable explanation.

The Universe Is a Fluke

We are simply incredibly lucky to find the universe the way it is. The universe and its laws were formed for no particular reason whatsoever. It was not formed for humans or intelligent beings. As David Hume wrote, "The life of man is of no greater importance to the universe than that of an oyster." If such intelligent beings exist, that is not of any importance. To people who accept this idea, we live in an absurd universe and need to accept this fact. If forced to think about it, such people would have to agree with the weak anthropic principle that if the universe was any other way we would not be able to exist. But they will not look into the implications of this. This stance is extremely antiphilosophical. The universe is the way it is and there is no reason for it. This response does not give any answers to the three questions we had about the structure in the universe. While this might be satisfying for people who do not think about the origins of the universe or of reason,[48] the rest of us, cursed with that strange desire to understand why the universe is the way it is, will continue looking for answers. To us, ignoring or denying a mystery does not make the mystery go away. We will just have to look further.

Out-of-Tune Universe

Some researchers scoff at the whole idea that this universe is fine-tuned for life in general and for intelligent life in particular. When we look at the cosmos, rather than seeing a place suitable for life, every star system that we point our telescopes at seems devoid of life. How can we say that the universe is waiting for life when there are billions of destructive supernovas, black holes, asteroids, and comets slamming into each other and into planets and stars? Despite years of looking, we have never found another planet that could sustain life. A further indication that the universe is not fine-tuned for intelligent life is that no one has ever visited us from any other place in the cosmos. Even within our solar system, there are no planets that would be able to support any type of intelligent life. If an astronaut stepped out of a spaceship, they would either instantly freeze to death or the sun will burn them in seconds flat.

What about Earth as a place perfectly suited for life? Let us take a closer look at our own beautiful planet. Two-thirds of our little blue planet is covered with water and (other than the supposed intelligence of dolphins), the sea does not seem very conducive to intelligent life. We are usually restricted to the dry surface of the Earth. Yet even in those parts, there are large swaths of land that are too high, too hot, too dry, or too cold to sustain long-term human life. Within environments capable of sustaining humans, there are constant tsunamis, volcanoes, earthquakes, hurricanes, mudslides, poisonous mushrooms, and lawyers, all of which make human life painfully fragile. There is an unending list of diseases, viruses, plagues, and deadly bacteria that have, over time, vanquished large segments of human societies. Perhaps the most destructive of all forces against human life is human nature itself, with its indomitable desire to murder and destroy its own species and environment. The list of reasons why the Earth is not perfectly suited for human life can go on and on.

Rather than seeing our universe as perfectly fine-tuned and propitious for intelligent life, such people see the universe as unsuited to the existence of sentient beings.[49] They would say that there is no reason to explain the supposed fact that the world is so ordered—because it is, in fact, not so ordered. This brings to light an even deeper mystery: if the universe is so inappropriate for the development of intelligent life, why did just such life develop anyway?[50]

Notice that this is only an answer to questions 2 and 3. It does not answer any of the deep questions.

Restrictive Definitions of Life

Others disparage the idea that the universe is fine-tuned for intelligent life by pointing out that our requirements for intelligent life are tailored to our lives, and this is too restrictive. They feel that the constants of nature and the laws of physics are not as restrictive as we think. Other forms of life would have emerged if the laws were different. Maybe life could have been created with other materials besides carbon. Some speculate that silicon life forms are a possibility. Scientists have recently found certain life forms that live in arsenic.[51] Marine biologists have been shocked to find some life forms living near active underwater volcanoes. Perhaps there are certain types of creatures made entirely of neutrons that live on the surface of a star. Scientists have formulated other ways of making more complicated elements in suns even with different physical constants.[52] Some people wonder if a computer virus with its uncanny ability to replicate, maintain homeostasis, and overcome all forms of security is not a type of life form. In fact, computer viruses even show signs of intelligence. To a certain extent, the assumption that the only life forms that can exist are the ones we are familiar with is a sign of a lack of imagination. The questions posed by the anthropic principle are answered by including more exotic types of life forms—the idea being that if we include all these possibilities, then redefined-life could have emerged with any of the other ways the universe was set up. Had the universe been any other way, the other life forms would have been amazed that the universe was their way and not any other way.

This response to the anthropic principle answers questions 2 and 3. There is no mystery why the universe is the way it is because it could be many different ways and (intelligent) life would have emerged. This solution again does not answer question 1 in our list of questions. Whether there is some weird type of life does not take away from the fact that there is a lot of structure in the universe. Why should that be?

Many Universes

One very popular explanation for the amazing fine-tuning of the universe is that our universe is just one of many universes that comprise something called a *multiverse*. Each of these universes has its own sets of laws and constants. In the vast majority of these universes the laws and constants are not propitious for life or intelligent life. Our universe is one of the lucky ones where intelligent life is a possibility.

Before you immediately discard the very idea of the existence of many different worlds, let us look at science's ever-expanding horizons. Throughout ancient and medieval times, people believed that our sun was the only sun in the universe. It is only in modern times that we have realized that our sun is just one of billions in the Milky Way galaxy. It was not too long ago that we learned that our galaxy is one of billions of other galaxies in the universe, each having billions of stars. People who believe in a multiverse are just taking this idea one step further. Maybe there are billions of other universes besides our universe, and we just cannot see them or get any empirical confirmation of their existence.

How does a multiverse help in explaining why our universe has such a life-sustaining structure? Imagine walking into a bingo hall and being the only player there. If your card wins the game, you must believe that this was miraculous. From all the random numbers to call, it happens to be your numbers that are called. This must be divine intervention. Similarly, if you walk into a random hotel room and this is the only room that has your size clothes in it, that would be a miracle. If this is the only universe in existence, then the fact that it is so perfectly made for us is somewhat miraculous. Now consider walking into a crowded bingo hall with many other players. One of them is going to win the game and jump up and scream "I won! It's a miracle!" To the winner, it is, in fact, a miracle. Why should she win and everyone else lose? But to you who are observing all of the players, and know that *someone* must win, it is not a miracle that *someone* wins. Similarly with the hotel rooms. It is not so miraculous if every room has clothes in it. Someone has to find the clothes that fit so well. Similarly with our universe. If there are many universes and some have laws that make intelligent life seem possible, then that is not so strange. We happen to find ourselves in one of those types of universes. Presumably the vast majority of the other universes in a multiverse are devoid of life. All these different universes will have different laws and different constants of nature. Only some of them will be "just right." It is not miraculous that the intelligent inhabitants of those universes that can support intelligent life jump up and scream, "It's a miracle!" It is expected of them.

The theory of a multiverse comes in many different flavors. Over the years, scientists have developed various theories explaining why there is a multiverse rather than a universe. Recently Brian Greene published an excellent book, *The Hidden Reality: Parallel Universes and the Deep Laws of the Cosmos*, where he describes nine different variations of the multiverse theme. I will briefly describe some of them in the coming paragraphs. A word of caution is in order. Some of these ideas are totally insane, to put it mildly. They take science fiction to new levels and come to very strange conclusions.

We first met the concept of a multiverse in section 7.2 when we discussed Hugh Everett's theory that every time there is a measurement, the universe splits into different universes. For every measurement, there are different possible outcomes and the universe splits into that many exact copies with each daughter universe having that particular outcome. Since there are millions of observations every second, there are going to be billions and billions of universes. It is not clear how these universes differ in their physical laws. Nor is it clear how these universes can have nonintelligent life if only observers can make measurements. Nevertheless, this was the first theory of a multiverse.

We met string theory several times on these pages. As we have seen, the idea that the universe is made out of very small strings solves many problems in physics. It turns out that string theory also predicts the existence of a multiverse. The strings wiggle around in different multidimensional spaces called *branes* (short for membranes) or *D-branes*. String theorists believe that there are many different branes and they can bang into each other and cause new universes with many different types of properties. There are so many types of universes that the multiverse is called the *string-theory landscape*. With so many universes, it is not surprising that some parts of this landscape have life in them.

The above two explanations for a multiverse posit many different universes, but all of them will be as improbable as ours. While the hypothesized multiverses answer the questions we asked, things seem a bit too random. Lee Smolin in his book *The Life of the Cosmos* has proposed an interesting model of the multiverse where universes with intelligent life are more probable than universes without life. He puts forward the idea that one universe can emerge from a collapsed black hole of another universe. This new universe can have many of its own black holes and in turn many baby universes. He further postulates that the laws of physics for the new daughter universes will be only slightly different than the laws for the parent universe. This mechanism gives the universe a certain aura of natural selection. The universes that have more black holes will have more daughter universes and are more likely to survive. Since black holes come from heavy suns where stellar nucleosynthesis can occur, there will be more universes with such heavy suns. In Smolin's multiverse, universes are evolutionary, coming closer and closer to being life-permitting.

Max Tegmark of MIT has a very interesting notion of a multiverse. He takes Plato and Pythagoras to an extreme and believes that the only thing that really exists is mathematics. Every type of mathematical structure that can exist, does exist. If the structure is coherent and within reason, then it exists. Some of those systems describe life-sustaining universes. Some mathematical structures even describe life forms that have some mental processes that we might call intelligent. And there are even mathematical structures that describe creatures that can ponder their own existence. We happen to find ourselves in a universe of the Tegmark multiverse where the mathematics is sophisticated enough for human beings. Tegmark explains why we don't see mathematics but we see trees, flowers, and humans. On the one hand, this theory clearly violates Occam's simplicity-of-ontology rule since *everything* exists. On the other hand, since there is no choosing what does and does not exist, there are actually fewer rules in such a multiverse. In other words, Tegmark's multiverse does satisfy the simplicity-of-hypothesis criterion. This intriguing idea deserves much more thought but is beyond the scope of this book.

To me, the most interesting versions of the multiverse are a consequence of some of the ideas we met in section 7.2 with our short exploration of quantum mechanics. One of the main ideas was that a property of an object is in a superposition of values until it is observed by a conscious being, and then it collapses to a single value. John Wheeler applies this concept to the entire universe. When the universes came into being, there were no human observers and so everything that existed was in a hazy superposition of values. But rather than thinking of this as a hazy superposition, think of it as a form of multiverse within our universe. Objects had many possible values in our single universe.[53] Each of these superpositions followed the laws of physics (or perhaps many different laws of physics) and continued on that path. The theory goes on to say that one of the many possible superpositions developed a complicated life or consciousness with the capacity to observe the surrounding universe, as in figure 8.8.[54] This observation (symbolized by the eye in the figure) collapsed the entire superposition into the one universe we know and love. The superposition that brought along consciousness caused the superposition to collapse. This theory is called the *participatory anthropic principle*. The observer participated in the creation of the universe and permits there to be observers. All the other superpositions that did not spawn sentient beings (they are without eyes in the figure) simply collapsed away. It is important to realize that if the participatory anthropic principle is true, not only would it be a legitimate explanation for the weak anthropic principle but it would actually satisfy the strong anthropic principle. The universe would stay in a hazy superposition unless there was an intelligent observer. Wheeler goes further with this theory. We saw with the delayed-choice quantum eraser experiment in section 7.2 that outcomes of experiments can change the past. More exactly, the outcome of an experiment depends on the total experiment. Perhaps we can say

Figure 8.8
The universe that produces an observer is observed. Figure by Hadassah Yanofsky.

that the universe's past depends on the existence and observation of human observers.[55]

We saw how a multiverse would help us answer the questions about why intelligent life exists in the universe. We also glimpsed several different types of multiverses and how they work. However, the idea of the multiverse is not without its critics. The most obvious objection to a multiverse is that there is no empirical evidence for any multiverse. We live in one universe and we only see one universe. This book is only being published in this universe (at least I have not received any royalties from any other universe), and there is no empirical evidence of any other universe. If they do exist where are they? What do they look like? We have simply postulated them because they help explain the existence of intelligent life, they help us with determinism (Everett's many-worlds interpretation), or their math says it is so (string theory), but that does not make it so. Yes, they lend support to the anthropic principle, but that does not make their existence an actual fact. Your financial problems would be solved if you won the lottery, but that does not mean that you, in fact, won the lottery.

There are other objections to a multiverse. All these concepts of a multiverse have laws that explain how the universes branch off each other and come into existence. These laws are not particular to one universe but are laws for the whole multiverse and are called *superlaws* or *metalaws*. Now we can ask a deeper question: Why are these superlaws set up so perfectly that some universes will produce intelligent life? The superlaws are created so that different universes will have different characteristics and some will be well suited for intelligent life. Why? We first asked why there is a structure in the universe that makes intelligent life possible. This was answered by positing the notion of a multiverse and saying that the reason there is intelligent life in our universe is because our universe is just one part of the vast multiverse. Now we are asking why there is a structure *in the multiverse* that can bring about intelligent life.[56]

Yet another criticism of a multiverse is that it is often invoked as an ad hoc alternative to the notion of an intelligent designer. In other words, an atheistic scientist would rather posit the existence of a multiverse than a deity of some sort. Neil Manson, a philosopher, called the concept of multiverse "the last resort for the desperate atheist."[57] In fact, the notion of a multiverse is just as unscientific as a deity. That does not mean they are equally possible or equally probable. It just means they are unobservable, unprovable, undeniable, and untestable. Many critics say that positing the existence of a multiverse demands the same leap of faith as most religions do.

Some scientists deny that the very concept of a multiverse is even science. A multiverse is neither empirical nor testable. If the different universes in a multiverse do not interact with each other, how can we even test if there are other universes? In other words, one of the deepest questions in all of science and perhaps existence is answered by many scientists with an answer that is, by definition, beyond the limits of science. But whether the concept of a multiverse is, in fact, science should not stop us from thinking about multiverses. We think about many topics that are not exactly science. They are interesting ideas and they just might explain our universe.

Symmetry

There is another appealing group of ideas that help explain the structure of the universe. The gist of these ideas is that any structure in the universe that we observe comes from the fact that we are observing the universe in a certain way.

In a sense, some of these ideas—like so many other ideas in philosophy—are a response to the problems raised by David Hume. In section 8.1 we saw that Hume's problem of induction was extremely vexing. He called into doubt the very notion of cause and effect, which is central to all of science. Immanuel Kant took these problems as a wakeup call. He tried to address the problems by saying that humans do not look at phenomena through clear glasses. According to Kant, we look at the world through colored glasses. We have preconceived notions that are built into us and help us understand and categorize all the phenomena we see. Notions like space, time, and causality are ideas that are part of us, and we use these ideas to make sense of the universe. These notions preexist in us and do not come from experience. With these hardwired notions, the universe looks the way it does. Without these notions, we would not be able to see the structure that we do see. For Kant, our view of the universe is influenced by our own mind and we cannot observe what is really out there "in itself" without these built-in notions. This is a step away from the traditional view of the relationship between human beings and the universe. The traditional view is that the universe is the way it is and we are viewing it that way. Kant is advocating the notion that our view of the universe is dependent on our perspective on it.

Einstein was also interested in how we view different phenomena. As we saw in section 7.2, he formulated relativity theory by insisting that the laws of physics should be the same regardless of how they are observed. Before Einstein, Galileo insisted that the laws of physics should remain the same as long as the observer is moving at a constant velocity. Einstein generalized this with special relativity, where he insisted that the laws of physics must remain fixed regardless of whether an observer is moving at a constant velocity or moving close to the speed of light. His general relativity further generalized this. He insisted that the laws should be the same even when changing velocity (i.e., accelerating or decelerating). This fact that the laws of physics must remain the same regardless of how the viewer observes them reflects a type of *symmetry*. Colloquially, we say that a room has symmetry if the room looks the same after the left and right sides are swapped. Scientists extend this notion of symmetry to the way laws of nature are described. The laws look the same if they are viewed from diverse perspectives. As science progresses, the notion of symmetry is playing an ever-increasing role.

Einstein actually did something more radical here. Before Einstein, most physicists would find and describe a law of physics and then go on and describe its properties such as its symmetries. In the case of relativity theory, *Einstein used symmetry to find the laws of physics*. He postulated that the law must satisfy these symmetries and then went on to describe that law that satisfies them. Anything that did not satisfy the symmetries could not be a law of physics. He was the first person to use symmetries as an important arbitrator or sieve that determines what is or is not a law of physics.

Emmy Noether (1882–1935) went further with these ideas. She focused on laws of physics called *conservation laws*. Such laws say that throughout a process or experiment a certain quantity does not change. Prominent examples include:

1. *Conservation of momentum.* This says that the total momentum of all the objects in a system will always remain the same. We see this, for example, when "breaking" the billiard balls on a pool table. In the beginning only the white ball is moving very fast toward the other balls. After the other balls are hit, they scatter in all different directions and at all different speeds. Conservation of momentum says that the speeds and directions all add up to the speed and direction of the original white ball.
2. *Conservation of angular momentum.* This law says that how and at what speed bodies spin must remain the same. A classic example of where such a conservation law is demonstrated is when an ice skater is spinning around and then brings her arms in toward her body. To preserve angular momentum, she will spin faster when her arms are pulled toward herself.
3. *Conservation of energy.* In short, this says that the type of energy in a system can change but the amount of energy must stay the same. For example, when you press on your brakes, the energy of the car moving turns into heat within the brake pads. At a dam, water drops from a high height and can rotate a turbine, which makes electricity.

Noether showed that each of these conservation laws corresponds to a certain symmetry of the system. The above three conservation laws come from the following three symmetries respectively:

1. *Symmetry of place.* This means that an experiment can be done in different places and the results will still be the same.
2. *Symmetry of orientation.* Independent of how an experiment is oriented, the results of the experiment will be the same.
3. *Symmetry of time.* Regardless of when an experiment is done, the results will be identical.

(We are not going to show how the conservation laws correspond to the symmetries.) Noether actually proved something far more general: she showed that *any* conservation law (of a certain type) will have a corresponding symmetry (of a certain type), and furthermore *any* symmetry (of a certain type) will have a corresponding conservation law (of a certain type).

Following Noether's theme, later researchers went further. Rather than looking for conservation laws, they looked for symmetries. This put symmetries at the center of their experiments and calculations. Physicists like John von Neumann (1903–1957) and Eugene Wigner formulated large parts of quantum mechanics in group theory, which is the mathematical language of symmetry. Many other branches of physics also use group theory in a fundamental way.

All these scientists are proposing that instead of humans seeing real structure in the universe, humans are acting as sieves. Scientists do not see the laws of physics; rather, what they select, they call science.

A modern exponent of these ideas is Victor J. Stenger. In a fascinating book titled *The Comprehensible Cosmos: Where Do the Laws of Physics Come From?* Stenger explains much of modern physics from the point of view of symmetry. He uses more sophisticated forms of symmetry than we have discussed to explain modern quantum mechanics, cosmology, quantum field theories, and all the other areas of contemporary physics. The book discusses local symmetries, global symmetries, gauge symmetries, and so on. All these different symmetries can be treated under the umbrella of something he calls *point of view invariance*. This notion says that regardless of how or when a phenomenon is observed, and regardless of how it is described, the laws of physics must be the same. Using these ideas as a sieve to determine the laws of physics, Stenger shows that the laws of physics are not "out there." They are just the way we look at the universe. In the preface he summarizes: "The laws of physics are simply restrictions on the ways physicists may draw the models they use to represent the behavior of matter." Again, what this means is the laws are those ways of describing the symmetries that we observe.

Probably the first to promote these ideas was Arthur Stanley Eddington. He was not only a world-class scientist but also a very deep philosopher. Some of his ideas also have an impact on several of the themes of this book.

Rather than just looking at the universe, Eddington turned to look at the scientist, asking "Who will observe the observers?"[58] He became an epistemologist to see how a scientist learns about the universe. Eddington's concept of science is something he called *selective subjectivism*. The laws are not objectively out there. Rather, they are subjectively chosen. The scientist selects certain phenomena on the basis of their symmetry and then calls the recurrent rules that describe such phenomena *laws of nature*. The laws can be expressed in mathematical language because that is the way we view the external world. Or as he says, "The mathematics is not there till we put it there." It is not the structure that we are looking at, rather it is the way we look at it the universe. Eddington finishes off his important book, *Space, Time and Gravitation*, with the following wonderful quote: "We have found a strange footprint on the shores of the unknown. We have devised profound theories, one after another, to account for its origins. At last, we have succeeded in reconstructing the creature that made the footprint. And lo! It is our own."

Before we leave Eddington, let us consider a very profound thought that is relevant to the theme of this book. When we talk about the limits of scientific reasoning, we must keep in mind how we are observing the universe. Eddington presents a fantastic analogy about a scientist who studies fish (an ichthyologist):

> Let us suppose that an ichthyologist is exploring the life of the ocean. He casts a net into the water and brings up a fishy assortment. Surveying his catch, he proceeds in the usual manner of a scientist to systematise what it reveals. He arrives at two generalisations:
>
> (1) No sea-creature is less than two inches long.
> (2) All sea-creatures have gills.
>
> These are both true of his catch, and he assumes tentatively that they will remain true however often he repeats it.
>
> In applying this analogy, the catch stands for the body of knowledge which constitutes physical science, and the net for the sensory and intellectual equipment which we use in obtaining it. The casting of the net corresponds to observation; for knowledge which has not been or could not be obtained by observation is not admitted into physical science.
>
> An onlooker may object that the first generalisation is wrong. "There are plenty of sea-creatures under two inches long, only your net is not adapted to catch them." The ichthyologist dismisses this objection contemptuously. "Anything uncatchable by my net is *ipso facto* outside the scope of ichthyological knowledge. In short, what my net can't catch isn't fish." Or—to translate the analogy—"If you are not simply guessing, you are claiming a knowledge of the physical universe discovered in some other way than by the methods of physical science, and admittedly unverifiable by such methods. You are a metaphysician. Bah!"[59]

Eddington is stressing that we should look at the size of the net that we cast to obtain our observations. In other words, the way we look at the universe is the way it will present itself to us. He goes on to point out that by looking at the net we will see that the information determined in (1) is more fundamental than the information determined in (2). After all, if we are using a net with two-inch holes, we won't catch a fish that is one inch. In contrast, the fact that all the sea creatures have gills, could be an unwarranted generalization of the fish that we have seen. There might very well be fish without gills out there. Eddington concludes that we learn more from looking at the way we are observing the universe than from actually observing it.

One can only speculate about how our scientific method can be adjusted so that we can see more of the universe. Continuing with the analogy of the fish scientists, what type of fish would we find if we cast a net that has one-inch holes? What are we missing by looking at the universe the way we do? What is out there?[60]

There are some problems with this solution to the issues raised by the anthropic principle. For one thing, there is simply a feeling that there is an immense underlying structure "out there" that does not care if it is observed by human or conscious creatures. There are vast tracts of both space and time that do not seem to have human or any other observers. Are we really to believe that when a certain part of the cosmos comes into view, it also comes into existence? As for time, there were long eons before observers came into being. Are we really to believe that there was no structure then? The feeling is that the laws are out there and we are viewing them.

Another, related difficulty is that these ideas go against the central conception in all of physics. There have always been laws of physics and states of physics. The laws determine the states and we study the states to learn more about the laws. Here we are proposing that the state of physics—the way the observer is observing—affects the laws of physics. This is a radical change. How do states, which are formulated by laws, change the laws? Actually, the last few paragraphs have said something even deeper: there are no laws at all!

The last problem I will mention is that not all of what we see satisfies the symmetries that we expect in our laws. Many features of the physical universe were formulated by something called *symmetry breaking*. This is a seemingly random process that takes some law of physics that has good symmetry properties and changes it into some other law that has less symmetry. Why some forms of symmetry should be preserved and not others is beyond us (for now) and seems totally random. While the properties of symmetry might shed light on some of the structure, they do not explain all of it.

A few years ago these ideas finally hit me. After I had given a lecture on the basics of quantum mechanics and the central role that the complex numbers play in it, a student asked the following fundamental question: "Why does the universe follow the complex numbers?" The question is excellent in its simplicity. Why the strange complex numbers and not the real numbers that we are used to? It took me some time to formulate the response that really gets to the crux of the issues. The answer is that the universe does not follow complex numbers. Instead, the universe does what it does! Human beings use complex numbers to help them understand this part of the universe called quantum mechanics. If mathematicians had not invented complex numbers, then physicists would have had a harder time understanding the world. The Earth and the sun do not look up Newton's famous formula that describes the attractive force between two bodies and determine the force between them. Rather, the formula is used by humans to understand what is going on. Again, we must emphasize, the universe does not function using complex numbers, Newton's formula, or any other law of nature. Rather, the universe works the way it does. It is humans who use the tools they have to understand the world.

We Do Not Know (Yet)

There is one other response to our fine-tuned universe: we simply do not know (yet). Science has progressed at a breathtaking speed over the past few centuries and we expect a lot from science. Nevertheless, no one has said that science will provide the answers to *all* the questions. At the moment, all of the above answers have a science-fiction feel to them and none of them are really scientifically satisfying. Some new evidence may arise to show that one of the above explanations for the anthropic phenomena is correct. In the future, scientists may find better answers than the ones given. Another possibility is that there is an infinite chain of explanations, one deeper than the next. One explanation might work for a while and then a deeper one is found that explains the previous one, and this goes on for eternity.[61] However, we must also accept the possibility that there will never be a satisfying explanation for these overlapping questions. In that case, we simply will never know. It takes a touch of humility to say we don't know, and perhaps such humility is warranted.

Let us speculate on why it is so hard to find a legitimate answer to the anthropic principle. There is something inherently strange about all the possible explanations for this principle. For most questions that can be asked about a system, an answer can be found within the system. For example, "Why is it raining today?" Answer: "The clouds are full and the temperature is correct." This is a question about the environment, with an answer *from* the environment. However, sometimes when a question is so fundamental, one must go outside the system to a deeper level. For instance, fundamental chemistry questions are answered within chemistry or at a deeper physics level. “Why does water boil?” Answer: “Because the fire under the pot increases the energy, etc.” This is a physics answer. Fundamental sociology questions like “Why did the people rebel?” can probably be answered at the deeper psychological level, etc. With our questions about the fine-tuned universe, we are asking fundamental questions that can only be answered by looking *outside* the universe. Why should the entire universe be a certain way? We are searching for an answer outside the universe. What is outside the universe? A deity? Many other universes? We are not used to answers that are outside the universe. Scientists want answers from within the universe and do not like traveling to regions beyond. Even for such fundamental questions about the entire universe, we like answers within the universe and are uncomfortable looking elsewhere. I suspect this discomfort will be with us for some time.

The existing evidence does not favor any particular response to our questions about the structure of the universe. We can speculate about all the possible definitions of life and all the possible multiverses, but we have no empirical proof that such life forms or universes actually exist. And so we come to the conclusion that we are in a state of ignorance about the ultimate reasons the universe is the way it is. The universe that created us to have reason and to question everything we see is very coy about revealing its deep secrets in response to these reasonable inquiries.

No matter what reason you accept for the anthropic principle, it is hard to avoid the eerie feeling that something strange and wonderful is going on here. The physicist Freeman Dyson said it in its clearest form: “I do not feel like an alien in this universe. The more I examine the universe and study the details of its architecture, the more evidence I find that the universe must in some sense have known that we were coming.”[61]

Further Reading

Section 8.1

There are many fine introductions to the philosophy of science. Some recommendations from easiest to hardest include Okasha 2002, Gorham 2009, Godfrey-Smith 2003, and Losee 2001. Our discussion has also gained much from Rescher 1999. Our presentation of the beauty and the end-of-science issues has borrowed much from the extremely readable Weinberg 1994. We cannot talk about the end-of-science issue without mentioning the controversial—but fun to read—Horgan 1996. Kuhn 1970 is very readable and interesting.

Section 8.2

Mickens 1990 is a collection of nineteen articles that discuss and give possible answers for Wigner's unreasonable effectiveness. Included is Wigner's original article. Many other articles and chapters deal with this issue, including chapter 27 of Klein 1981. Burtt 1932 offers a fascinating historical tour of the ever-increasing role of mathematics in the sciences. One must also include Pickering 1984 for an intriguing perspective.

Section 8.3

The anthropic principle is discussed in many books, including the classic Barrow and Tipler 1986 and Davies 1982, 2008. The second Davies book is a popular book that covers every aspect of the topic and is very deep. Gibbon and Rees 1989 is another popular account of the anthropic principle. There is also a very nice summary article found online by P. C. W. Davies titled "Where Do the Laws of Physics Come From?" (2012).

For more about the multiverse, see Greene 2011. Also see Carr 2007, which contains twenty-eight up-to-date scholarly articles from every different angle. See also Deutsch 1997 as well as Max Tegmark's papers on his web page, http://space.mit.edu/home/tegmark. For a basic introduction to string theory see Musser 2008.

For more about the role of symmetry in formulating the laws of physics, see Eddington 1958, Kilmister 1994, Weyl 1952, Stenger 2006, Cook 1994, and Lederman and Hill 2004.

9 Mathematical Obstructions

As soon as I understood the principles, I relinquished for ever the pursuit of the mathematics; nor can I lament that I desisted, before my mind was hardened by the habit of rigid demonstration, so destructive of the finer feelings of moral evidence, which must, however, determine the actions and opinions of our lives.[1]
—Edward Gibbon (1737–1794)

I had a feeling once about Mathematics, that I saw it all—Depth beyond depth was revealed to me—the Byss and the Abyss. I saw, as one might see the transit of Venus—or even the Lord Mayor's Show, a quantity passing through infinity and changing its sign from plus to minus. I saw exactly how it happened and why the tergiversation was inevitable: and how the one step involved all the others. It was like politics. But it was after dinner and I let it go![2]
—Winston Churchill (1874–1965)

The New Mathematics: standard mathematics has recently been rendered obsolete by the discovery that for years we have been writing the numeral five backward. This has led to reevaluation of counting as a method of getting from one to ten. Students are taught advanced concepts of Boolean algebra, and formerly unsolvable equations are dealt with by threats of reprisals.[3]
—Woody Allen

Mathematics is the epitome of pure reason. Much of science is based on mathematics and logic. In a sense, mathematics is the language of reason. It is the most successful of all human inventions. And yet, as we will see, it too has its limitations.

Section 9.1 introduces some limitations that were dealt with during the classical Greek era. Section 9.2 contains a short description of Galois theory, which is an entire field of modern mathematics dedicated to showing that certain problems cannot be solved with the usual methods. In section 9.3 we return to our discussion of solving problems using computers and show that several problems in mathematics are unsolvable by computation—be it mechanical computation or human computation. Section 9.4 discusses some self-referential aspects of logic including Gödel's famous incompleteness theorem. The chapter closes with a discussion of some technical and philosophical aspects of logic and axiom systems.

9.1 Classical Limits

It was clear to the followers of Pythagoras that Hippasus, a former member of their club, had to be killed. He must be thrown into the sea lest he continue to blaspheme and reveal all of their secrets. Hippasus was being totally irrational!

One of the more interesting characters in the ancient Greek world was Pythagoras of Samos. He lived in the sixth century BC and had a major impact on the ancient and the medieval worlds. Pythagoras—a philosopher, mystic, music connoisseur, and religious leader—was one of the first to study mathematics for its own sake rather than for its applications. This makes him one of the world's first *pure* mathematicians. He also discovered relationships between musical harmonies and mathematics. Pythagoras had an immense influence on the progress of philosophy and mathematics.

One of the central dogmas for Pythagoras and his legion of followers was that the entire world was governed and described by whole numbers or ratios of whole numbers. Such ratios were the only numbers thought to be sane or "rational." In other words, they did not believe in what we now call *irrational* numbers. To them, rational numbers were the only numbers that existed.

All was well with this worldview until a student of Pythagoras named Hippasus of Metapontum came along. Hippasus contemplated a common square whose sides are of length 1, as in figure 9.1.

By Pythagoras's famous theorem for a right triangle we have

$$x^2 + y^2 = z^2$$

or

$$\sqrt{x^2 + y^2} = z$$

In the square in figure 9.1, both x and y have length 1. That implies that the diagonal, z, has length $\sqrt{2}$. This was fine with everyone, but Hippasus then proceeded to prove that this commonly occurring number, the square root of 2, is not a rational number. This easily described number that is found everywhere was the first known example of an irrational number.

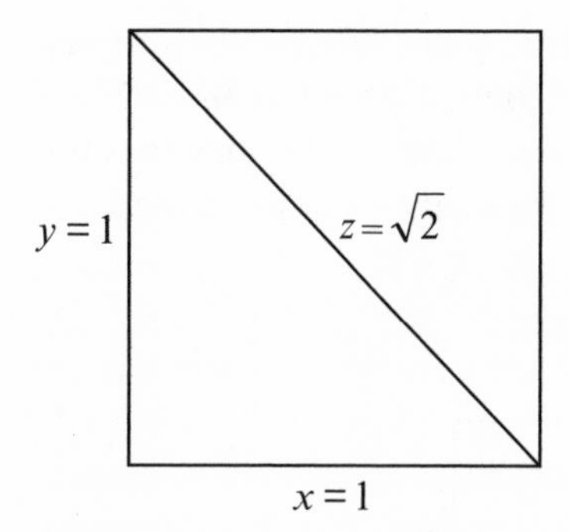

Figure 9.1
A square with an irrational hypotenuse

This was a shocking result to his contemporaries. In a sense, this was the first violation of a limitation for the mathematics of that era.[4]

Legend has it that Pythagoras and his followers were upset with Hippasus's finding. They feared that he might reveal his discovery to other people and hence show the inadequacies of their faith. The brotherhood did the only thing they thought they could do: they took Hippasus out to sea and threw him overboard, hoping he would take his secret to a watery grave.

It is not known what proof Hippasus used to show that the square root of 2 is an irrational number. However, a very elegant proof exists that is worthy of our consideration. The proof is geometric and does not contain many of those nasty equations that make people unhappy. It is a proof by contradiction, with which we are already familiar. If we assume (wrongly) that the square root of 2 is a rational number, then we are going to find a contradiction:

The square root of 2 is rational $\Rightarrow$ contradiction.

If $\sqrt{2}$ were a rational number, then there would be two positive whole numbers such that their ratio is the square root of 2. Let us assume that the two smallest such numbers are a and b. That is,

$$\sqrt{2} = \frac{a}{b}$$

Squaring both sides of this equation gives us

$$2 = \frac{a^2}{b^2}$$

Multiplying both sides by b^2 gives us

$$2b^2 = a^2.$$

Let us look at this equation from a geometric point of view. This means that there is a large square whose sides are length a (that has area a^2) and two small squares whose sides are length b (each has area b^2) such that the sum of the areas of the two small squares is exactly the area of the large square, as seen in figure 9.2.

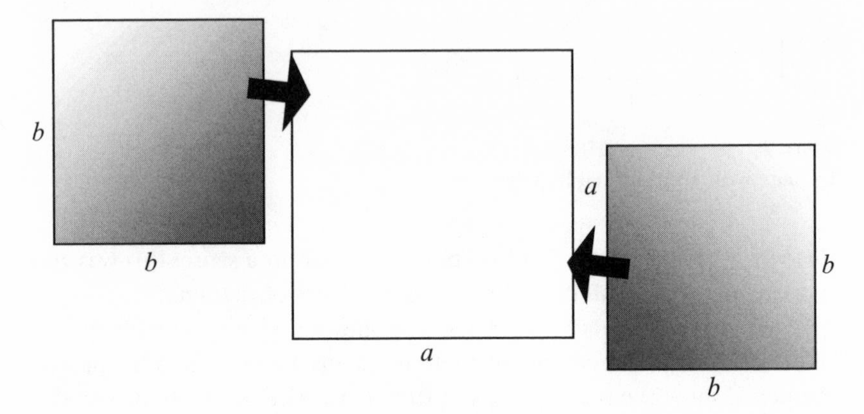

Figure 9.2
The smallest two squares that fit into one

Furthermore, if we assume that a and b are the smallest such numbers, there are no smaller squares that make this true.

When the two smaller squares are placed into the larger one, there must be some overlap and two missing parts, as in figure 9.3.

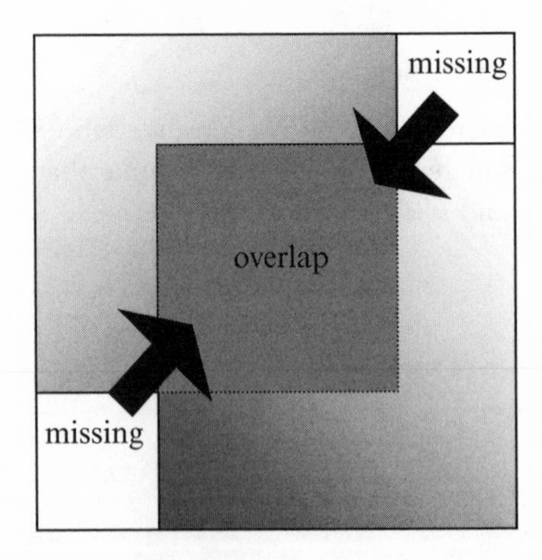

Figure 9.3
Still smaller squares

There are two problems in figure 9.3: there is an overlap that shows we counted an area twice and there are two smaller regions of the a box that are not covered by either of the two b boxes. In order for the areas of the larger square to equal the sum of the two smaller ones, the two missing parts must collectively have the same area as the overlap—that is, the two lesser (missing) squares must fit into the larger (overlap) square. These are necessarily smaller than our original squares. But wait! We assumed that the squares we started with (of size a and b) were the smallest ones such that the two lesser ones fit into the larger one, but now we've found smaller ones. This is a contradiction. There must be something wrong with our assumption that such numbers exist. Conclusion: the square root of 2 is not a rational number. It is an irrational number.

Since classical times, many other math problems have also defied human reason. These problems continued to be pursued through medieval times and were finally shown to be unsolvable in modern times.

The ancient Greeks approached mathematics with a geometric slant. Their methods continue to be employed today in high school geometry classes. For the classical Greeks, mathematics was all about constructing geometric objects using a straightedge and a compass. If a shape could be constructed in this manner, the Greeks were able to deal with it, and if not, the shape was considered unreasonable. The constructions of shapes were easy to describe: given two points, one can use a straightedge to draw a line connecting the points. Given a point that is the center and another point that indicates the radius, one can use the compass to make a circle.[5] Iterating these two operations can lead to many different shapes. We, of course, are interested in what geometric objects *cannot* be constructed.

Three of the most famous construction problems from classical times are depicted in figure 9.4.

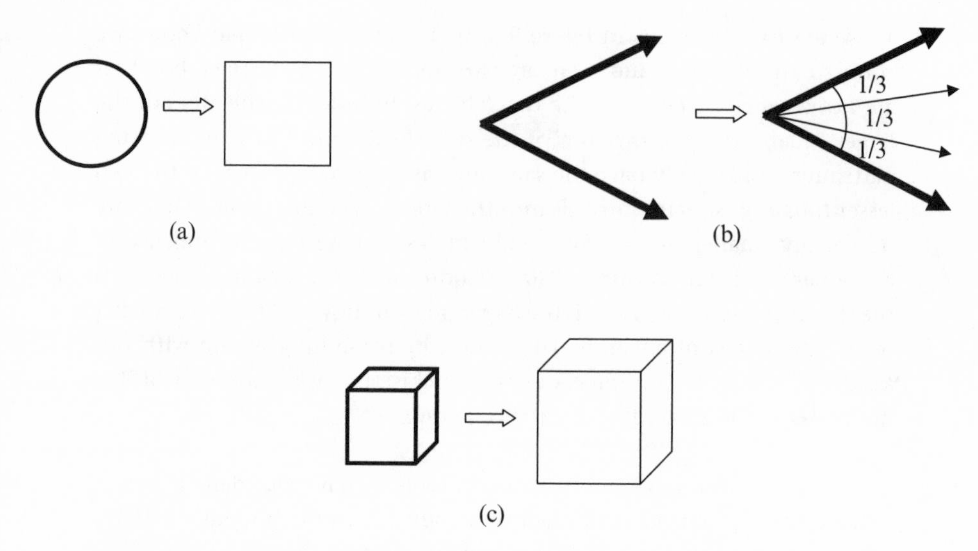

Figure 9.4
(a) Squaring a circle, (b) trisecting an angle, and (c) doubling a cube

The first problem (a) is called *squaring the circle*. Given a circle with a certain area, use a straightedge and a compass to construct a square that has the same area as the circle. In order for someone to perform such a construction, they would need to construct a line segment comparable to the number π. Another famous problem (b) is to take a given angle and split it into three equal parts. This problem is called *trisecting an angle*. The third problem (c) is called *doubling the cube*, and entails taking a cube of a certain size and forming another cube twice its volume. To construct such a cube, one needs to construct a line segment proportional to the cube root of 2. We will soon see that all three of these constructions are impossible using a straightedge and a compass.

The Greeks declared that any number for which a line segment of that size can be constructed using a straightedge and a compass is called a *constructible number* (also called a *Euclidean number*). All whole positive numbers can be constructed. It can be shown that we can multiply and divide using those instruments, hence all rational numbers are constructible. There are, however irrational numbers that are also constructible. For example, if you make a square whose sides are length 1, then the diagonal is the square root of 2, which can be constructed.

Modern mathematicians have defined a larger class of numbers called *algebraic numbers*. These are numbers that are solutions to polynomial equations of the form

$$a_n x^n + a_{n-1} x^{n-1} + \cdots + a_2 x^2 + a_1 x + a_0 = 0$$

where all the coefficients are whole numbers. Since every rational number a/b is the solution to the equation

$$bx - a = 0,$$

every rational number is algebraic. It is known that every constructible number is algebraic. However, there are more algebraic numbers. For example, the cube root of 2 is not constructible, but it is algebraic because it is the solution to

$$x^3 - 2 = 0.$$

However, not every real number is an algebraic number. Real numbers that are not algebraic are called *transcendental numbers*. Such numbers are not solutions to an algebraic equation. In a sense these numbers cannot be described by the usual algebraic operations. If a number is transcendental, then it is not algebraic and definitely not constructible. It turns out that it is very hard to prove that a number is a transcendental number. It was not until 1844 that mathematicians proved that any transcendental number exists. Then, in 1882, Ferdinand von Lindemann (1852–1939) proved that π is transcendental. This shows that π cannot be constructed, and hence it is impossible to square the circle using a straightedge and a compass. It has also been proved that it is impossible to trisect an angle and double a cube.

Very few numbers are known to be transcendental. One might be tempted to say that since there are so few known examples, few exist. But a short counting argument shows this premise is totally wrong. Consider the hierarchy of types of numbers in figure 9.5.

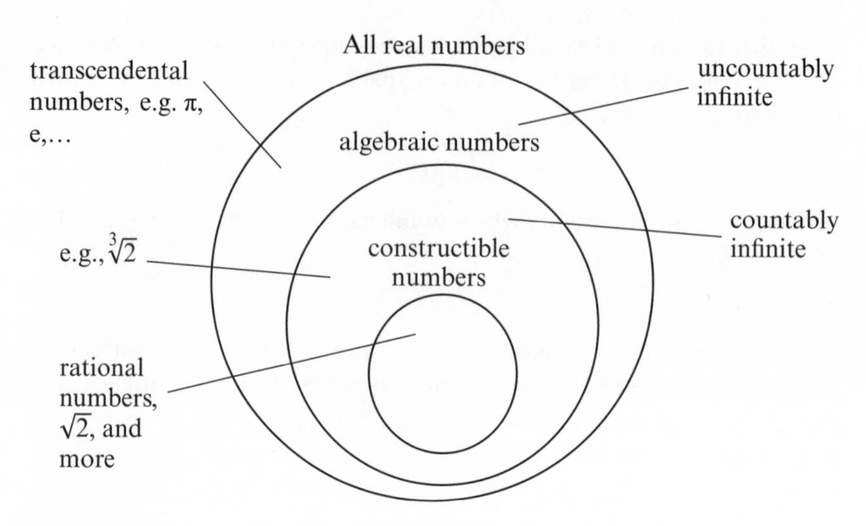

Figure 9.5
Different types of real numbers

Every algebraic number has some integer polynomial equation for which it is a solution. Using a clever counting technique (like the zigzag theorem or the necklace theorem that we saw in section 4.2), one can show that the set of integer polynomial equations is countably infinite. This shows that there is only a countably infinite number of algebraic numbers. Therefore, since there is an uncountably infinite number of real numbers, the numbers that are transcendental real numbers are uncountably infinite. Another way of looking at this is that the numbers we can describe with the usual operations—algebraic numbers—are countably infinite, but there are vastly more numbers that cannot be described by the usual operations. From this we can conclude that the quantity of shapes and numbers that cannot be constructed is vastly larger than the quantity of shapes and numbers that can be constructed.

9.2 Galois Theory

Paris. The night of May 29, 1832. A young man was frantically writing a long letter. He had to write quickly because there was much to say and he knew he would be killed the next day. The letter contained a summation of his mathematical research, and he wanted to put it all on paper before it was too late. He finished the letter with a plea to his friend to "ask [world-famous mathematicians] Jacobi or Gauss publicly to give their opinion, not as to the truth, but as to the importance of these theorems. Later there will be, I hope, some people who will find it to their advantage to decipher all this mess."[6] The following day he fought a duel of honor for the love of a woman, and as he predicted, he was mortally wounded. Brought to the local hospital, he survived one more day. Supposedly, his last words were to his brother: "Don't cry, Alfred! I need all my courage to die at twenty."[7] The young man's name was Évariste Galois and his work will always be a major part of modern mathematics.

What was in this letter? Hermann Weyl (1885–1955), one of the greatest mathematicians of the twentieth century, wrote that "this letter, if judged by the novelty and profundity of ideas it contains, is perhaps the most substantial piece of writing in the whole literature of mankind."[8] Perhaps Weyl was being a little bombastic in his assessment. Nevertheless, Galois's work contains ideas that are essential for modern mathematics and physics. What can a twenty-year-old possibly have to say that is so important?

Born in 1811, during post–French Revolutionary fervor, Galois had a short and tragic life. His father was the mayor of a small town outside Paris who committed suicide after a bitter political dispute. Évariste was a passionate and complicated youth who was not an easy person to deal with. At a young age he became obsessed with mathematics to the exclusion of his other studies. He failed to get into the École Polytechnique, the most prestigious school in France. He eventually was admitted to a second-tier school, but his brilliance was mostly misunderstood by his teachers. Galois submitted two articles for publication and both were supposedly lost by the editors. At some point he became involved with radical political groups, which led to his expulsion from school. It is not known who the other player was in the fatal duel. The identity of the woman the duel was fought over is also unknown. One can only speculate about the other works this young genius might have accomplished if his life had not been tragically cut short.

Galois's work dealt with solving polynomial equations. Before we can understand his contributions, we have to study some history. Consider the following simple equation:

$$ax + b = 0.$$

Such an equation is called a "linear" equation, and most ninth-grade students know how to solve for x:

$$x = -b/a.$$

More complicated, "quadratic" equations take the following form:

$$ax^2 + bx + c = 0.$$

The solutions to such equations were known in ancient times and are still taught to high school students using the "quadratic formula." There are actually two solutions:

$$x_1 = \frac{-b + \sqrt{b^2 - 4ac}}{2a}$$

and

$$x_2 = \frac{-b - \sqrt{b^2 - 4ac}}{2a}.$$

Note these formulas use addition, subtraction, multiplication, division, squares, and square roots.

What about "cubic" equations such as

$$ax^3 + bx^2 + cx + d = 0?$$

Are there standard formulas to solve such equations? In the sixteenth century Gerolamo Cardano[9] showed that there are three solutions and that they are given with fairly complicated formulas.[10] The solutions employ the usual operations, including the square root and cube root.

Pressing on, we can ask about a "quartic" equation:

$$ax^4 + bx^3 + cx^2 + dx + e = 0.$$

Lodovico Ferrari (1522–1565) and Niccolò Fontana Tartaglia (1499–1557) found solutions for such problems. Many readers would be anxious if we actually wrote down the four formulas for the four possible solutions. Rather than write them, let's just say that the "quartic formula" uses the usual operations, square roots, cube roots, and fourth roots.

What about a "quintic" equation?

$$ax^5 + bx^4 + cx^3 + dx^2 + ex + f = 0.$$

Here things get more interesting. One would imagine that there are "quintic formulas" composed of the usual operations and all root operations up to the fifth power. This is not true! There are no such formulas! In the early nineteenth century, Paolo Ruffini (1765–1822) and Niels Henrik Abel (1802–1829)[11] proved that no such general formulas utilizing the usual operations and root operations exist. This means that there will never be a simple formula that provides the solutions for every single a, b, c, d, e, and f in a quintic equation. This is another clear example of a limitation of mathematics.

In general, the problem is unsolvable. Nevertheless, there are solutions for certain quintic equations that are easy to find. For example,

$$x^5 - 1 = 0$$

has a solution at $x = 1$.

This is the heart of Galois's work. He was able to use the coefficients of a given quintic equation to determine if the equation would be solvable with the usual operations. To this end, Galois introduced the notion of a *group*. A group is a mathematical structure that models the idea of symmetry. Galois showed how to associate a group with every equation. With these symmetries, he was able to determine if the given quintic equation would be solvable by means of the normal operations. Once his work for solving quintics was understood, it was used in many other areas of mathematics and science.

The idea of describing symmetries involved a major revolution in modern mathematics, chemistry, and physics. Much of modern mathematics and the sciences studies different forms of symmetry and hence different types of groups. It is from this point of view that we can finally understand Weyl's statement about the importance of Galois's letter: modern mathematics and science extensively use concepts that Galois introduced.

We would be in way over our heads if we were to get into the nitty-gritty of how *Galois theory* actually works. Suffice it to say that first the symmetries of a mathematical or physical system are described. With this in place, researchers make sure that the symmetries are preserved with different operations or physical laws. The fact that a system cannot violate its symmetries can be seen as limitations of the system.

The classical unsolvable problems of constructions with straightedge and compass that we met in the last section can all be proved unsolvable using Galois theory. An additional problem we have not discussed asks whether a polygon for which each edge has the same length—called a *regular polygon*—can be constructed. An equilateral triangle and a square can be constructed. What about a pentagon? Or an arbitrary n-sided regular polygon? Galois theory tells us exactly which n-sided regular polygons can be constructed using a straightedge and a compass. So if

$$n = 3, 4, 5, 6, 8, 10, 12, 15, 16, 17, 20, 24, \ldots, 257, \ldots \text{ or } 65{,}537, \ldots$$

then an n-sided regular polygon is constructible. By contrast, if

$n = 7, 9, 11, 13, 14, 18, 19, 21, 22, 23,$ or $25, \ldots$

then an n-sided regular polygon cannot be constructed.

A fun place to see the limitations that Galois theory describes is in the classic children's fifteen-piece puzzle. This well-known puzzle has fifteen little pieces on a four-by-four grid. One is permitted to move the pieces one at a time. The goal is to get the pieces in order, as in the right-hand diagram in figure 9.6. However, there are some starting configurations that cannot lead to the ordered configuration. When the 14 and 15 are simply swapped, there is no way to get to the ordered configuration. There is also no way to get from the ordered configuration back to the original.

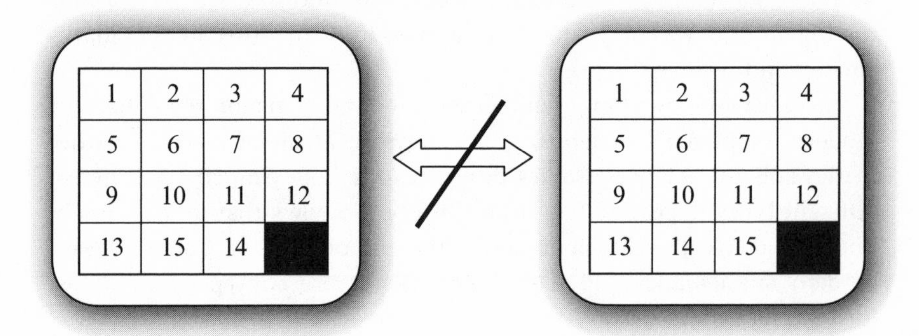

Figure 9.6
An impossible manipulation

In fact, there are fifteen factorial (15!) possible configurations. Exactly half of them are called "even permutations," while the other half are called "odd permutations." The usual movements of pieces have a certain symmetry that takes even permutations to even permutations and odd permutations to odd permutations. These are the symmetries of the system. The fact that one of the configurations in figure 9.6 is an even permutation and the other is an odd permutation ensures that no number of legal moves will get us from one configuration to the other.

Another fun place to observe an impossibility related to Galois theory involves Rubik's Cube. Take a completed Rubik's Cube and (unnaturally) twist one of the corners. Mix it up more and then let a poor unsuspecting friend (who has not read this book) try to solve the puzzle. It cannot be done. One twist makes it impossible to solve, no matter how many moves are performed.

In summary, the Galois theory of equations shows the inherent limitations of the usual operations of multiplication, division, exponentiation, and roots in solving equations. Mathematicians have, over the years, developed other techniques that use calculus and infinitary methods to solve some of these problems. So Galois theory demonstrates that some problems cannot be solved *using certain methods*. Similarly, all n-sided regular polygons can be constructed if you are permitted to measure exact sizes with the straightedge. The fifteen-piece puzzle can be easily solved by removing the pieces from the puzzle and putting them back correctly. A Rubik's Cube can always be solved by cheating—that is, by taking it apart and putting it back in order. These are all simple tricks for getting around the mathematical limitations described by Galois theory.

9.3 Harder Than Halting

Imagine you land a job working for a construction contractor helping customers design their future dream kitchens. Everything is going fine until the wife of a millionaire walks in and wants to change her kitchen floor. She does not want the usual pattern of squares. She wants something totally different. She wants to only use circles. You show her that circles will not work since this will leave gaps that cannot be filled (as in figure 9.7).

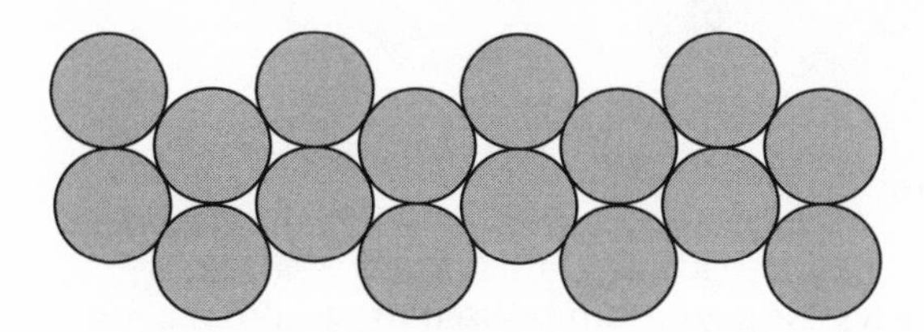

Figure 9.7
Circles are not good for tiling.

What about pentagons (figure 9.8)?

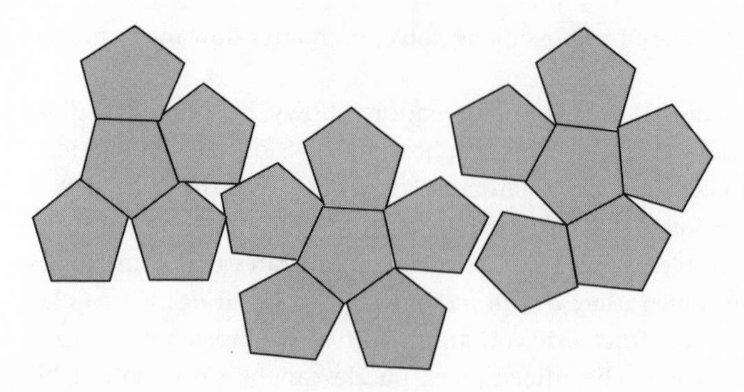

Figure 9.8
Pentagons are not good for tiling.

Pentagons will also not work, but you try to sell her on hexagons (figure 9.9).

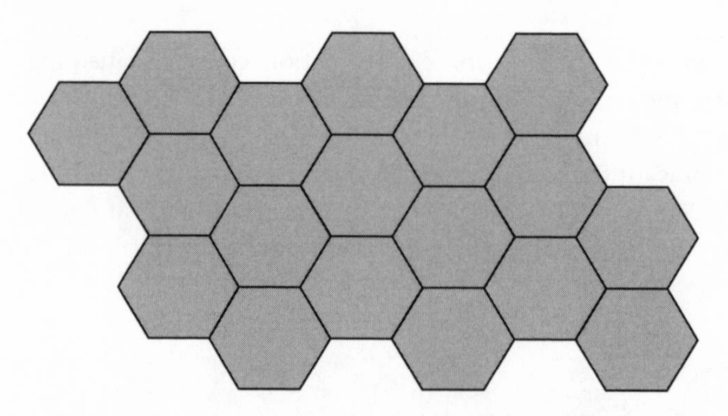

Figure 9.9
Hexagons are good for tiling.

There are no empty spaces. Hexagons can be used to tile a room. Some shapes work, and some do not work. Figure 9.10 has two other tilings that use a single shape.[12]

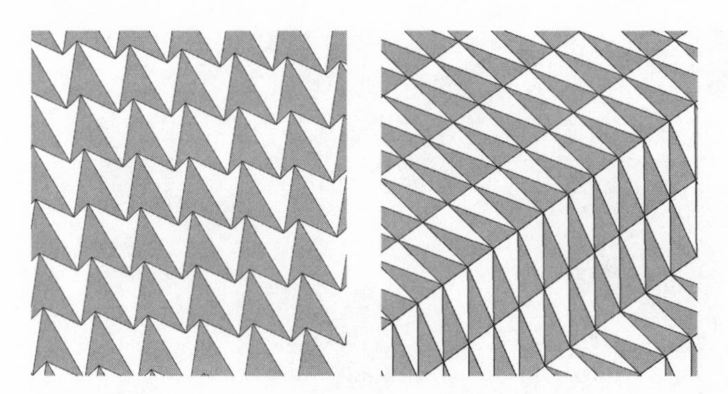

Figure 9.10
Other tilings with a single shape

There are obviously many more different shapes that can tile without leaving any gaps. The famous Dutch artist M. C. Escher (1898–1972) made some great etchings of strange shapes that fit together perfectly to tile without gaps.

Consider the weird shape in figure 9.11 called the Myers shape.

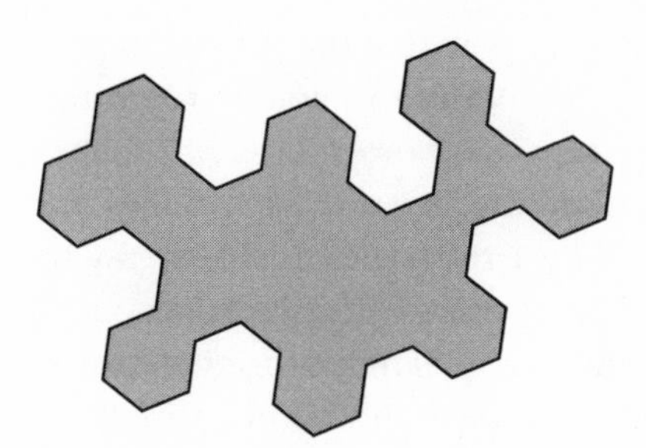

Figure 9.11
The Myers shape

One might think that there is no way such a strange tile can cover a floor without leaving gaps. But in fact it is perfectly legitimate. As figure 9.12 shows, it can be done.

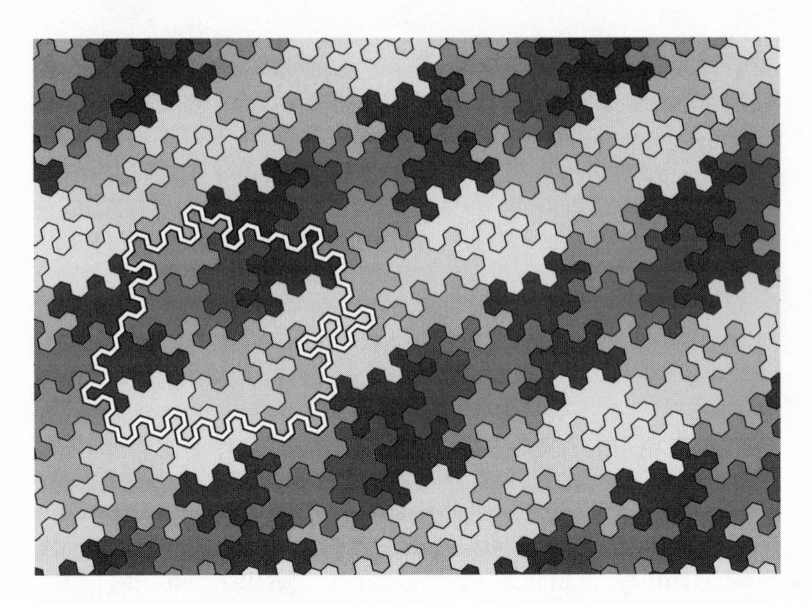

Figure 9.12
A tiling using the Myers shape

Most of the shapes that we have seen tile in a way that the patterns repeat themselves. Such tilings are called *periodic*. Within a periodic tiling the same patterns show up over and over. There are, however, certain shapes that have tilings that do not have a repeating pattern. Such tilings are called *nonperiodic*. Consider the simple 2-by-1 rectangle. It is very easy to make periodic tilings using this shape. However, we can also make nonperiodic tilings with the rectangle. Putting the rectangles together makes squares, which can be placed vertically or horizontally as in figure 9.13. Since any pattern can be made like this, it is easy to make nonperiodic patterns.[13]

One can go further with nonperiodic patterns. There are certain sets of shapes that when you tile with them they are *never* periodic. In other words, the only patterns that can be made with these shapes are nonperiodic patterns. Such shapes are called *aperiodic tiles*. Roger Penrose discovered two such sets of shapes. One pair of shapes are "kites" and "darts" and the other pair are called rhombuses (see figure 9.14).

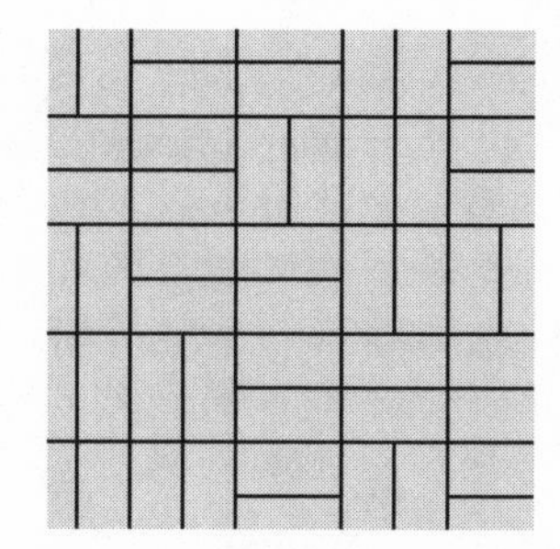

Figure 9.13
Two examples of nonperiodic tilings

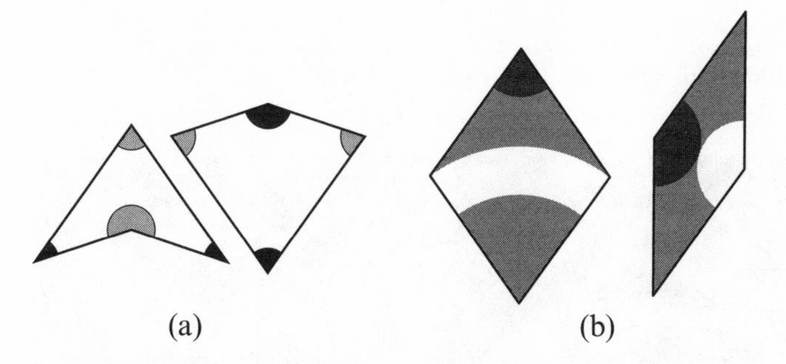

Figure 9.14
Penrose tiles: (a) kite and dart and (b) rhombuses

These shapes have different colors. When the shapes are matched up such that the colors match, the patterns formed will not be periodic. Figures 9.15 and 9.16 are examples of such nonperiodic tilings.

Let us go back to your job helping people tile their kitchen floors. It would be wonderful if there were some way of entering different sets of shapes into a computer that would be able to tell if those shapes are able to tile a large floor (don't worry about the ends) without gaps. We call this task of accepting shapes and determining if they are good for tiling the *Tiling Problem*. A computer that could solve the Tiling Problem would answer "No" to circles and pentagons, while answering "Yes" to squares, triangles, hexagons, Myers shape, Penrose's kite and dart, and Penrose's rhombuses. This computer program would be a tremendous help in your job.

Alas, no such computer program can ever exist! In the mid-1960s Robert Berger proved that no computer exists that can solve the Tiling Problem.

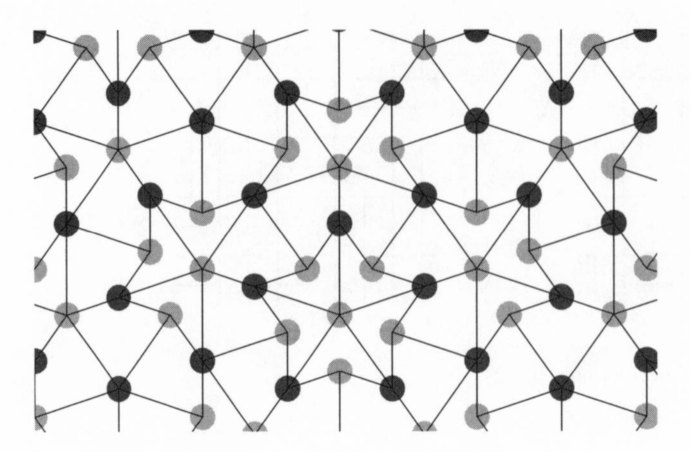

Figure 9.15
A nonperiodic tiling with kites and darts

Figure 9.16
A nonperiodic tiling with rhombuses

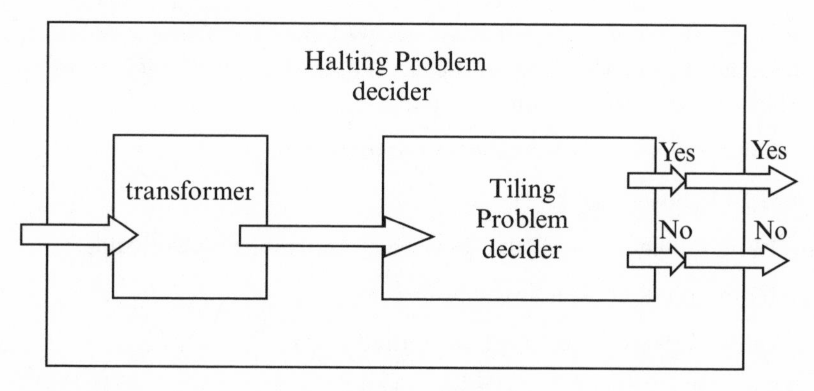

Figure 9.17
A reduction of the Halting Problem to another problem

He proved this by showing that this problem is harder than the Halting Problem, which we met in chapter 6. Remember that the Halting Problem asks whether a computer program will eventually halt or go into an infinite loop. As we saw, it is impossible for any computer to solve the Halting Problem. Since the Tiling Problem is even harder, it too cannot be solved by any computer.

In detail, one can transform any computational process into a set of shapes such that those shapes can tile any plane if and only if the computational process will never halt. That is, the fact that the shapes tile the room is equivalent to the computational process going into an infinite loop. (We saw such transformations in section 6.3.) We might envision this transformation, or reduction, of one problem into another as in figure 9.17.

In the diagram, a program enters on the left and is then transformed into a set of shapes. Assume (wrongly) that there is a computer that can determine whether a set of shapes can be a good tiling. Then we would have a way of deciding if any program will go into an infinite loop or will halt. Since we already know that there is no possible way to solve the Halting Problem, we know that there is no possible way of deciding the Tiling Problem.

Decision problems like the Tiling Problem, which a computer can never solve, are called *undecidable problems*. Although these problems are clearly defined and have objective answers, there is no way any computer can ever solve them.

It is worth emphasizing that we showed that the Tiling Problem is undecidable by piggybacking on the fact that the Halting Problem is undecidable. In section 6.2 we showed that

The Halting Problem is decidable $\Rightarrow$ contradiction.

Figure 9.17 shows that

The Tiling Problem is decidable $\Rightarrow$ the Halting Problem is decidable.

Combining these implications tells us that

The Tiling Problem is decidable $\Rightarrow$ contradiction

and hence the Tiling Problem is undecidable.

Deciding whether a specific set of shapes can tile a floor is only one of many problems that are harder than the Halting Problem and hence undecidable. There are many more such problems. I will examine two of them.

We saw in the last section that Ruffini and Abel proved that there is no general method for solving a polynomial equation with one variable that is to the fifth or higher power. Here is a related problem: given a polynomial equation with integer coefficients and any number of variables, determine if integer solutions to the equation exist. Equations where we only permit integer solutions are called *Diophantine equations*. Also, we are not looking for the solution to the equation. Rather, we are interested in an algorithm that will decide if an integer solution exists. The challenge to find such a procedure was one of the hard problems that David Hilbert presented to the mathematical world at the beginning of the twentieth century and that came to be known as *Hilbert's Tenth Problem*.

Some examples:

$$x^2 + y^3 = 134$$

has a solution: $x = 3$ and $y = 5$. In contrast, it is easy to see that the equation

$$x^2 - 2 = 0$$

does not have an integer solution. What about a complicated equation like

$$x^4y^3z^7 - 23x^5y^2 + 45x^2 = 231?$$

Is there a solution? I simply do not know. It would be nice if there were a computer program that would solve Hilbert's Tenth Problem. Such a program would accept a Diophantine equation as input and reveal if any integer solutions exist.

In 1970, a twenty-three-year-old Russian mathematician named Yuri Matiyasevic—building on earlier work of Martin Davis, Hilary Putnam, and Julia Robinson—finally proved that no computer program that solves this problem can exist. That is, it is undecidable to tell if any given Diophantine equation has a solution.

Without getting into the nitty-gritty details of the proof, let us try to provide an intuition. Basically the mathematicians showed that for a given computation, one can devise a Diophantine equation such that the computation will halt if and only if the equation has integer solutions. In terms of figure 9.17, the inner decider would be a Diophantine equation decider. If there were a mechanical way to decide if the Diophantine equation has a solution, then there would be a mechanical way to decide the halting problem. In other words, Hilbert's tenth problem is harder than the halting problem. Alas, no method exists to solve the halting problem and hence no method exists to solve Hilbert's tenth problem.

Another major problem shown to be undecidable is called the *word problem for groups*. Groups are mathematical structures that were first formulated by Galois, whom we met in the last section. These structures express symmetries and show up everywhere . . . including your bedroom. To get a feel for what groups are all about, let's step into the bedroom for a few minutes and meditate on a mattress. Proper care of a mattress demands that it be reoriented every few months. There are actually three basic ways of reorienting a mattress, as depicted in figure 9.18. A mattress can be flipped along its long edge (L), flipped along its short edge (S), or simply rotated (R).

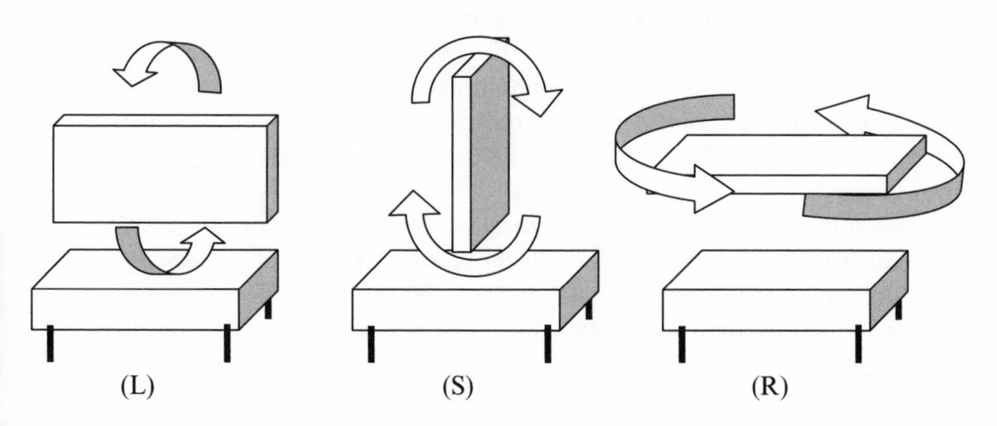

Figure 9.18
Three basic ways to reorient a mattress

These different actions bring the mattress back to its normal position and express a symmetry of the mattress. There are still more reorientations that one can do with the mattress. If S denotes flipping the mattress clockwise on the short edge, let S′ denote flipping it counterclockwise on the short edge. There will similarly be reorientations L′ and R′. Given these different reorientations, we can combine them by first performing one action and then another. For example, first flip the mattress along the long edge (L) and then perform a rotation (R′). We denote this combined action as LR′.

Consider some reorientations. It is obvious that LL means just flipping it twice along the long edge. This will return the mattress to its original orientation. Similarly, R′R will rotate the mattress counterclockwise and clockwise to bring it back to its original orientation. Slightly less obvious is that LRS will also bring the mattress back to its original orientation. (Do not strain yourself with a heavy mattress just to discover mathematical truths. The facts can easily be proved using a light book.) LRL will not bring the mattress back to its original position. After a few minutes, you can see that

SR′R′RRRLLL′SRLLL

also brings you back to the original position. What about

R′SLR′R′L′SL′RLS′S′L′S′L′SRLS′L′R′LSR?

Given enough time and patience, you can actually sit down and determine if the mattress will go back to its original position. Such a problem, where you are given a sequence of operations or a word in the group and asked whether doing these operations will return you to the original position, is called a *Word Problem* for the group. It turns out that the word problem for this trouble-free group of mattress orientations is decidable. That is, one can sit down and write an algorithm that indicates if the given word gets back to the original orientation. Is this true for other groups?

In the mid-1950s, Russian mathematician Pyotr Novikov (1901–1975) and American mathematician William W. Boone (1920–1983) proved that there are certain groups for which there are no algorithms or computers that can solve their Word Problems. For these groups the Word Problem is undecidable.

Once a few problems have been shown to be undecidable, many others can also be shown to be undecidable. To show that a new problem is undecidable, one has to show that an undecidable problem can be transformed into it. One might say that a problem that is harder than a problem that is already known to be harder than the Halting problem is harder than the Halting Problem. In particular, since group theory is at the core of large chunks of modern mathematics and physics, there are a lot of problems within those areas that are undecidable.

Since there are many aspects of modern physics that use group theory in a fundamental way and there are many problems in group theory that are undecidable, there will be many aspects of modern physics that are undecidable. There are decision problems regarding the physical world that no computer could ever answer. One aspect of science is the ability to predict or reveal certain facts about the physical universe. This undecidability makes predictability impossible. However, we must be cautious here. While it is true that many physical theories can be expressed in the language of group theory (and other mathematical structures that have undecidable problems associated with them), if these mathematical structures were the only way of expressing these theories, then we would be right in saying that these physical theories are undecidable. However, there might be other ways of expressing certain physical theories that are not plagued with such undecidable problems.

A case in point is geometry. An elementary school child knows that one needs to do a lot of basic arithmetic to get results in geometry. As we will see in the next section, basic arithmetic has certain limitations. One might conjecture that since there are limitations in basic arithmetic and geometry is expressed using basic arithmetic, there are certain limitations on geometry. However, this is not true! There are ways of expressing basic facts of geometry without using arithmetic and hence geometry does not have these limitations.[14]

Of course, one might object and say that just because a problem is undecidable does not mean that it is beyond human capabilities or beyond reason. Even if a computer cannot solve the problem, there may be other ways of solving it. Perhaps human beings can solve certain problems that a computer cannot. That brings us back to the question of the nature of the human mind. Is it more powerful than a computer or not? Can a human being perform actions that a computer cannot? We have already discussed this question in section 6.5 while pondering the Halting Problem. Although we did not come to any firm conclusions in that section, whichever answers one arrives at for the Halting Problem will be the same for the undecidable problems discussed in this section.

9.4 Logic

Logic is the language of reason. Its roots reach back to ancient Greece, where philosophers realized that certain forms of reasoning used common constructions. To understand the structure of reason, they studied these recurring patterns. In the nineteenth century, mathematicians and logicians turned to the structure of proofs in mathematics and studied their patterns. They set up axiom systems to put mathematics on a firm foundation.

An Italian logician named Giuseppe Peano (1858–1932) created a simple axiom system to describe the basic arithmetic of natural numbers. This system came to be known as *Peano Arithmetic*. The axioms of this system are as follows:

1. There is a natural number 0.
2. Every natural number, a, has a successor denoted $a + 1$.
3. There is no natural number whose successor is 0.
4. Distinct natural numbers have distinct successors: if $a \neq b$, then $a + 1 \neq b + 1$.
5. If the number 0 has a certain property, and any natural number a has the property implies that $a + 1$ also has the property, then every natural number has the property.[15]

Hilbert and others showed that these few axioms describe most of the properties of the natural numbers.

Logicians then went further by using symbols to characterize these logical systems. Axiom 2, for example, can be written as

$$\forall a \exists b \; b = a + 1$$

and axiom 4 can be written in symbols as

$$\forall a \forall b \; (a \neq b) \rightarrow (a + 1 \neq b + 1).$$

Using such symbols, every statement and proof in basic arithmetic can then be turned into sequences of symbols. We refer to this conversion of mathematics into symbolic statements as *symbolization*.

Kurt Gödel, impressed by the power of self-reference, showed that mathematics can also have self-reference. By converting the symbolic language into mathematical language, he was able to complete the loop and have mathematical statements talk about themselves (see figure 9.19). Gödel's method gives a number to every logical symbol, statement, and proof. Since the logical statements that were about numbers also have numbers, we have mathematics dealing with mathematics, or numbers talking about

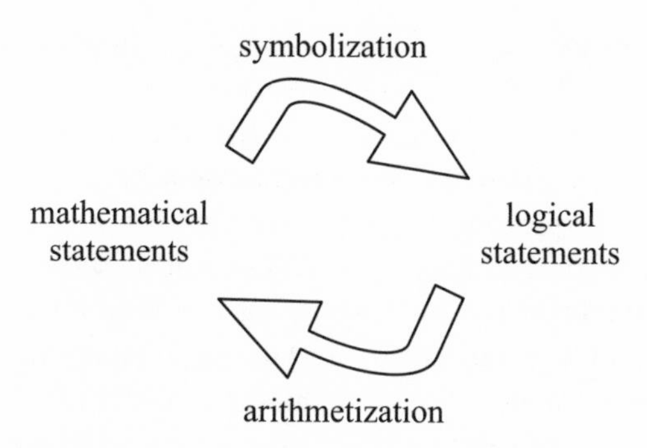

Figure 9.19
Getting mathematics to be self-referential

numbers. The logician Emil Post (1897–1954) summed it up nicely with the statement that "Symbolic Logic may be said to be Mathematics become self-conscious."[16]

This process of converting logical symbols, statements, and proofs into numbers is called *arithmetization.* While it is not important for us to know exactly how Gödel assigned numbers to logical structures, we can sketch some of his ideas. Since there are a finite number of symbols, we can assign each symbol a number: to every symbol x, we can assign a number "x." So for example "$\rightarrow$" = 1, "$\exists$" = 2, "V" = 3, etc. Once every symbol is given a number, then a statement, which is a sequence of symbols, can be assigned a unique number by making each symbol an exponent of a prime number and multiplying them. For example,

$$\forall a \exists b\ b = a + 1$$

will be assigned the number

$$2^{``\forall"}3^{``a"}5^{``\exists"}7^{``b"}11^{``b"}13^{``="}17^{``a"}19^{``+"}21^{``1"}$$

where the base numbers are the prime numbers in order. This number is unique in the sense that distinct formulas will be given distinct numbers. We can take this arithmetization or encoding of symbols to the level of proof. Since a logical proof consists of a sequence of statements, we can assign a unique number to every proof. Moreover, these encodings are "mechanical" in the sense that given a symbol, statement, or proof, we can easily find the number that corresponds to it. Similarly, given a number, we can mechanically find the logical symbol, statement, or proof that corresponds to it.[17]

The number generated for even the smallest proof will be humongous. For larger statements or proofs, these numbers will quickly become greater than the number of particles in the universe. This need not concern us, for we are never going to run out of natural numbers. Besides, our goal is not really to calculate these numbers. The simple idea is that every proof can be given a number and that makes self-reference possible. Exactly which number is generated is uninteresting.

We are dealing with basic arithmetic and we would like to describe certain sets of numbers in a symbolic way. Predicates are like functions that accept numbers as input and then output true or false depending on whether the numbers satisfy that property. For example, the property that determines if a number is even or not can be written as

$$\text{Even}(x) \equiv (\exists y)(2y=x).$$

This predicate is true for x if there is a natural number y such that 2 times y is x. There are predicates with two variables. The predicate that determines if x divides y evenly is given as

$$\text{Divide}(x,y) \equiv (\exists z)(xz=y).$$

That is, x divides y evenly if a z exists such that x times z equals y. One can use predicates to formulate other predicates. For example, the predicate that determines if a number is prime or not is

$$\text{Prime}(x) \equiv x>1 \wedge (\forall y)(\text{Divide}(y,x) \rightarrow (y=1 \vee y=x)).$$

In English, this says that x is a prime number if x is greater than 1, and if any number divides x then that number is either 1 or x itself.

The final ingredient that we will need in order to have self-reference is something called a *fixed-point machine*. This is a way of turning any predicate into a self-referential statement. For any predicate $F(x)$ that depends on one number, there is a logical statement C such that C is going to be equivalent to $F(\text{“}C\text{”})$. In symbols this is

$$C \equiv F(\text{“C”}).$$

In words, C is a statement that says

“This logical sentence has property $F(x)$.”

or

“I have property $F(x)$.”

C is called a “fixed point” because $F(x)$ is thought of as a function and generally the output of a function is different from the input. Here the input is “C” and the output is equivalent to the input. It is “fixed.” This means that the sentence is going to talk about itself. It would take us too far afield to show exactly how this fixed-point machine works and how C is actually constructed. Suffice it to say that the fixed-point machine works very similarly to the diagonalization proofs done in chapters 4 and 6. All of these are ways of describing self-reference.

Now let's use our fixed-point machine to make some interesting logical statements.

Tarski's Theorem

If our goal is to get a formula analogous to the liar paradox using math and logic, then let us assume that a predicate $Truth(x)$ exists that expresses

$Truth(x) \equiv x$ is the Gödel number of a true statement in Peano Arithmetic.

Use this supposed predicate and form:

$F(x) \equiv \sim Truth\ (x)$.

In other words, $F(x)$ will be true when x is the number of a statement that is not true. Place this $F(x)$ into our fixed-point machine and get a logical sentence T such that

$T \equiv F(\text{“}T\text{”}) = \sim Truth\ (\text{“}T\text{”})$.

T says that T is not true. Or in other words, we formed the logical sentence that says

"This logical sentence is not true."

or

"I am false."

Such a sentence is called a *Tarski sentence*. Let's analyze this logical sentence. If this sentence is true, then since it states that it is false, it is false. On the other hand, if the sentence is false, then since it states that it is false, it is, in fact, true. We have a genuine contradiction. Such contradictions are not permitted in the exact world of mathematics and logic! What went wrong? The only part that was at all troublesome was the assumption that a logical predicate $Truth(x)$ exists.

The existence of $Truth(x) \Rightarrow$ contradiction.

Since we do not permit contradictions, we must have that $Truth(x)$ cannot exist. That is, we just proved a limitation of basic arithmetic: it cannot determine when its own statements are true or not. For purely logical sentences without arithmetic, we determine the truth or falsity of a statement with a truth table. But here we are not dealing with pure logic. Here we are dealing with mathematics and logic. And for that, it is not possible to use a truth table. Self-reference has handed us a limitation.

The Tarski statement is analogous to the famous liar paradox. Whereas the liar paradox uses English sentences, this uses exact logical sentences. We might be flippant about the meaning of English sentences and say that the liar paradox is like many other English sentences that are meaningless and/or contradictory. But we are going to have to be more careful about logic.

Gödel's First Incompleteness Theorem

We come now to one of the most celebrated theorems in twentieth-century mathematics. It expresses a shocking limitation of mathematics and logic.

Rather than examining truth, as in Tarski's theorem, let's examine provability. Let $Prove(y, x)$ be a predicate that is true whenever

"y is the Gödel number of a proof of a statement whose Gödel number is x."

Since our language is so exact and our encoding is so mechanical, this predicate actually exists. Unlike the *Truth* predicate, which does not exist, one can actually sit down and describe the *Prove* predicate. In broad strokes, the predicate must look at the number y and determine that it is the number of a proof. It then must verify that this proof actually proves the statement whose number is x. It would be extremely painful to go through all the steps, but it is done in many logic textbooks. Suffice it to say that, unlike the *Truth* predicate, the *Prove* predicate actually exists.

With the *Prove* predicate in hand, we can form the predicate

$$F(x) \equiv (\forall y) \sim Prove(y, x).$$

That is, $F(x)$ means that every single number y is *not* a proof of the statement x. In other words, $F(x)$ is true if and only if no proof exists for the logical statement whose Gödel number is x. A fixed point for this $F(x)$ is a sentence G such that

$$G \equiv F(\text{“}G\text{”}) = (\forall y) \sim Prove(y, \text{“}G\text{”}).$$

G says that every single number y is not a proof of G. In English G says

"This logical statement is not provable"

or

"I am unprovable."

Such a sentence is called a *Gödel sentence*. Let's analyze this statement. If G were provable, then since the statement says it is not provable, we proved a false statement. A good logical system like Peano Arithmetic cannot prove false statements. Let's assume then that the Gödel statement is *not* provable. That is exactly what the sentence says and hence makes it true. So the statement is true but not provable.[18]

We just showed that

The Gödel sentence is provable $\Rightarrow$ contradiction.

We can conclude that the Gödel sentence is unprovable and hence true. But we have to be a little careful here. It could be (although I assure you it is not) that Peano Arithmetic is inconsistent and can prove anything, including the Gödel sentence. While I am firmly convinced that Peano Arithmetic is consistent, we still have to state this assumption very carefully:

"If Peano Arithmetic is consistent, then the Gödel sentence is (unprovable and) true."

We will revisit this statement in the next section.

Gödel's amazing result deserves much reflection. The "obvious" belief before Gödel came along was that for simple systems like arithmetic, whatever was true was also provable. That is, if a statement is true there must be some proof of it. This is shown in the left-hand diagram in figure 9.20.

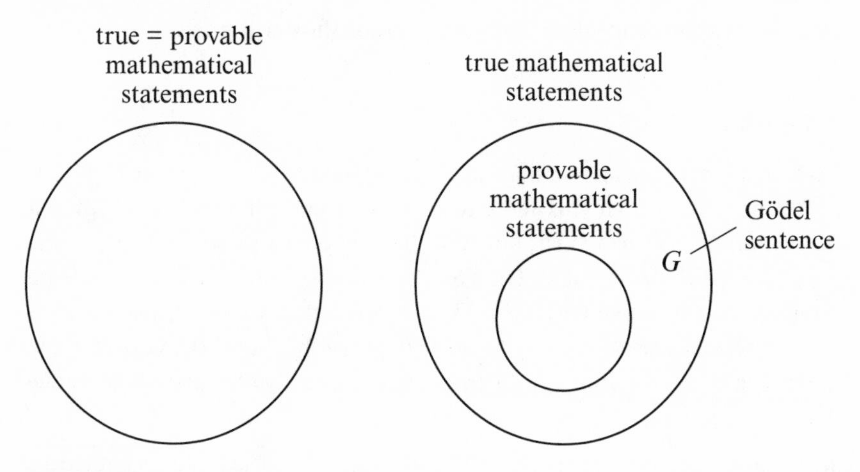

Figure 9.20
The "obvious" view and the correct view

Gödel showed that this view was wrong. There are sentences like G that are true but not provable. (We will see at the end of this section that G is not the only statement of this type.)

Parikh's Theorem

Rohit Parikh took Gödel's ideas further by pointing out some of the stranger aspects of the nature of proof. Consider the following predicate:

$Prooflen(m, x) \equiv m$ is the length (in symbols) of a proof of the statement whose Gödel number is x.

For a given m and x, there is no problem determining if this predicate is true or false: simply go through all proofs of length m (there are a large but finite number of them) and see if any of the proofs end with a statement whose Gödel number is x. If such a proof exists, then the predicate is true. Otherwise, it is false.

Now, let n be a large number, and consider the predicate:

$F(x) \equiv \sim(\exists m < n\ Prooflen(m, x))$.

This predicate is true if there is no proof of x whose length is shorter than n. Applying the fixed-point machine to $F(x)$ gives us a fixed point P such that

$P \equiv F(\text{“}P\text{”}) \equiv \sim(\exists m < n\ Prooflen(m, \text{“}P\text{”}))$.

That is, P is equivalent to the sentence that says it is false that there exists an $m<n$ such that m is the length of a proof of P. In other words, P says

"This logical sentence does not have a proof shorter than n."

or

"I do not have a short proof."

We call such a logical sentence a *Parikh sentence*.

Let us determine if this sentence is true or false. If P were false then a (short) proof of P *does* exist. But how can there be a proof of a false statement within a consistent system? So the sentence is not false and must be true. As we saw above with Gödel's incompleteness theorem, just because a statement is true, does not mean it is provable. Now let's consider the following relatively short proof that a (long) proof of the Parikh sentence exists:

If the Parikh sentence does not have a proof, then in particular it does not have a short proof. Then we can easily check all proofs less than n and see that none of them prove P. Summing up: if the sentence cannot be proved, then we can prove it.

This proof can be formulated in Peano Arithmetic and is fairly short. A Parikh sentence is an example of a sentence that has a very long proof, but a short proof of the fact that it has such a long proof. Strange but true!

Löb's Paradox

For our final use of the fixed-point machine we are going to go off the deep end and prove that every logical sentence—no matter how crazy—is provable. In particular, we will show that, as we long suspected, the moon *is* made of green cheese!

We saw that if A is a logical statement then "A" is the Gödel number that corresponds to A. One of the requirements of the arithmetization process is that we can also go the other way: given a number x that corresponds to a logical statement, we will refer to its corresponding logical statement as $|x|$. If we start with a logical sentence A, look at its number "A", and then look at the number's logical sentence $|"A"|$, we return to the same logical sentence A. In symbols $|"A"|=A$.

Let M (for moon) be any logical sentence. We will prove that M is always true.

Consider the predicate

$$F(x) \equiv |x| \rightarrow M.$$

$F(x)$ is true only if the logical sentence that corresponds to the number x implies M. Use the fixed-point machine on this predicate. A fixed point looks like

$$L \equiv F(\text{"}L\text{"}) = (|\text{"}L\text{"}| \rightarrow M) = (L \rightarrow M) .$$

In other words, L says

"This logical sentence implies M"

and is called a *Löb sentence*. Assume, for a moment, that L is true. Then since L is the same as $L \rightarrow M$, our assumption implies that $L \rightarrow M$ is also true. Since both L and $L \rightarrow M$ are true, then, by modus ponens, M is true. So, by assuming that L is true, we have proved that M is true. In other words, we have proved that L implies M and hence $L \rightarrow M$ is true. But $L \rightarrow M$ is equivalent to L. So L is true. We conclude that M—which stands for the fact that the moon is made of green cheese—is true.

We just proved that the moon is made of green cheese. This is ludicrous! What went wrong? The problem arises because we did not put a restriction on the formulas $F(x)$ for which we are permitted to use the fixed-point machine. We saw in section 4.4 that in order to avoid Russell's paradox of set theory, we had to restrict the axiom of comprehension. We also saw in section 2.1 that some researchers insist that to avoid linguistic paradoxes we must adhere to a hierarchy of permitted English sentences. We must stay away from certain English sentences to steer clear of contradictions. In a similar way, here we must restrict the fixed-point machine in order to avoid proving false statements. Such a restriction might seem strange because the proof that the fixed-point machine works seems applicable to all $F(x)$. But restrict we must lest we go beyond the bounds of reason.

Although we have stated all these limitations in the language of Peano Arithmetic, such phenomena can be described in much more general terms. Peano Arithmetic has all the usual arithmetic operations to code and decode logical statements as numbers. However, there are many other systems that allow one to talk about statements as numbers and numbers as statements. Once any such encoding is possible, we are going to have self-reference and hence similar limitations. It is worthwhile to state Gödel's incompleteness theorem in this general form:

In any "nice" logical system that has enough structure so that it can encode and decode its own statements, there will be self-reference and hence limitations. In particular, statements exist that are true but unprovable.

There are systems of logic that are very "weak" in that they cannot encode their own statements. These systems will not have self-referential limitations and hence Gödel's incompleteness theorem will not apply to them. Such systems are called *complete* because every statement that is true has a proof. An example of such a complete system is *Presburger Arithmetic,* which is a logical system that only deals with addition and does not have enough power to deal with multiplication and other operations. Another example is a certain logical system that deals with basic geometry, which is also not powerful enough to encode its own statements. There is something a bit paradoxical here. Systems that are so weak that they cannot encode their own statements have the strength that everything that is true is provable. In contrast, a system that has the strength to encode its own statements has the weakness that it has true but unprovable statements.

In this section we have described several true but unprovable mathematical statements. One might think that only these few "pathological" mathematical statements are problematic and that most "regular" true mathematical statements can be proved. A little counting argument will show us that this is not so.[19]

Consider all the subsets of the natural numbers. As we saw in section 4.3, there are uncountably infinite such subsets. If S is a subset of the natural numbers and x a natural number, then exactly one of the following two mathematical facts is true:

x is an element of S or x is not an element of S.

Since there are uncountably infinite subsets of the natural numbers, such true facts about numbers are uncountably infinite. This is all about mathematical facts—not about what can be stated. A mathematical statement is a mathematical fact that can be put into symbols. We saw above that arithmetization is a correspondence between mathematical statements and the natural numbers. This implies that there are countably infinite mathematical statements. Hence there are massively more mathematical facts than mathematical statements.

Gödel's theorem takes this further. The entire purpose of his incompleteness theorem is to show that the set of provable mathematical statements is a proper subset of all mathematical statements, as in figure 9.21.

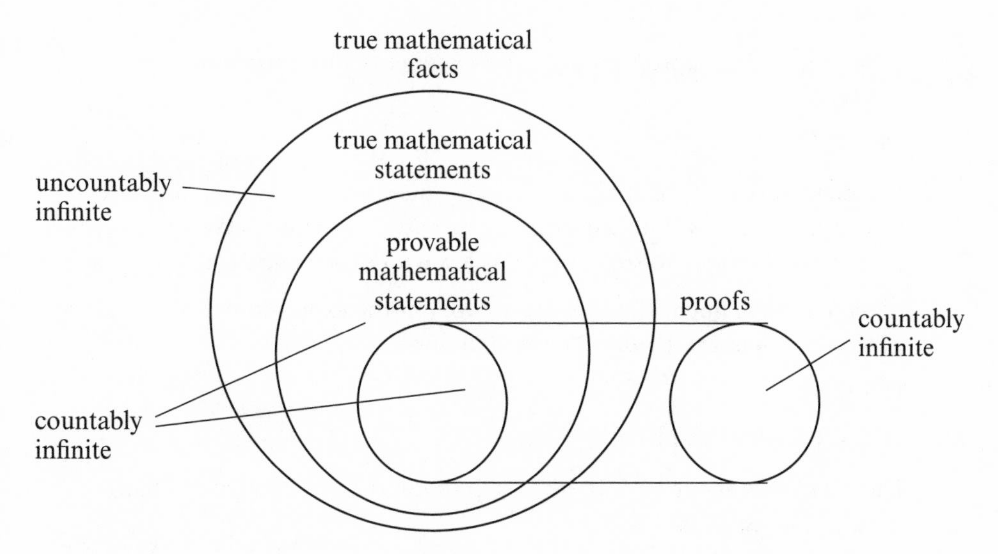

Figure 9.21
True facts, true statements, and provable statements

Since there is a natural number for every proof, the set of provable mathematical statements is also countably infinite. Cristian S. Calude, together with several others, recently proved that even within the set of all true mathematical statements, the ones that are provable are rarities. We conclude that although mathematical proofs can prove a countably infinite number of true statements, there are immensely more true mathematical statements and facts that are unprovable and, hence, beyond rational thought.[20]

Gödel's shocking theorem was that a true but unprovable sentence exists in Peano Arithmetic. Although the Gödel sentence was the first example of such a sentence, there is a feeling that it is a bit "contrived" or "unnatural" and not "real mathematics." As we showed, there are countably infinite statements that are also true but unprovable in Peano Arithmetic. It would be nice to see more such statements that are "natural" and look like "real mathematics." One such statement is *Goodstein's theorem*.

First some background. Any number can be written as a sum of powers of 2. For example,

- $19 = 2^4 + 2^1 + 1$
- $87 = 2^6 + 2^4 + 2^1 + 1$
- $266 = 2^8 + 2^3 + 2^1$

Let's look at 266 more carefully. The exponents can also be written as powers of 2:

$$266 = 2^{2^{2+1}} + 2^{2+1} + 2^1$$

The number is now totally expressed with numbers of 2 or less. This way of writing numbers is called "hereditary base-2 notation." We are now going to perform a two-step process as a way of expressing 266:

- Step (a): Increment the base from 2 to 3 in the expression.
- Step (b): Subtract 1 from the whole number.

This gives us

$$(3^{3^{3+1}} + 3^{3+1} + 3^1) - 1 = (3^{3^{3+1}} + 3^{3+1} + 2).$$

This number is about 10^{38}. The two-step process can be iterated by changing all 3s to 4s:

$$(4^{4^{4+1}} + 4^{4+1} + 2) - 1 = 4^{4^{4+1}} + 4^{4+1} + 1 \approx 10^{616}.$$

And once again:

$$(5^{5^{5+1}} + 5^{5+1} + 1) - 1 = 5^{5^{5+1}} + 5^{5+1} \approx 10^{10,000}.$$

Each time, we are incrementing the base by 1 and subtracting 1. As you can see, the numbers increase wildly. You could imagine that if you continue this process, the numbers will simply continue to grow and grow. Or would they? In 1944, English mathematician Reuben Goodstein (1912–1985) proved the following remarkable theorem:

Take any number, write it in hereditary base 2 notation and then iterate the following two-step procedure: (a) change all the n's in hereditary base n notation to $n + 1$'s and (b) subtract 1. Eventually the result will hit . . . zero!

That is, instead of the number getting bigger and bigger, it eventually—after a very long time—will start going down and will eventually go all the way down to zero. This is shocking. Of the two steps in our procedure, one step massively increases the number and the other step simply subtracts 1. Our intuition tells us that this sequence of numbers will simply get bigger and bigger. Our intuition is wrong! The numbers will "eventually" start getting smaller. The number of iterations this will demand is immense, but it will, nevertheless happen. Goodstein proved this amazing theorem using infinitary methods and the full power of set theory. In 1982, Laurie Kirby and Jeff Paris demonstrated that this theorem can only be proved using such infinitary methods and that although the theorem can be stated in the language of Peano Arithmetic, it cannot be proved in that system. The numbers simply get too big for that system. So Goodstein's theorem, like Gödel's sentence, is true but unprovable in Peano Arithmetic.

9.5 Axioms and Independence

Let's move on and use Gödel's first incompleteness theorem to obtain more results. We found that the Gödel sentence,

"This logical statement is not provable."

is unprovable and hence true. We actually proved that

"If Peano Arithmetic is consistent, then the Gödel sentence is true."

This fact can be formalized and proved within Peano Arithmetic. That means we can write a proof of what we did in the last section within the language of Peano Arithmetic. This proved statement is written as the following implication:

"Peano Arithmetic is consistent" $\rightarrow$ "The Gödel sentence is true."

Assume we can prove that "Peano Arithmetic is consistent" within Peano Arithmetic. Then we would have the following deduction in Peano Arithmetic:

"Peano Arithmetic is consistent"
"Peano Arithmetic is consistent" → "The Gödel sentence is true."
___
"The Gödel sentence is true."

That would be a proof in Peano Arithmetic that the Gödel sentence is true. But Gödel's sentence says that no such proof can exist! There must be something wrong with our assumption. The only statement we assumed was that Peano Arithmetic can prove its own consistency. This leads us to Gödel's second incompleteness theorem: *Peano Arithmetic* cannot prove its own consistency. That is, using basic arithmetic one cannot prove that basic arithmetic is consistent.

It is important to notice that Gödel's second incompleteness theorem was proved piggybacking on Gödel's first incompleteness theorem. The first theorem is the implication:

The Gödel sentence is provable in Peano Arithmetic ⇒ contradiction.

In this section, we showed the implication:

"Peano Arithmetic is consistent" is provable in Peano Arithmetic

⇒ The Gödel sentence is provable in Peano Arithmetic.

Combining these two implications gives us

"Peano Arithmetic is consistent" is provable in Peano Arithmetic ⇒ contradiction.

We conclude that "Peano Arithmetic is consistent" is *not* provable in Peano Arithmetic.

We just showed that the statement "Peano Arithmetic is consistent" cannot be proved in Peano Arithmetic. But is the statement true? Can it possibly be false? Are we really to believe that arithmetic is not consistent? Will someone one day prove that 2 + 2 is not 4? Fear not, dear reader. In 1935, Gerhard Gentzen (1909–1945) proved that the stronger axiom system of Zermelo-Fraenkel set theory with choice (ZFC) *does* prove the consistency of arithmetic. In detail, since we can interpret arithmetic within ZFC, and since that system has the ability to deal with the more powerful ideas of infinity, it can prove the consistency of Peano Arithmetic. In other words, a simple "finitistic" proof of the consistency of arithmetic does not exist but there is an "infinitary" proof of it. Gentzen proved that if ZFC is consistent, then so is Peano Arithmetic.[21] There is one small problem: Who says ZFC is consistent?

Gödel's second incompleteness theorem actually says something much more. There is nothing special about Peano Arithmetic in our discussion. We said that Gödel's first incompleteness theorem was true for any axiom system that can encode its own statements. By using the same reasoning we used at the beginning of this section, we can also extend Gödel's second incompleteness theorem to any axiom system that can encode its own statements. ZFC is such a system. Hence, by Gödel's second incompleteness theorem, ZFC cannot prove its own consistency.

This is rather troubling! Most of modern mathematics can be formulated in ZFC. It would be nice to know that the axiom system of ZFC is consistent and that no contradictions can be found. Alas, no such proof exists within ZFC. Researchers have constructed other, more powerful axiom systems that can prove the consistency of ZFC. These systems consist of the ZFC axioms and potent "axioms of infinities." These new axioms are not used in the typical mathematics. They have an unnatural, ad hoc feel to them. Unlike the usual axioms of ZFC, it is hard to tell if these other axioms are true or not. They are not self-evident. But, even with these new powerful axioms, we do not have a system that can prove its own consistency. One always has to go to a stronger system.

In the last section we saw that Goodstein's theorem is a "natural," "uncontrived" mathematical result that is beyond the ability of Peano Arithmetic to prove but is nevertheless true and provable in ZFC. Analogous to that result, Harvey Friedman has come up with several "natural," "uncontrived" mathematical statements that are beyond the ability of ZFC to prove but are nevertheless true in stronger systems.

Figure 9.22 summarizes some of our findings.

Peano Arithmetic cannot prove the consistency of Peano Arithmetic (Con_{PA}). ZFC can prove the consistency of Peano Arithmetic but cannot prove the consistency of ZFC (Con_{ZFC}). We can always build stronger systems that can prove the consistency of ZFC, but these stronger systems will also not be able to prove their own consistency. We can go on forever. This conforms to one of the central themes of this book: any self-referential system, no matter how powerful, is somewhat limited.

Do not worry that modern mathematics or even basic arithmetic is inconsistent. I assure you that they are consistent. Arithmetic has been around for millennia and no one has ever found an inconsistency. There are millions of people working in modern mathematics and no inconsistency has ever been found. Nevertheless, it is unnerving that proving the

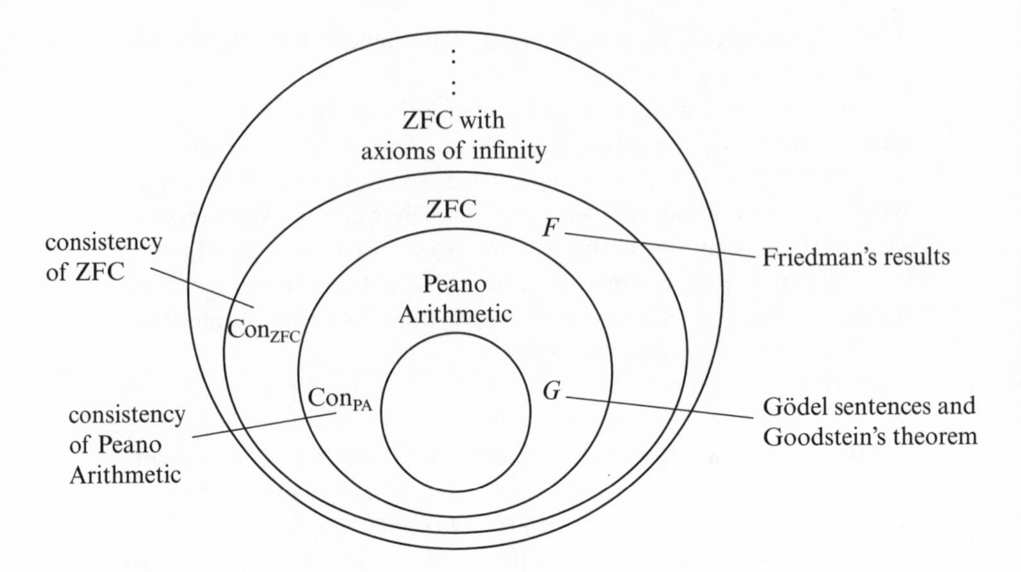

Figure 9.22
A hierarchy of axiom systems

consistency of the logical systems that are at the foundation of mathematics and science is beyond the bounds of reason.

In a sense, one can view Gödel's theorems and their consequences as a criticism of the axiomatic method. Mathematicians have always looked for axioms in order to find the fewest and simplest statements that imply all the other statements. But Gödel has shown us that no matter how many axioms are added, there will always be something missing. That is, there will always be statements that are true but cannot be proved using those axioms. Much of mathematics can be done without axioms. The vast majority of mathematicians do not work with axiom systems. They simply follow the rules they have learned. It is logicians and set theorists who keep their eyes on what assumptions are used. Even Cantor, the founder of modern set theory, did not prove his theorems using axioms. Rather, the axioms of Zermelo Fraenkel set theory were formulated *after* Cantor did his work. Cantor simply used his intuition.

Another criticism of the axiom system is the way that set theorists seemingly add axioms to make stronger and stronger systems beyond ZFC. The usual axioms of Peano Arithmetic and ZFC seem obvious (or as Thomas Jefferson and his colleagues wrote, "We hold these truths to be self-evident"). The Peano Arithmetic axiom that distinct numbers have distinct successors is obvious to any five-year-old. The ZFC axiom that given a set, its powerset also exists, is obvious. In contrast, certain axioms of infinity that set theorists work with are not self-evident or obvious. They are added because they don't seem to contradict the other axioms and are useful in proving certain theorems. This may feel to some like cheating: we are simply adding axioms that we find useful.

On the other hand, without axioms where would we be? We might stumble into contradictions. Although no sufficiently strong axiom system is complete, with axioms we can at least see the hierarchy of different systems. We can say that ZFC is more powerful than Peano Arithmetic because we can compare their axioms. The power of mathematics is that it is firmly based on axioms and not based on mere human intuitions.

There is something slightly disingenuous about the mathematical limitations discussed in this chapter: they are not really limitations, but merely stumbling blocks. There are ways of getting around them. For every limitation *of a mathematical system*, some larger, more powerful system exists that overcomes the limitations:

- We saw that it is impossible to trisect an angle *with a straightedge and a compass*; however it is very easy to perform this task with a simple measuring stick. Measure the angle and divide it by three. With the same amount of ease, we can similarly square the circle and double the cube.
- Galois theory determines when certain equations are unsolvable *using the operations of addition, multiplication, division, and roots*. However, there are calculus-based methods that can always solve such equations.
- Gödel's first incompleteness theorem says that certain statements are true but unprovable *in some finite arithmetical system*. But we saw that the same statements are provable in stronger systems.
- Gödel's second incompleteness theorem says that the consistency of a finite arithmetical system is unprovable *within that system*. Gentzen showed that the consistency is provable in the stronger system of ZFC.
- The consistency of ZFC (as the continuum hypothesis and the axiom of choice that we met in section 4.4) is not provable *within ZFC*. However, there are larger systems in which it is provable.
- For any system larger than ZFC, there are statements that are not provable *in that system*, but there will always be larger systems.

We call these limitations *relative limitations*, as opposed to *absolute limitations* that can never be solved in any mathematical system. Are there any known absolute limitations? That is, are there any mathematical statements that are true but no human being can ever show that they are true no matter how much ingenuity is applied? Note that even if such a statement did exist, we would never be able to prove or know that it was such a statement since to prove a statement is unprovable but true would mean that we proved it is true. So an absolutely unsolvable problem is a problem that we can never prove, and we can never know that we can never prove. This makes the question of their existence somewhat metaphysical since whether or not they exist is something we can never know.

In contrast to mathematical limitations, the computational limitations discussed in chapter 6 and in section 9.3 are, in a sense, absolute. We showed that no computer or mechanistic method can solve certain problems. There are no known better machines that can solve these problems. Even quantum computers, when they come into existence, cannot do more than what regular computers can do. (Their power is their speed at doing what regular computers already can do, not that they can solve unsolvable problems.) The absoluteness of computational limitations comes from the fact that we really know what computers are all about.[22] This is in contrast to our limited knowledge of the nature of human consciousness and reason.

Gödel summarized the past two paragraphs by positing that either (a) absolutely unsolvable problems exist or (b) the human mind with its seemingly unlimited capacity for ingenuity "infinitely surpasses the powers of any finite machine." In other words, if the human mind is just a machine, then it will have the same limitations as machines and finite systems. In that case, (a) will be true and the human mind will also have problems that are absolutely unsolvable. By contrast, if the human mind can always solve any problem, then (b) will be true and the mind must be stronger than any computer or finite axiom system that has limitations.

Let us (somewhat hesitatingly) accept these two choices as the only possible options. Which of these alternatives should we believe? It occurs to me that as we learn more and more about the human brain and as cognitive science continues to make progress in understanding how the mind works, we will have no choice but to accept that (a) the brain/mind works in a mechanical way. The human brain is probably the most complicated machine in the entire universe and we are hundreds of years away from actually understanding how the human brain works. Nevertheless, from what we have seen of the workings of the brain, there does not seem to be any mysterious process that would give the human mind powers to "infinitely surpass" a finite physical process. It seems plausible to accept that there are absolute limitations on the abilities of the human brain, as there are absolute limitations on any computing device and any physical machine. Even in the world of pure mathematics, there are limitations of reason.

Further Reading

Section 9.1

The beautiful proof of the irrationality of the square root of 2 was first shown to me by Leon Ehrenpreis. He heard it from Stanley Tenenbaum, who seems to have developed it himself. Eves 1976, Kramer 1970, and Kline 1980 have more about classical limitations.

Section 9.2

A fascinating biography of Galois can be found in chapter 20 of Bell 1937. While many parts of Bell's book have been shown to be stretching the truth, it is still fun to read. Stewart 2003 is a good place to start learning about Galois theory. For math students, good places to learn about Galois theory include chapter 15 of Birkhoff and Mac Lane 1957 and chapter 4 of Jacobson 1985.

Section 9.3

Goodman-Strauss 2010, 2011, are two beautiful articles with many pictures on tilings. The basics on undecidable problems are included in any textbook on theoretical computer science, such as chapter 6 of Cutland 1980. A nice presentation of Hilbert's tenth problem can be found in Davis and Hersh 1973. The use of mattress reorientations as a way of introducing group theory was shamelessly taken from Hayes 2008. A further discussion of undecidability in physics can be found in Pour-El and Richards 1989.

Section 9.4

The theorems of Tarski and Gödel are covered in any logic book—for example, in section 3.5 of Mendelson 1997, chapters 2 and 7 of Manin 2010, or chapter 17 of Boolos, Burgess, and Jeffrey 2002. A readable and complete proof is included in Van Heijenoort 1967. For a short, modern approach and the latest developments, see Davis 2006, Bunch 1982, and Hofstadter 1979. The amazing result about lengths of proofs can be found in Parikh 1971, 496. All the results are also included in Yanofsky 2003.

Goodstein's theorem was taken from Kirby and Paris 1982.

Section 9.5

You can read about some of the implications of Gödel's theorem in Kline 1980. However, be mindful of coming to wild conclusions with regard to Gödel's incompleteness theorems. Much nonsense has been published on the topic. Torkel 2005 and Berto 2009 provide antidotes.

10 Beyond Reason

The eternal silence of these infinite spaces fills me with dread.[1]

—Blaise Pascal

At the present time I seem to be thinking rationally again in the style that is characteristic of scientists. However this is not entirely a matter of joy as if someone returned from physical disability to good physical health. One aspect of this is that rationality of thought imposes a limit on a person's concept of his relation to the cosmos.[2]

—John Nash

The meaning of the world is the separation of wish and fact. Wish is a force as applied to thinking beings, to realize something. A fulfilled wish is a union of wish and fact. The meaning of the whole world is the separation and the union of fact and wish.[3]

—Kurt Gödel

We have come to the end of our journey. It is time to sum up some of our findings and try to make sense of our explorations. In section 10.1 I categorize the different types of limitations that we have considered. Section 10.2 discusses the definition of reason. I conclude by looking at what, if anything, is beyond the limitations of reason.

10.1 Summing Up

Every chapter of this book has discussed a different subject and its limitations. There are, however, other ways to categorize the myriad limitations we have found. Here I give another classification of four types of limitations on reason.

Physical Limitations

The simplest type of limitation is one that shows reason does not permit a certain physical object or physical process to exist. The very first limitation we met (in chapter 1) was with the chessboard and dominoes. This is an example of a physical process that cannot exist. There is no way to place the dominoes on the chessboard with the two black corners removed. The barber paradox of section 2.2 also demonstrates that a certain isolated village with a particular rule cannot exist. The same section discussed a reference book that cannot exist. At the end of section 3.2 we saw that the time-traveler paradox shows either that time travel is impossible, or that even if it were possible, some actions by the time traveler will not be permitted. The universe simply will not permit a contradiction-causing process. All of chapter 6 and section 9.3 show that certain physical computers or algorithmic processes cannot exist. There are tasks that simply cannot be performed in this world. And finally in section 7.2, we talked about the possibility that quantum mechanics is inherently nondeterministic. In that case, no physical process can anticipate a quantum mechanical outcome. In all these examples we see that the physical universe is constrained by the dictates of reason.

Mental-Construct Limitations

A second, more subtle type of limitation states that a certain idea or mental construct cannot exist. I consider language a mental construct that is used to describe a mental state or a part of the universe. When discussing the liar paradox in section 2.1, I showed that certain sentences are neither true nor false. If a sentence is true then it is false, and if it is false then it is true. The mind cannot give such sentences any meaning. Similarly for other linguistic paradoxes—like the heterological paradox of section 2.2 as well as the interesting-number paradox, Berry's paradox, and Richard's paradox of section 3.2—Zeno's paradoxes of section 3.2 deserve some thought. They are not physical limitations since the slacker will get to the door and Achilles will win the race. Rather, they demonstrate a problem with the descriptions of certain actions. The descriptions are faulty because they seen to demand an infinite process. In contrast, the actions are perfectly legitimate. Zeno's paradoxes show that there are problems with certain mental and linguistic descriptions of basic movements. Similarly, our discussion of vagueness in section 3.3 showed a limitation of this type. Deciding whether a certain collection is a heap, and whether someone is considered bald, is a mental and/or a linguistic problem. We showed there were certain problems with such vague predicates. The inability to prove or disprove statements like the continuum hypothesis and the axiom of choice (section 4.4) demonstrates limitations of our logical ability. Similarly, most of the mathematical limitations discussed in chapter 9 are limitations of mental constructs. Human beings like their mathematics free of contradictions. Systems with contradictions do not have the right to be called mathematics and are ignored by researchers. Mathematicians avoid axioms and definitions that will bring about contradictions.

Several times throughout the book, I have had to restrain myself from taking an obvious step so as not to cross the limits of reason. A limitation of a mental construct was needed so as to avoid contradictions.

- In section 2.1 we saw that it is wrong to demand that every declarative sentence is either true or false. If we were to make such a demand, then we would have to say a liar sentence is true or false and get into contradictions.
- In section 2.2 we saw that there are phrases (e.g., "not true about itself") or even words (e.g., *heterological*) that—regardless of their obvious meaning—cannot be given meaning.
- In section 3.3, we saw that we had to restrict the use of the logical law of modus ponens when there are vague terms involved for fear of proving false statements.
- In section 4.4, we saw that we are constrained from making the obvious assumption that for every property there is a set of objects that satisfy that property. We simply cannot say that for fear of tripping on a predicament like Russell's paradox.
- In section 9.4, we saw that we must restrict our use of the logical fixed-point machine for fear of Löb's paradox.

In all these cases, we have an obvious logical step to take. But we realize that we are at the edge of the precipice and we restrict ourselves so as not to fall into the world of contradictions. Notice that all these restrictions are of mental constructs, not physical objects.

There is a slight exit strategy for natural language and for some mental constructs. As I stressed in chapter 1, we as humans are not bothered by having some contradictions in our everyday language, nor are we perturbed by contradictions in our mind. So when we meet a limitation of natural language and of certain mental constructs, we might simply ignore the limitation and have a contradiction. However, when our language is used to describe the physical universe (science) or made to describe mathematics, we do not have the luxury of having contradictions. We must keep such mental constructs and language contradiction-free so as to describe the contradiction-free physical universe.

Practical Limitations

Another type of limitation that we have met is of a slightly less fundamental character. They are not limitations that show it is impossible for something (physical or mental) to exist. Rather, they show that it is extremely impractical for something to exist. That is, it is impossible to make some prediction or find some solution in a normal amount of time or with a normal amount of resources. Chapter 5 is concerned with certain solvable computer problems that demand trillions of centuries to solve. While computers can theoretically solve such problems, for all practical purposes it is beyond human ability. Section 7.1 discussed chaos theory, which shows that systems with extreme sensitivity to initial conditions are essentially impractical to solve. In terms of the butterfly effect, while theoretically we might be able to keep track of all butterflies in Brazil to predict tornadoes in Texas, it is simply impractical to do so. These and several other practical limitations were discussed in this book.

Limitations of Intuition

We have seen that our naive intuition is somewhat flawed, reflecting more an error than a limitation. Our basic intuitions about the universe around us were challenged many times and the usual view was shown to be wrong. In section 3.1 our discussion of the ship of Theseus showed that objects do not really have an existence *as that object* (outside of a mind). We saw that this is not only a problem with physical objects and people but also with institutions and concepts. Chapter 4 showed that our naive intuitions about infinities are somewhat problematic. Section 3.4, where we met the Monty Hall problem, indicated that our concept of knowing needs adjusting.

Another point we have seen, especially in quantum mechanics (section 7.2) and in relativity theory (section 7.3), is that the observer plays a major role in the observed universe. There is a naive belief that the world is objective and external to the observer. This naive belief says that we can learn about this external world without changing it. The belief needs to be updated. The truth is that our view of the external world depends on how it is observed. The results of our experiments depend on what types of experiments are employed. The answers to our questions depend on what questions are asked and how the questions are posed. This places the conscious mind of the observer in a more central position in the study of the universe. Scientists are not outside the universe looking in. Rather, they are part of the universe and trying to make sense of it. They are part of the phenomena they are studying. It is hard to separate the experimenter from the experiment. That is, the universe is the ultimate self-referential system: the universe uses scientists to study itself.[4] We have seen several other counterintuitive ideas in both quantum mechanics and relativity theory.

There are also many questions about the sciences and the physical world for which our intuitions are in need of adjustments (chapter 8). The very nature of science and its relationship with mathematics, the universe, and the mind is open. The big questions of how and why we perceive structure and order in the universe are far from being resolved.

The various limitations were found or proved with different methodologies. However, there are some similar patterns that are worth highlighting.

Many of the limitations that we found were simply byproducts of self-reference. Once a system has the ability to talk about itself and deal with its own properties, there will be limitations of the system. Table 10.1 provides a list of some diverse areas where we found self-reference.

These diverse areas all have self-reference and have paradoxical situations and limitations.[5] We can also classify these self-referentail limitations into the four types of limitations listed in the beginning of this section. The situations in sections 2.2, 3.2, 5.2, and 7.1 are about physical limitations. The situations in sections 4.3, 4.4, and 9.4 are about limitations of mental constructs (math, set theory, and logic), and these limitations show that such exact mental constructs are not permissible. Sections 2.1 and 3.4 deal with human language and beliefs where contradictions are commonplace. What is remarkable is the common scheme of all the self-referential paradoxes: they all negate some basic aspect of themselves. This scheme seems to be a fundamental facet of reason.[6]

Once there is a limitation (from a self-referential paradox or in any other way), it is not hard to find other limitations by looking at reductions. One limitation can piggyback on another. In fact, a careful reading of chapter 6 and of section 9.3 will demonstrate that only one problem was shown to be computationally unsolvable: the Halting Problem. All the other problems were simply shown to be reductions of the Halting Problem. From a single limitation one can go on to build an entire edifice of limitations. We wonder if perhaps every type of limitation somehow comes

Table 10.1
Self-referential systems

Section	Topic	Object of self-reference	Consequences
2.1	Language	A sentence that *negates its own* truth.	There are declarative sentences that are neither true nor false.
2.2	Barber paradox	A villager who *negates his own* village's rule about shaving.	Such a village with this rule cannot exist.
3.2	Time-travel paradoxes	An event that *negates its own* existence.	Time travel (with the freedom to perform any action) is impossible.
3.4	Beliefs	A belief that *negates its own* truth.	There are declarative thoughts that are neither true nor false.
4.3	Infinity	A subset of the natural numbers that *negates its own* existence in a purported correspondence between the natural numbers and the powerset of natural numbers.	No such correspondence can possibly exist (i.e., the infinity of the powerset of natural numbers is higher than the set of natural numbers).
4.4	Russell's paradox	A set that *negates its own* rule of only containing sets that do not contain themselves.	No such set can exist.
6.2	Turing's halting problem	A program that *negates its own* "haltingness." The program will give the wrong answer when asked if it will stop or go into an infinite loop.	No such program can exist. No halting program can exist.
7.1	Predicting	A prediction that *negates its own* future.	Perfectly predicting the future is logically impossible.
9.4	Gödel's first incompleteness theorem	A logical statement that *negates its own* provability.	The statement is unprovable and hence true. The set of provable statements is a proper subset of true statements.

Table 10.2
Describable and indescribable

Section	Countably infinite	Uncountably infinite
6.3	Solvable computer problems	Unsolvable computer problems
7.1	Describable phenomena	Indescribable phenomena
9.1	Algebraic numbers	Transcendental numbers
9.4	(Provable) mathematical statements	Mathematical facts

from some form of self-referential limitation or a reduction from such a limitation.

Another common theme we have seen is the distinction between what is describable and what is indescribable.[7] By the very nature of language, what can be described is countably infinite. In contrast, what actually exists "out there" is uncountably infinite. Table 10.2 reminds us of some of the differences we have encountered.

The immense difference between countably and uncountably infinite exemplifies that what can be captured by language and formal reasoning is minuscule compared to all that exists. As Isaac Newton is quoted as saying, "What we know is a drop, what we don't know is an ocean."[8]

10.2 Defining Reason

At the end of chapter 1, I asked for a definition of reason. An entire book on what is the limits of reason begs for a definition of what is within the boundaries of reason. Why is it that some processes are reasonable and some are not? Why is it reasonable to check your blood pressure and not reasonable to check your horoscope? Why is it reasonable to agree with the chemists of today and to ignore the alchemists of centuries ago?

Over the centuries, philosophers have proposed many different definitions of reason and discussed their properties. Philosophers have also spilled much ink distinguishing between reason and related terms such as *intelligence, intellect, rationality, understanding,* and *wisdom*. There have been many attempts as well to distinguish human reason from animal reasoning and computer reasoning.[9] A summary of all the various theories and opinions on the topic is beyond the scope of this book.

However, there is a property that most thinkers agree reason has: one cannot use reason to derive contradictions and false facts. Throughout this book these two pitfalls were studiously avoided. Whenever we assumed some idea and derived a contradiction, we knew immediately that the assumption was wrong. The universe simply does not permit contradictions. A property cannot be both true and false. Similarly, if we derived a fact that is false, we knew that some assumption or reasoning process was wrong. Our reasoning would be worthless if we derived false facts. The facts stand and bad reasoning must be avoided.[10]

Our quest for a definition of reason can now be focused. What reasoning processes will ensure that we do not reach a contradiction or derive a false fact? Can we characterize the types of processes that lead to such bad conclusions? Which reasoning processes avoid such dreadful results?

Rather than giving an exact characterization of the processes that avoid contradictions and falsehoods, let us make this our very definition of reason: *Reason is the set of processes or methodologies that do not lead to contradictions and falsehoods.* Any process that will not lead us to a contradiction or falsehood is legitimate reason. Any time we arrive at a contradiction or falsehood, we know we have stepped over the bounds of reason and are being irrational. We have defined reason by examining its boundaries. What exactly will ensure that we don't step over these boundaries is hard to determine, but when we get to a contradiction or a direct falsehood, we know that we went too far. While we cannot provide exact rules on what is reasonable, we can tell when a process is wrong. Although this might sound like a strange definition, it works.

Since we do not know what will and will not cause one to err, our definition of reason is somewhat time dependent. What was once considered reasonable, could, in the future, be shown to cause contradictions. In fact, throughout history, there have been many times that something was considered part of science and only later turned out to be false. Some prominent examples include the following:

- Ether was a substance that light waves were supposed to travel through. Scientists only realized in the early twentieth century that this substance could not be detected.
- Chemists of yore believed in a substance called phlogiston that evaporates when an object is burned. It took many years until chemists realized that there was no such substance.
- Phrenology, the belief that different human traits can be learned by looking at physical characteristics of skulls and brains, was considered good science well into the twentieth century.
- Humorism and spontaneous generation were once considered legitimate science.

In contrast, there are many times that some idea has been considered silly and as time progresses it becomes part of reason and science:

- In classical Greece, the universe was considered to have always existed. It is only relatively recently that science realized that the universe came into existence within some finite time.
- The germ theory of Louis Pasteur and Ignaz Semmelweis was ignored when it was introduced. Now their ideas are taught to preschoolers to induce them to wash their hands.
- Negative and imaginary numbers were considered strange curiosities when they were first introduced. It took centuries for people to realize they actually had meaning.

This list can be extended for pages and pages. The point I am making is that there are no exact rules to determine when some idea or process is part of reason or beyond its boundaries.[11]

One might protest that empirical evidence should always be used to determine the truth of an idea. While this might be true, it is not always so simple. For example, Copernicus and Galileo said the Earth moves around the stationary sun. Pope Urban VIII said that the sun travels around the stationary Earth. As you look around the seemingly stationary Earth it feels like the pope was right. We now know that the pope and his empirical evidence were wrong. Galileo had done experiments showing that since the Earth is moving at a constant speed, we cannot feel it. As he said, one must look at *all* the empirical evidence to see that although the Earth appears not to move, "Nevertheless it moves."[12]

One problem with differentiating reason and nonreason is that our intuition sometimes fails us. Usually when something goes against our intuitions we deem it false. This is not always justified. For example, one of our most basic intuitions is that objects are in one place at one time. We learn this simple lesson as a young toddler. However, quantum mechanics with its shocking doctrine of superposition has shown us that this simple intuition is false. Another obvious intuition is that objects have fixed lengths and processes last set times. Relativity theory has shown us that this simple intuition is false. What other intuitions that we currently have will turn out to be false?

Since the limits of reason are not fixed, there are ideas whose status is still in limbo. There are many ideas in contemporary science that have neither been shown to be true nor false. Some examples are dark matter, dark energy, multiverses, string theory, the Higgs boson, and supersymmetry. All these ideas have many proponents and might be true, but we are not sure yet. Are they within the bounds of reason or not?

With this definition of reason in hand, we can perhaps answer some of the questions posed in chapter 1 about the nature of reason.

Why is one assumption more reasonable than another? Many times in the book when dealing with limitations, we have encountered conflicting reasonable assumptions. For example, it is intuitive to assert that

(a) Proper subsets of a set have fewer elements than the original set.

However, in our chapter on infinity, we found that

(b) Some proper subsets are the same size as the original set.

Which of these assertions is the right one? As I showed in chapter 4, we can get into contradictions if we do not follow the dictates of set theory, which assert (b). Since we want to avoid contradictions in reasonable processes, (b) is accepted and (a) is ignored. As counterintuitive as it sounds, (b) is correct and (a) is incorrect. Bad intuitions are not feared. Only contradictions are feared.

Why is it reasonable to check your blood pressure and not check your horoscope? Very simple: it is a basic fact that a healthy life depends on having good blood pressure. In contrast, there is no reason to check your horoscope because its predictions have been shown to be different from observable facts. Hence paying attention to your horoscope is beyond the bounds of reason.

There might be a process that is somewhat counterintuitive. We do not know if this process will lead us to correct predictions or to contradictions. Is this process reasonable? Perhaps yes and perhaps no. One thing we know for sure is that if a process leads to a false fact or a contradiction, then it is not reason. This gives a hierarchy of reasonable processes. At the very core is our fear of contradictions and false facts. That will be our final criterion for any reasonable process.

What should we believe in and what should we discard? Human beings are inherently gullible. People accept many ideas that are beyond reason. We want to believe our horoscope. We want to believe that by popping some pill we will lose weight without doing any exercise or curtailing our caloric intake. We want to believe that we will look as good as the model when we wear that outfit. But, alas, it is false. The human mind is susceptible to false ideas. When we step outside the bounds of reason, we need to be vigilant about what to accept as truth. By being cognizant of our gullibility, we are taking the first steps in our defense.[13]

As time progresses and we gain more experience with reason, science, and mathematics, we learn more about the boundaries of reason. We learn more about our faulty intuitions, and about processes that will make false predictions and contradictions. Thus the boundaries and definition of reason become more apparent.

10.3 Peering Beyond

After seeing the limitations of reason and defining it, we can wonder what is beyond reason. What methods can we use when reason fails us? How can we learn information that reason will not reveal to us? We should be cautious about overstepping the boundaries of reason. It is dangerous to use other methods outside of reason to understand the world around us or to build new technology. What was said at the beginning of chapter 1 must be reiterated: reason is the only methodology that improves our well-being.

• Jonas Salk did not find a cure for polio using intuition. He used reason and the scientific method.
• Imagination was not used to get humans to the moon. Technology based on reason and science was employed.
• World hunger will not be solved using feelings of love and warmth. Rather, genetically modified crops and ammonium nitrate fertilizers will help feed the world.

Reason is necessary for all these achievements. Even though we have highlighted many limitations of reason, still we should be hesitant to go outside of reason.

What is beyond the bounds of reason? This book has discussed certain boundaries that we are not to exceed for fear of contradictions and falsehoods. But what happens if we peer beyond the boundaries like the searcher in the old wood engraving shown in figure 10.1?[14] What are we missing by staying within the bounds of reason? What is out there?

Figure 10.1
The searcher peering beyond the bounds

Let me include a caveat about the many spatial metaphors in this book. The phrase "outer limits" is used as if it is a place. The word *beyond* is used as if there is a geographic boundary with two sides. We have "boundaries" that we are not permitted to "step over." Are we "peering" into a place? Such spatial phrases can lead to error. There are no places and there are no walls. By visualizing a wall, we may incorrectly assume that it is equally legitimate to investigate both sides of the wall. This is simply not so. As beautiful as the woodcut in figure 10.1 is, it is not real. Do not mistake the metaphor for reality.[15]

Nevertheless, the question of what is beyond reason does have a certain legitimacy. Reason is a methodology of learning and applying information. This book has shown that there are limitations on our ability to gather and use such information. However, the information exists. Some examples of such hidden information include the following:

• There is a shortest route for the Traveling Salesman Problem. We will not be able to acquire this knowledge, but it does exist.

• Chaos theory has shown us that we will never be able to predict the future of a chaotic system. Nevertheless, a future for the system does exist.
• We might not be able to determine whether a computer program will eventually halt, but the program will halt or go on for eternity.
• Although, by definition, we cannot prove that Gödel's sentence is true, nevertheless it is true.

In all these cases, the information exists[16] and is "really out there" but we have no way of knowing what that information is. What other methods are there for obtaining such information? We have the right to speculate.[17]

Although it is nice to speculate, unfortunately I do not think it is possible to say anything intelligent on this topic. Our definition of reason is an approach that avoids contradictions or false facts. Anything intelligible about the information beyond such boundaries must be a guess and hence there is nothing intelligent we can really say. We might be able to guess the information or receive the information in some other way. Someone might give us the information and we might simply accept it as true. Such methods of obtaining information are not within the purview of this book and, as with anything outside the bounds of reason, we have to accept the fact that contradictions and falsehoods might be before us. There might be much information out there but without the right tools for ascertaining that information, we are forced into silence.

Imagine being given a metal box and told that it is impossible to learn the contents of the box. You can try to drill it, burn it, x-ray it, shake it, break it, and so on, but you will never be successful at opening it or knowing its contents. It is true that in the box there might be an expensive jewel or useless sand. The box could also contain a piece of paper with the number 42 on it. We must also entertain the possibility that there is nothing inside the metal box. It could simply be an empty box. We will never know. This is similar to the point I am making. We are given the bounds of reason and are told we cannot go beyond those bounds without getting into the world of contradictions or inaccuracies. There might be some type of knowledge or information outside the bounds of reason that we cannot come to know. Nevertheless, as with the box, we must realize that we will simply never know. We might feel or intuit some information. We might have wishes about what is beyond the boundary. But we must tread with great trepidation for fear of the specter of contradictions and falsehoods.

One should not be too burdened by the bounds of reason and rationality. There is no cause to be despondent because we are unable to see beyond the bounds outlined in this book. *We human beings already live beyond reason.* The world humans inhabit is not the cold, heartless world of reason, logic, mathematics, and science. Our minds do not live in a world of stones, carbon-based life forms, and molecules following habitual laws of physics. Rather, we all have feelings and emotions that are not dictated by reason and logic. We have a sense of beauty, wonder, ethics, and values that are beyond reason and defy rational explanation. We appreciate beautiful art and music for no logical reason. While contemplating a mountain range, we are full of awe and wonder. We try to avoid performing deeds that are wrong even though they might be beneficial to us. The time we spend with our loved ones is treasured even though there is no logical necessity for it. We feel pain when we are distant from our loved ones. Our decisions are not made on the basis of logic and reason. Instead we use aesthetics, practical experience, moral inclinations, gut impulses, emotions, intuitions, and feelings. In this sense, every one of us already transcends the bounds of reason.

To some extent, the explanation for why human beings have such a hard time getting along with each other is that we all have desires and values that are beyond objective logic and reason. If we were all strictly logical like a computer or *Star Trek*'s Spock, we would all be on the same page and never argue with each other. This diversity of will makes life interesting. The cacophony of various human desires gives color to our relationships with other people. It also gives us the feeling that everyone *else* is crazy and irrational. Of course, everyone else has similar feelings about us. We all have irrational desires and wills that control us in ways that are different from others.

Not only do we have this nonreasonable part of our psyche, but this irrational component is our most important component. It is what gets us out of bed in the morning. It is our motivation and our will. There is no logical reason to do anything. Reason and logic tell us what is and in some cases they can tell us what will be. These tools can be used to help us get what we want. But they do not tell us what to want or what ought to be. Only will and desire tell us that.[18] Unless love, desire, music, and art exist, our world has no meaning. Real life has importance only when it includes ethics, values, and beauty. Will and desire are fundamental, while reason is a tool for that will and desire. Reason is a powerful—but nevertheless limited—tool.

Further Reading

Many interesting philosophy books discuss related topics—for example, Eddington 1958, Fogelin 2003, Priest 2003, and Rescher 1999, 2009. There is also a nice BBC documentary that covers some of our topics called *Dangerous Knowledge* (http://www.dailymotion.com/video/xdoe8u_dangerous-knowledge-1-5_shortfilms).

Notes

Preface

1. Popper 2002, 38.

Chapter 1

1. From the preface to Kant 1969. The original is *Die menschliche Vernunft hat das besondere Schicksal in einer Gattung ihrer Erkenntnisse: daß sie durch Fragen belästigt wird, die sie nicht abweisen kann; denn sie sind ihr durch die Natur der Vernunft selbst aufgegeben, die sie aber auch nicht beantworten kann; denn sie übersteigen alles Vermögen der menschlichen Vernunft.*

2. It is not clear if Einstein ever actually said this. However, Horgan (1996, 83) quotes a similar statement by John Archibald Wheeler: "As the island of our knowledge grows, so does the shore of our ignorance." Friedrich Nietzsche uses the same metaphor in *The Birth of Tragedy* (2000, 97): "But Science, spurred by its powerful illusion, speeds irresistibly toward its limits where its optimism, concealed in the essence of logic, suffers shipwreck. For the periphery of the circle of science has an infinite number of points; and while there is no telling how this circle could ever be surveyed completely, noble and gifted men nevertheless reach, e'er half their time, and inevitably, such boundary points on the periphery from which one gazes into what defies illumination."

3. This statement needs a little justification. One must distinguish between craft or technique, which does build on itself, in contrast to art and creativity, which do not build on themselves. In fact, creativity demands that the art be *different* from previous generations. It would be hard to claim that literature progresses when it can be argued that the greatest literature was written centuries ago by writers like Dante and Shakespeare. The Holocaust, genocides, and the wars of the twentieth century are counterexamples to any claims of an improvement with regard to human morality.

4. This chessboard-and-dominoes puzzle was cribbed from Gardner 1994, where it is called the "Mutilated Chessboard." The puzzle is, however, much older.

5. From Quine 1966, 3.

6. I am not equating human thought and human language. The latter is far more organized, coherent, and codified than the former. Whereas human thought does not have to be intelligible to any mind other than the thinker—in fact, it usually is not—human language is an attempt at making human thought understandable to other human minds. This is true for spoken language and even more so for written language. For the written word, even more codification and organization are required. A great piece of writing will be codified and clear enough for many other minds to appreciate. In contrast, literary theorists describe written language that "descends" to the level of human thought as "stream of consciousness." Examples of such literary works are James Joyce's *Finnegans Wake* and T. S. Eliot's *The Love Song of J. Alfred Prufrock*. Most people find these works unreadable. The relationship between human thought and language was dealt with by the Russian psychologist Lev Vygotsky in his book *Thought and Language* and in the later works of Wittgenstein. Nevertheless, both thought and language are prone to contradictions.

7. Einstein 1936.

Chapter 2

1. *Wovon man nicht sprechen kann, darüber muß man schweigen.*

2. Alas, I was unable to ascertain if Yogi Berra actually said this.

3. Some analysis shows that Epimenides' declaration is not really a paradox. For one thing, we are subtly assuming that every sentence that a liar utters is a lie. This is false. A liar is someone who lied at least once. We have all lied at least once in our lifetime and hence we are all liars.

Furthermore, there is something wrong with the logic in deriving a contradiction. Assume for a moment that Epimenides is telling the truth. This would imply that he is a liar and that the sentence is false. But a false sentence is not a contradiction. In contrast, assume that Epimenides' sentence is false. That means that not all Cretans are liars and that a Cretan exists who is not a liar. Such a pious truth-teller can be anyone on the island. (What if Epimenides were the only person on the entire island?) If this truth-teller were Epimenides, then he is telling the truth and the sentence is true. That would be a contradiction. However, the truth-teller need not be Epimenides and could be someone else on the island. So there would be no contradiction if we simply accept that Epimenides was stating a falsehood.

There is one last interesting idea to point out. We have determined that Epimenides' statement cannot be true and must be false. From this we logically conclude that someone on the island must be a truth-teller. This demonstrates the power of language and logic: from the fact that Epimenides made his statement, we conclude a fact about someone else's piety.

Despite these problems with Epimenides paradox we will see that there are other similar linguistic paradoxes that are bona fide, red-blooded paradoxes. It turns out that the classic example of a paradox is not really a paradox at all. How paradoxical!

4. The feminist response to this little quandary is that the barber's wife shaves the barber. This is just one of the many arduous tasks that she has had to perform over the centuries without getting any credit. She has pulled mankind from the abyss of contradiction!

5. We will meet this paradox again in section 4.4.

6. Hardy 1999, 12.

7. Michael Barr pointed out to me that 1729 is the first number only if you restrict it to positive whole numbers. If you permit negative whole numbers, then $91 = 6^3 + (-5)^3 = 4^3 + 3^3$.

8. We will meet vague terms again in section 3.3.

9. Richard's paradox will be much better understood after reading section 4.3.

Chapter 3

1. Similarly, the USS *Constitution* has been docked in the port of Boston for almost 200 years. The USS *Intrepid* is docked in the port of New York City.

2. Aristotle posits that there are four main "causes" of objects: the material cause (what it is made of), the formal cause (its shape), the efficient cause (who/what made it), and the final cause (its purpose). Each of these causes somehow accounts for what the object is. In this paragraph, I showed how we can make changes to all four "causes" of the ship of Theseus. Nevertheless, despite these changes, many people would consider the ship unchanged.

3. However, we can perform the same analysis with atoms. Is a carbon atom still a carbon atom if it loses one of its electrons? How about if it loses a neutron? What if it forms a chemical bond with another atom? Even atoms do not exist as atoms.

4. I am invoking Occam's razor as a criticism of (extreme) Platonism.

5. What Kant called *ding an sich* or thing-in-itself.

6. It is beyond the purview of this book to discuss Eastern philosophy. However, some of these ideas figure prominently in classical Indian and Chinese philosophy. It might be said that one of the primary purposes of meditation and "nullification

of the self" is the ability to see beyond the classifications of objects. Without classifications and names, there are no distinctions between concepts or objects. Reality then takes the form of the oneness and unity that is so central to the mystical traditions.

7. This is actually one of the simplest theorems to prove and worthy of taking a minute to actually demonstrate it. Let us say we do not know what the sum is and call it x:

$x = 1/2 + 1/4 + 1/8 + 1/16 + 1/32 + \ldots$

Again, we do not know what x is but, by naming it, we can manipulate it. Consider $1/2\,x$. By the distributive property of arithmetic we know that

$1/2\,(a + b + c) = 1/2\,a + 1/2\,b + 1/2\,c.$

This is not only true for three numbers but for infinitely many numbers. And so we have that

$$1/2\,x = (1/2)\,1/2 + (1/2)\,1/4 + (1/2)\,1/8 + (1/2)\,1/16 + \ldots$$
$$= 1/4 + 1/8 + 1/16 + 1/32 + \ldots.$$

Subtracting $1/2\,x$ from x gives us

$x - 1/2\,x = 1/2\,x$

or

$x - 1/2\,x = (1/2 + 1/4 + 1/8 + 1/16 + \ldots) - (1/4 + 1/8 + 1/16 + 1/32 + \ldots) = 1/2.$

This gives us

$1/2\,x = 1/2$

or $x = 1$. And we are done.

8. We will meet and explain some of these concepts in section 7.2.

9. Aristotle wrote: "Besides, a view which asserts atomic bodies must needs come into conflict with the mathematical sciences, in addition to invalidating many common opinions and apparent data and sense perception" (*De Caelo*, 303a21).

10. If we assume the world is discrete, the mathematics needed to build rockets and bridges is far more complicated than calculus. Perhaps calculus is simply an easy approximation of the true mathematics that has to be done to concretely model the discrete world in which we live.

11. Chapter 7 discusses them in greater depth.

12. We are all time travelers: we constantly travel *forward* in time.

13. It should be noted that if I had been able to go back to something similar to the Continental Congress, this would only be confusing to me. For my part, I know that I was not there at the original Congress, and now I will be there. However, to everyone else, given that I changed history by being there, it will not be strange.

14. This was a theme in the 1985 movie *Back to the Future.*

15. Rucker 1982, 168.

16. Even *pornography* is a vague term. As Supreme Court Justice Potter Stewart has said, he cannot define pornography "but I know it when I see it."

17. One usually says that two animals belong to the same species if they can mate with each other. However, there is a major problem with this definition: animal A might be able to breed with animal B, making them the same species. At the same time, animal B might be able to breed with animal C, rendering them of the same species. However, animal A might not be able to breed with animal C, making them different species. This lack of transitivity of identity for species is similar to other problems we saw with the problem of identity in section 3.1.

18. This explains why this section is in this chapter, not in the chapter on linguistic paradoxes (chapter 2).

19. Our old friend Zeno also has a version of this paradox. Some even find sorites-type arguments in the Bible (see Genesis 18:23–33).

20. Rohit Parikh also wrote a very interesting paper on this topic: Parikh 1994.

Chapter 4

1. From Pascal, *Pensées* (267): *La dernière démarche de la raison est de reconnaître qu'il y a une infinité de choses qui la surpassent, Elle n'est que faible si elle ne va jusqu'à connaître cela.*

2. We will see more of this in section 9.1.

3. The set (0,1) should not be thought of as the ordered pair of numbers 0 and 1. Rather it denotes the interval of all real numbers between the number 0 and the number 1.

4. Cantor announced this earth-shattering result in a letter to his friend Richard Dedekind (1831–1916). The letter was dated December 7, 1873, and that date can be taken as the beginning of modern set theory.

5. In section 2.3, I presented Richard's paradox about English phrases that describe real numbers. It should now be clear that this diagonalization proof was in mind when Richard's paradox was conjured up.

6. There are many other axiom systems besides Zermelo-Fraenkel set theory, but most of them are known to be just as powerful as Zermelo-Fraenkel set theory. That means that whatever can be proved using Zermelo-Fraenkel set theory can be proved using the other systems and vice versa. Thus, I will only discuss Zermelo-Fraenkel set theory.

7. Nicholas Bourbaki, "Aujord'hui qu'il est possible, logiquement parlant, de faire deriver toute la mathématique actuelle d'une source unique, la Théorie des Ensembles" (*Théorie des ensembles* (Paris, 1954), 4).

8. One of the leading number theorists of the last century, André Weil, summed it up nicely: "God exists since mathematics is consistent, and the Devil exists since we cannot prove it."

9. I am going to ignore Washington, D.C., and all the anomalies that make that place so detrimental to rational thought.

10. Note that the infinitesimal mathematical points of one ball can easily be put into correspondence with the infinitesimal mathematical points of the two balls, just as the natural numbers can be put into correspondence with the positive and negative integers. However, if the ball were made out of atoms, such a correspondence would not be possible. There are only a finite number of atoms in any ball and a finite number cannot be put into correspondence with twice that size. Once again, as we did with Zeno's paradox, we must ask whether we are justified in using (infinite) mathematics as a model for the real physical world.

11. There are many different schools of thought, and I am doing an injustice by lumping them all into one group. Large tomes are written defending nominalism of type x against nominalism of type y, and so on. However all of these different schools of thought share a dislike for abstract entities.

12. In fact, this is done in geometry. We will see in section 8.2 that Euclid's fifth axiom is independent of the other nine axioms. Rather than restricting themselves to one axiom system, mathematicians study both. They look at the nine axioms with the truth of the fifth axiom and get Euclidean geometry; at the same time, they also study the nine axioms with the falsehood of the fifth axiom and get non-Euclidean geometry. This still leaves us with some open questions. In geometry, the two systems correspond to different backgrounds. Euclidean geometry is the geometry of the flat surface. Non-Euclidean geometry describes curved and twisted surfaces. What is the analogy in set theory? What does the system with choice correspond to? What does the system without choice correspond to? After all, this is set theory where we are talking about collections of objects. How could there be variability about collections of objects? Another possible solution to all our questions is to regress and abandon all axiomatic systems. After all, Cantor did his magnificent work without the benefit of axioms. However, without axioms, we might return to having paradoxes and contradictions. More about this at the end of section 9.5.

13. We will discuss these questions in more depth in chapter 8.

14. I found this quote from Paul Cohen on the Internet (but could not find a source for it): "The notion of a set is too vague for the continuum hypothesis to have a positive or negative answer." That means that we cannot come to an answer about

the continuum hypothesis because the notion of a set is not well defined. This leads to the following question: What could possibly be vague about the obvious concept of a collection of objects?

Chapter 5

1. From Al-Daif 2000, 4.

2. There are 2.5 million people in Brooklyn. Due to the speed and efficiency of the web, phone books have really fallen out of use.

3. On the island is the tomb of Königsberg's most famous resident, Immanuel Kant.

4. In fact, only 120 routes have to be checked since it does not matter which city we designate as the first. In other words, for n cities, there are only $(n - 1)!$ possible routes.

5. A certain Internet company wanted to use the vastness of this number as their name. Legend has it that they misspelled the word. It's a shame they did not have Google to check their spelling.

6. We are being a bit wishy-washy about the exact definition of **NP**. To give the real definition would take us a little too far afield. However, most examples of NP problems are, in fact, either 2^n or $n!$. It is important to know that NP does not stand for "*N*on*P*olynomial." Rather, it stands for "*N*ondeterministic *P*olynomial." This means that the problems can be solved in polynomial time, if a nondeterministic machine performs the task. A nondeterministic machine is essentially a machine that makes a lot of guesses at the same time. Alas, nondeterministic machines do not exist and we cannot use nonexistent machines to solve our problems . . .

7. We will meet these ideas in depth in section 7.2.

8. In essence, what Cook and Levin did was to show that for every NP problem and for every input to that problem, one can write down a (very long) logical expression that mimics the actions of the computer finding a solution. If such a solution can be found by the computer, the logical expression will be satisfiable. If no such solution can be found, the logical expression will not be satisfiable.

9. There are those who believe that neither of these statements can be proved with the axioms of modern mathematics. They believe that the **P** =? **NP** question is "independent" of the axioms of mathematics. I discussed such independent statements in section 4.4 and will discuss them again in section 9.5.

10. I feel obligated to warn you that one of the seven Millennium Problems proposed, the Poincaré conjecture, has already been solved by Grigori Perelman. So there is now more pressure to solve one of the remaining problems. Do it fast!

11. Since the traveler is not permitted to visit any city twice, we must assume that in her trip from vertex c to e she is flying over vertex a.

12. I have never seen this approximation algorithm in the literature. This could be because, in general, it is not very good at getting to the solution.

Chapter 6

1. Mac or Linux users might find these concepts hopelessly abstract. Good for them.

2. Interestingly enough, Microsoft uses this proposed solution to the Halting Problem. When a program runs for a longish period of time, a "Not Responding" message is put on the top bar of the open window. One is then supposed to "break out" of the alleged infinite loop. Unfortunately, it is not clear that the program is really in an infinite loop. It might simply be going on longer than the Windows operating system expects.

3. It is easy to see this correspondence between programs and numbers: every program is stored in a computer as a unique series of zeros and ones. This sequence is simply a very large binary number. A typical program might be stored as millions of zeros and ones. The associated number will be astronomically large, but that need not bother us.

4. Other than 42 being the answer to the ultimate questions of life, the universe, and everything, there is really no reason why this number is special.

5. A very similar concept was presented in the last chapter. However, since I do not assume knowledge of that chapter, I undertake it again.

6. Currently, the hardest problem in mathematics is called the *Riemann hypothesis*. It is one of the seven Millennium Problems that the Clay Institute set up in 2000. If you solve this problem, you will get $1,000,000. The problem is somewhat harder to state than the Goldbach conjecture, so I will not delve into it. However, for the cognoscenti, just as there exists a program to look for a counterexample to the Goldbach conjecture, so too can a program exists to systematically search for zeros of the zeta function whose real part is not 1/2.

7. Formally, they showed that this question is independent of Zermelo-Fraenkel set theory. There is more discussion of such independence in sections 4.4 and 9.5.

8. Bierce, 1906/2010, 21.

9. Can human beings determine many facts about their own mind? What about our subconscious? Psychology is the study by human minds about human minds and especially their subconscious. Do psychologists have a deeper insight into themselves than others? One must give a resounding No! to this question.

Chapter 7

1. Poincaré 2010, 75.

2. From Alanis Morissette's song *Ironic*.

3. Pierre-Simon Laplace, introduction to *A Philosophical Essay on Probabilities*. The original passage is as follows: "Nous devons donc envisager l'état présent de Fin il vers, comme l'effet de son état antérieur, et comme la cause de celui qui va suivre. Une intelligence qui, pour un instant donné, connaîtrait toutes les forces dont la nature est animée, et la situation respective des êtres qui la composent, si d'ailleurs elle était assez vaste pour soumettre ces données à l'analyze, embrasserait dans la même formule les mouvemens des plus grands corps de l'imivers et ceux du plus léger atome: rien ne serait incertain pour elle, et l'avenir comme le passé, serait présent à ses yeux."

4. In fact, there has been some fascinating work done on laboratory coin tossing. See Diaconis, Holmes, and Montgomery 2011.

5. If complex numbers are unfamiliar, simply consider c to be a pair of real numbers $<c_1, c_2>$ and z to be a pair of real numbers $<z_1, z_2>$. Then iterate by simply taking $<z_1, z_2>$ to $<z_1^2 - z_2^2 + c_1, 2z_1z_2 + c_2>$.

6. Some subtlety is involved when dealing with the exact time and simultaneity.

7. One of the earliest sources seems to be Lucian of Samosata (125–180 AD) in his *The Sale of Lives*. The *Oxford English Dictionary* under the word crocodilite says it is from ancient Egypt. For more sources, see Von Prantl 1855, 493.

8. It should be noted that precisely because quantum mechanics is the *only* known system that seems random, there are physicists who believe that even this system must also be deterministic. As we will see, such researchers believe that although it seems random, there are hidden variables that determine the behavior of the system. Some actually provide (not simple) formulas that make predictions about quantum mechanics. Although such researchers include the likes of David Bohm and Albert Einstein, they are not considered part of the orthodoxy of modern physicists who believe that quantum mechanics is, in fact, random. If hidden variables for quantum mechanics do exist, then all known physical systems are deterministic.

There is, however, another side of the story. While I am taking the usual approach in presenting other physical systems besides quantum mechanics as deterministic, that is slightly disingenuous. *All* the laws of physics are—to the extent that they are complete—nondeterministic. The laws of electricity are based on quantum electromagnetism, which is a quantum system and hence nondeterministic. The laws of fluid dynamics are based on quantum mechanics and statistical mechanics. Hence, they too are nondeterministic. Gravity, classical mechanics, and general relativity might seem deterministic, but in fact, they are not real laws because they do not take quantum mechanics into account. Such laws are incomplete. When scientists finally do formulate quantum gravity, it will take quantum mechanics into account and be nondeterministic.

The two paragraphs mean that there are two options: (a) If there are hidden variables for quantum mechanics then all the laws of physics are deterministic. (b) If there are no hidden variables for quantum mechanics, then all the laws of physics are nondeterministic. Is our universe random or not?

9. This is similar to the discussion about solvable computer problems at the end of section 6.3. We will meet similar arguments about mathematics in sections 9.1 and 9.4.

10. Chapter 5 of Rescher 2009 actually has several proofs that there are "More Facts Than Truths." Rescher also says this disparity between what can be expressed and what actually exists shows that nominalism is false. We don't go there.

11. Goethe (1749–1832) said it poetically: "Let us seek to fathom those things that are fathomable and reserve those things which are unfathomable for reverence in quietude."

12. Feynman 1963, vol. 3, 1-1.

13. In the language of section 3.3, before the measurement, there is no epistemic vagueness, rather there is ontological vagueness.

14. Quoted in Pais 1979, 907.

15. Heisenberg 2007, 129.

16. In an uncanny way, the quantum physicists of the last century anticipated current sensibilities. They were very careful with living creatures and only performed a thought experiment on Schrödinger's cat. Thus quantum physics advanced, and no lives were lost.

17. This is actually slightly different from the original EPR experiment. In that experiment they measured position and momentum. I am discussing David Bohm's version of the experiment in which spin is the phenomenon examined.

18. Newton's formula that characterizes the force between two objects (which we met in the last section) also has a feeling of nonlocality to it. The force is strong when the two objects are close to each other. The larger the distance between two objects, the weaker the force. Nevertheless, as large as the distance is, there is still a force. In real life what this says is that even though a grain of sand is very small, and the moon is very far away, by moving a grain of sand on Earth you are affecting the gravitational force between the Earth and the moon. This change is extremely minuscule but it still exists. However, there are two major differences between gravity and entanglement. For one, gravity works at the speed of light. That is, it

takes time for any changes on Earth to affect the moon. In contrast, entanglement is instantaneous. A second major difference is that the gravitational force fades as the two objects get further apart. In contrast, the entanglement phenomenon remains as powerful and as instantaneous, regardless of whether the two objects are five feet apart or five million light-years apart. This makes entanglement much stranger than gravity.

19. Einstein expressed his discomfort with measurements affecting distant objects in a letter to Max Born. He started by describing some characteristics of physics: "It is further characteristic of these physical objects that they are thought of as arranged in a space-time continuum. An essential aspect of this arrangement of things in physics is that they lay claim, at a certain time, to an existence independent of one another, provided these objects 'are situated in different parts of space.' Unless one makes this kind of assumption about the independence of the existence (the 'being-thus') of objects which are far apart from one another in space—which stems in the first place from everyday thinking—physical thinking in the familiar sense would not be possible. It is also hard to see any way of formulating and testing the laws of physics unless one makes a clear distinction of this kind. . . . The following idea characterizes the relative independence of objects (A and B) far apart in space: external influence on A has no direct influence on B; this is known as the 'principle of contiguity.' . . . If this axiom were to be completely abolished, the idea of the existence of (quasi-) enclosed systems, and thereby the postulation of laws which can be empirically checked in the accepted sense, would become impossible" (Born 1971, 170–171).

20. We have already seen this with the double-slit experiment and more emphatically with the Kochen-Specker experiment. Although the Kochen-Specker result is cleaner and does not require two particles to prove that superposition is a fact, Bell's theorem was published three years earlier than the Kochen-Specker theorem. Furthermore, Bell's results were shown to be true experimentally, which had a huge impact on the physics world. In contrast, the Kochen-Specker theorem has been largely ignored by experimentalists until recently.

21. The inequality that I am describing is not actually Bell's original formulation. Rather, I am describing a variation of Bell's theorem that is due to Bernard d'Espagnat. The results of this formulation are the same as those of Bell's original.

22. This is one of the most detailed theorems I am going to discuss in this book. More than one reading is required for a complete understanding. Press on!

23. "Spukhafte Fernwirkung."

24. I am really simplifying the actual experiment here. The real experiment has to do with turning the diagonal filter on and off and is done with entangled particles. For simplicity's sake, I am describing the spirit of the experiment.

25. People have free will if their actions are not predetermined by what happened in the past. In other words, they act for no other reason than that this is what they want. (Whatever *they* means. As I stressed in chapter 1, human beings are full of conflicting ideas and desires. Which is the real person—the one who wants the cake or the one who wants to lose weight?) This is not a simple idea. After all, if I help an old lady cross the street because my mother used to tell me that I should do that, am I performing a freewill action or am I governed by previous programming by my mother? What if my mother had not told me to take such actions? Another question: If someone puts a gun to my head and tells me to perform a bad deed, and I do perform the deed, am I exercising free will or is the criminal forcing me to do it? After all, I do not need to perform the deed. At what point does a freewill act become an act of randomness? None of these questions have easy answers.

26. In 2006, John H. Conway and Simon B. Kochen published what they called the "Free Will Theorem." This theorem is based on an experiment that is a combination of the EPR and Kochen-Specker experiments. Whereas the Kochen-Specker experiment concerns one particle, the Conway-Kochen result depends on two spin-1 particles that are entangled. Two different observers are making measurements on the two particles. They claim that if a human being has free will, then so do the particles. There is some controversy as to whether this result was actually proved. Regardless, I believe I have duplicated their result from the delayed-choice quantum erasure experiment.

27. It is not clear how an experimenter's free will is impeded by the fact that a photon has knowledge of what freewill choice the experimenter will make. Even if the experimenter had knowledge of future choices, does that imply a lack of free will to choose? Free will is about control of actions, not about knowledge of actions.

28. There is one possibility that I did not mention. Maybe particles do have free will and the experimenter's decision on whether to pull away the diagonal polarization filter is somehow determined by the particle's decision to go into a superposition or not. That is, the particle controls the human observer. This, of course, is ludicrous. Nevertheless, an important experiment by Benjamin Libet (1916–2007) is worth mentioning. He found that certain regions at the back of the brain were excited seconds before people became aware of making certain decisions. In other words, there is a place in the brain that is controlling us and telling us what to want and what to do. For more, see part III of the excellent Nørretranders 1998. Recently neurologists have taken Libet's experiments much further.

29. This is similar to a game of bingo in which the winner of a round jumps up and screams that it is a miracle that she won. To all the other players who did not win, this is clearly not a miracle. However, to the "external" viewer—who knows that *someone* will win the game and scream out that this is a miracle—this is expected and very deterministic.

30. In mathematical language, one only has to deal with unitary operators and not with Hermitian operators.

31. Birkhoff and von Neumann (1936).

32. Deutsch 1997, 4–6.

33. From the end of the second part of Hume's *Dialogues Concerning Natural Religion* (Hume 1988, 19).

34. Bohr 1935, 702.

35. Exactly how far can we go with this process? One might argue that we cannot measure anything smaller than an atom and that should be the ideal measuring unit. In other words, count how many atoms are passed from the top of Norway to the bottom. This will make the coast of Norway a very large but finite number. If we were to go beyond an atom, we could get into a situation similar to that of the Mandelbrot set discussed in section 7.1. The boundary of that mathematical object does not have an atomic level where we must stop measuring. Rather, it has infinite complexity. It can be proved that the border of the Mandelbrot set is infinitely long but the area it surrounds is finite. What about Norway?

36. Galileo 1953, 186–187.

37. Galileo used this fact in his defense of Copernicus's view that the sun—not the Earth—is the center of the universe (heliocentric). People who believed that the Earth was the center of the universe (geocentric) argued that the Earth is not moving because we do not feel it moving. When we throw a ball up in the air, it lands in our hand, not where the Earth was. This seems counter to the idea that the Earth is moving. Galileo defended Copernicus by pointing out that we cannot feel movement if the movement is not accelerating.

38. The passenger is looking out the window and knows that he is moving toward the front of the train. He might reason that he is moving toward the front light and that is why he sees it first. But he is also aware of Einstein's postulate that the speed of light is measured as constant regardless of whether he is moving toward it or away from it.

39. Obviously one would lose weight simply from the exertion of climbing Mount Everest.

Chapter 8

1. Weinberg 1994, 259.

2. We must wonder what would happen if we did find a pink swan. We might say that it is not, in fact, a swan. After all, all swans are white, so this pink bird must be a different species. To what extent is being white part of the definition of being

a swan? If it is in the definition, then without any observations we can safely say "all swans are—by definition—white." This is all theoretical and not real. A quick Internet search confirms that there are, indeed, black swans and white ravens.

3. Wheeler and Zurek 1984, 195.

4. Hume 1955, 51.

5. In symbols: $\forall x(Raven(x) \rightarrow Black(x))$.

6. This is simply the contrapositive: $\forall x(NotBlack\ (x) \rightarrow NotRaven\ (x))$.

7. The idea can be found about a hundred years before the birth of William of Occam in Maimonides's *Guide for the Perplexed.* While discussing the different possible motions of the sun, he writes: "He will, besides, endeavour to find such an hypothesis which would require the least complicated motion and the least number of spheres: he will therefore prefer an hypothesis which would explain all the phenomena of the stars by means of three spheres to an hypothesis which would require four spheres" (Maimonides 1881, part 2, chap. 11, 1904). There are also similar statements in Aristotle.

8. We saw this idea in our discussion of quantum mechanics in section 7.2. We will see it again when dealing with the anthropic principle in section 8.3.

9. Dirac 1963, 47.

10. Criticizing such a view, Einstein is quoted as saying, "If you are out to describe the truth, leave elegance to the tailor." (Something similar was said earlier by Ludwig Boltzmann.)

11. Steven Weinberg brings down two examples of seemingly beautiful theories that did not live up to expectations. One is an early version of Watson and Crick's theory of DNA, and another is an early theory of Kepler's describing the distance of the planets from the sun. Both of these theories would really be beautiful if they worked. But, alas, they are false.

12. Weinberg 1994, 162–164.

13. Russell 2009, 67. This quote is taken slightly out of context. Russell was criticizing Eddington's type of metaphysics, where unity and wholeness play a role.

14. For an interesting history of the ever-increasing importance of mathematics in physics, see Burtt 1932.

15. I go much deeper into the nontrivial relationship of mathematics and physics in the next section.

16. With apologies to the related biological theory developed by Niles Eldredge and Stephen Jay Gould.

17. Kuhn 1987.

18. One must be careful when writing about Kuhn. His book was hijacked by many different philosophers, who took some of his ideas to the extreme. Kuhn spent much time trying to clarify his views and protesting against some of the ideas that seem to be consequences of his writings. Over time, he also modified and transformed his ideas. I will not go into the details of who said what at what time, or what was really meant. Rather, I am going to pose questions about the limitations of reason based on some of the ideas initiated by Kuhn.

19. In the literature, the example of finding the source of the Nile is attributed to Steven Weinberg in Weinberg 1994. In fact, he illustrates this point with the example of finding the North Pole (pp. 231–232). He illustrates another point with the example of finding the source of the Nile (p. 61).

20. Kant 1949, section 57, p. 122. George Bernard Shaw famously toasted Albert Einstein by saying, "Science is always wrong. It never solves a problem without creating ten more."

21. When exactly is "very soon"? It has been more than a decade and a half since John Horgan (1996) published his famous book predicting an imminent end to science. However, I do not know anyone who thinks this prediction has come true. Is science closer to ending now than in 1996? Making a prediction that science will end "soon" without providing a time table is not a falsifiable prediction. *Soon* is simply not a word that can be pinned down. How can we tell if the prediction is wrong?

22. The literature is full of very passionate ideas about this topic. Somehow researchers "know" the answers to these questions. Unfortunately this humble writer simply does not "know" the answers.

23. Many other topics in the philosophy of science impinge on the limitations of science. For example, philosophers discuss the existence of the laws of nature. To what extent are the laws real as opposed to being simply patterns of observations or socially constructed ideas? What are some of the practical limitations of science? To what extent is science influenced by the social structures of the scientists? See Rescher 1978 for more on the practical limitations of science. See Rescher 1999 for many other issues pertaining to the theoretical and philosophical limitations of science. There are many topics in classical epistemology that also have definite implications for the limitations of science. For example, philosophers ask how we can prove that we are not "brains in a vat" being fed stimuli. Another interesting philosophical thought is solipsism, the belief that one's mind is the only mind in existence. (Solipsists usually express shock that others do not accept their ideas.) Philosophers take this even further to "solipsism of the present moment," which is the belief that one's mind is the only mind in existence and that existence only began five minutes ago. In other words, even one's memories are of recent origin. While these crazy ideas are clearly false, there are no logical or reason-based proofs to show that they are false. We leave such ideas for other writers.

24. "La filosofia ´e scritta in questo grandissimo libro che continuamente ci sta aperto innanzi a gli occhi (io dico l'universo), ma non si puo intendere se prima non s'impara a intender la lingua, e conoscer i caratteri, ne' quali ´e scritto. Egli´e scritto in lingua matematica, e i caratteri sono triangoli, cerchi, ed altre figure geometriche, senza i quali mezi e impossibile a intenderne umanamente parola; senza questi e un aggirarsi vanamente per un'oscuro laberinto" (Galileo Galilei, *Opere Il Saggiatore*, 171).

25. Dirac 1963.

26. Einstein 1921.

27. Dirac 1982, 603.

28. Dirac 1939, 122.

29. Figure by Hadassah Yanofsky.

30. Whewell 1858, vol. 1, 311.

31. Einstein 1921.

32. We will meet group theory in much more detail in sections 9.2 and 9.3.

33. Weinberg 1994, 157.

34. Message of Pope Benedict XVI on the occasion of the international congress "From Galileo's Telescope to Evolutionary Cosmology: Science, Philosophy and Theology in Dialogue," 2009, http://www.vatican.va/holy_father/benedict_xvi/messages/pont-messages/2009/documents/hf_ben-xvi_mes_20091126_fisichella-telescopio_en.html.

35. Quoted in Bell 1937, 16.

36. Gardner 2005.

37. Even in physics, one can question the necessity of mathematics. Physics is about understanding causes and effects. Mathematics in physics is used to describe the exact *quantity* of causes and effects. My dissertation advisor, Alex Heller, reported a personal conversation with the great American physicist Richard Feynman, who said something along the lines of, "Physics is written in the language of mathematics. If we did not have mathematics, physics would not have progressed as much as it did and would be behind where it is now . . . by about fifteen minutes." On the web, I found the following story: Feynman remarked in a lecture that "if all of mathematics disappeared, physics would be set back exactly one week." The mathematician Mark Kac replied, "Precisely the week in which God created the world."

38. One might restrict the problems that sociology deals with and get phenomena suitable for contemporary mathematics. This is a central motivation of Rohit Parikh's social software project.

39. Isaac Asimov's classic *Foundation* series is based on this idea.

40. On the young physicist's boards are diagrams and equations of particle interactions. On the young mathematician's board are diagrams of objects with strange names like "cobordisms," "cohomology," and "homotopy functors."

41. Sidney Morgenbesser (1921–2004) is quoted as replying with this comment: "If there were nothing you'd still be complaining!" Woody Allen is quoted as saying, "What if everything is an illusion and nothing exists? In that case, I definitely overpaid for my carpet."

42. One can see the questions addressed in the last section as a further layer of these mysteries. Consider in particular the following question:

Question 4: Why does this intelligence that is capable of understanding the structure of the universe use the language of mathematics to describe this structure?

Why is the language of mathematics so perfectly suited to describing the laws of physics? Permutations and combinations of mathematical operations that form equations and inequalities become the laws of nature. I dealt with this issue in the last section.

43. Perhaps we are being a bit presumptuous in calling our species "intelligent." After all, this species has waged numerous inane wars where millions of their own were slaughtered. As a whole, this species spends trillions of hours a year watching insipid television shows. And "intelligent" is not the right name for a species that invented spam e-mails and encourages narcissistic pastimes like Facebook. Nevertheless, over the millennia, this species produced many shining lights that make us worthy of the lofty title: Blaise Pascal, Isaac Newton, David Hume, Marie Curie, Albert Einstein, Arthur Stanley Eddington, Emmy Noether, Andrew Lloyd Webber, Meryl Streep, and, of course, tiramisu.

44. In symbols, we can state this as $[(A \vee B) \wedge (\sim B)] \rightarrow A$. This rule is a common part of logic and is called a *disjunctive syllogism.*

45. Freud was actually the first to point out this three-pronged attack on human beings in "A Difficulty in the Path of Psycho-Analysis" (1917).

46. If we are to follow the participatory anthropic principle (which we will meet in a few pages), then conscious human observers are the actual *cause* of the universe being the way it is.

47. One must separate two types of deities that are used for such explanations. There are the personal deities of revelation, who want to see the human drama unfold for whatever reasons. In contrast, there are the impersonal deities of the philosophers. Such deities neither reveal themselves nor demand anything of humans. These two types of deities should not be confused with each other. As Pascal famously wrote, "'God of Abraham, God of Isaac, God of Jacob'—not of the

philosophers and of the learned" ("DIEU d'Abraham, DIEU d'Isaac, DIEU de Jacob' non des philosophes et des savants"). In general, the impersonal god is equated with nature or perhaps "Nature." A trendier New Age name would be something like "Cosmic Consciousness." However, most philosophers and theologians who discuss the impersonal deity prefer the name "God" so as to invoke the awe and reverence historically associated with that title. It is very unclear how the existence of the impersonal deity answers any of the questions about why the universe is the way it is.

48. J. D. Salinger describes the mother-in-law of a character in his fantastic novel *Raise High the Roof Beams, Carpenters* (2001) as follows: "A person deprived, for life, of any understanding or taste for the main current of poetry that flows through things, all things. She might as well be dead, and yet she goes on living, stopping off at delicatessens, seeing her analyst, consuming a novel every night, putting on her girdle, plotting for Muriel's health and prosperity. I love her. I find her unimaginably brave." This character clearly does not care about the implications of a fine-tuned universe or the anthropic principle.

49. Some people use the fact that the universe is not propitious for intelligent life to explain the *Fermi paradox*. This paradox asks why no intelligent beings from the billions of stars within each of the billions of galaxies have visited us (other than for some short visits in the episodes of the *X-Files*). There are many suggested answers to this mystery. In 2002, Stephen Webb published a book titled *If the Universe Is Teeming with Aliens . . . Where Is Everybody? Fifty Solutions to Fermi's Paradox and the Problem of Extraterrestrial Life*. The fiftieth solution—and the one Webb prefers—is that this universe did not generate other intelligent life forms and we are all alone. The first solution in the book, reportedly given by the physicist Leó Szilárd, is that "they are already here among us: they just call themselves Hungarians." See Webb 2002, 28.

50. Perhaps we can say that the universe is against having intelligent life and that the chances of having intelligent life are, say, 0.0000001 percent. We, therefore, only see intelligent life in 0.0000001 percent of the universe.

51. There is a bit of controversy in the literature about the veracity of this discovery. Originally it was thought that these life forms live off arsenic. It is now believed that they only live in arsenic. I am grateful to Jolly Mathen for pointing this out.

52. See Weinberg 1994, 221.

53. In fact, most authors do not take this as a multiverse theory.

54. Eye of Rivka Yanofsky. Figure by Hadassah Yanofsky.

55. Warning: long-term concentration about combining the participatory anthropic principle and the delayed-choice quantum eraser experiment can cause feelings of mysticism and madness.

56. A proponent of a multiverse theory might contend that at some point we must stop asking questions. They would say we are permitted to ask about a universe and its properties but we are not permitted to ask why the multiverse has the structure it has. Such questions would get us into an infinite regression or are nonsense. This argument is a page out of the playbook of medieval theologians who argued that everything must have a cause and the cause of the universe is a deity; however, one is not permitted to inquire about the cause of the deity. Such restrictions on permitted questions are unappealing. As long as we have intelligence, we must go on asking questions.

57. Manson 2003, 18.

58. Eddington 1939, 21.

59. Eddington 1958, 16.

60. We will meet these questions again in section 10.3.

61. Dyson 1979, 250.

Chapter 9

1. Gibbon 2001, 142.

2. Churchill 1996, 27.

3. Allen 1993, 62.

4. The difference between rational numbers and irrational numbers is not only a topic for ancient Greek philosophy and religion. We can have a similar discussion today. We imagine the world is somehow described and governed by real (and complex) numbers. The table is 5.82252932 . . . feet long; the temperature is 67.19153228 . . . . degrees Fahrenheit; the time it will take for the ball to land is 5.83245 . . . seconds. However, the human mind cannot retain an arbitrary real (or complex) number. Our brains are finite machines and can only deal with whole numbers or the ratio of two whole numbers (i.e., rational numbers). Similarly computers are limited in the types of numbers they can hold. This disparity between the "real world" and what we can know about the "real world" is a genuine limitation of reason.

There are at least two ways of getting around this limitation. First, we can say that the real world is discrete and thus suitable for rational numbers. As we have seen, quantum mechanics (section 7.2) and the wisdom of Zeno (section 3.2) assure us that the universe is discrete and that there is no information beyond Plank's length, Plank's energy, and Plank's time. Thus rational numbers are all that is needed to describe and understand the "real world." Another way of bridging this separation between the human/computer capacity and the "real world" is to realize that although we cannot hold arbitrary real numbers in our minds, we nevertheless deal with them. I cannot hold all the digits of pi and e in my head, but I can describe these numbers perfectly. Pi is simply the ratio of the circumference of a perfect circle (something that only exists in the mind, not in the real world) to its diameter. Besides, to describe arbitrary real numbers, I have ways of generating as many digits as I need. In that sense, there is a way of describing real numbers. I am certain the final opinions on this topic have not been uttered.

5. One can almost see Euclid's first four axioms of geometry (see section 8.2) from these two constructions.

6. "Tu prieras publiquement Jacobi ou Gauss de donner leur avis, non sur la vérité, mais sur l'importance des théorèmes. Après cela, il y aura, j'espere, des gens qui trouveront leur profit à déchiffrer tout ce gâchis."

7. "Ne pleure pas, Alfred! J'ai besoin de tout mon courage pour mourir à vingt ans."

8. Weyl 1952, 138.

9. We met this genius in the chapter 8. He invented complex numbers and was also one of the founders of probability theory. Despite a terribly tragic life he nevertheless made tremendous accomplishments. See section 5.5 of Penrose 1994.

10. Just for fun, here is one of the formulas used to find a solution:

$$x_1 = -\frac{b}{3a} - \frac{1}{3a}\sqrt[3]{\frac{2b^3 - 9abc + 27a^2d + \sqrt{(2b^3 - 9abc + 27a^2d)^2 - 4(b^2 - 3ac)^3}}{2}}$$
$$-\frac{1}{3a}\sqrt[3]{\frac{2b^3 - 9abc + 27a^2d - \sqrt{(2b^3 - 9abc + 27a^2d)^2 - 4(b^2 - 3ac)^3}}{2}}$$

It is obvious why such a cubic formula is not taught in high school!

11. Note that Abel also died young. He, too, lived a tragic life and died in extreme poverty.

12. I thank Chaim Goodman-Strauss for figures 9.10 through 9.16.

13. I am indebted to Chaim Goodman-Strauss for this clever example.

14. In technical terms, although basic arithmetic is *incomplete*, basic geometry was proven by Hilbert to be *complete*.

15. Basically this is induction, which we glimpsed in chapter 3 when we talked about heaps.

16. Post 1941, in Davis 2004, 343n12.

17. This is very similar to what we did in chapter 6. There we assigned a unique number to every program. Here we assign a unique number to every proof. The goal is also similar: self-reference. With programs, we wanted programs that deal with numbers to have numbers so that programs can deal with programs. Here we want

mathematical statements that deal with numbers to have numbers so that mathematical statements can deal with mathematical statements.

18. I am ignoring some details here. There was a slight complexity in Gödel's result concerning consistency and something called omega consistency (written ω-consistency). In 1936, John Barkley Rosser (1907–1989) modified Gödel's sentence to get what is now called a *Gödel-Rosser sentence*, showing the result for which we were aiming.

19. At the end of section 6.3 I showed that although computers can perform a countably infinite number of tasks, an uncountably infinite number of tasks remain that computers cannot perform. At the end of section 7.1 I presented an argument that science can only deal with a countably infinite number of phenomena, while an uncountably infinite number of phenomena remain that cannot be described by science. Here I present a mathematical analogy to those results.

20. Along these lines of thought, Leopold Löwenheim (1878–1957) formulated what has come to be known as the *downward Löwenheim-Skolem theorem*. This deep theorem states that what is described cannot be more complicated than the language used to describe it. In detail, statements in mathematics are written with a finite set of symbols. Using these symbols, there is a countably infinite number of possible statements. Now consider a system with an uncountably infinite number of elements. The downward Löwenheim-Skolem theorem states that if there is a consistent way of using a language to talk about such a system, then that language might very well be talking about a system with only a countably infinite number of elements. That is, the axioms might be intended for discussing something uncountably infinite, but we really cannot show that it is has more than a countably infinite number of elements. This gives us a severe limitation on what we can describe.

21. An interesting result is worth contemplating. As I have said, Peano Arithmetic is weaker than ZFC. In other words, whatever can be proven in Peano Arithmetic can surely be proved in ZFC. But how much weaker is it? It was shown that if you take the axioms of ZFC and leave out the axiom of infinity, then the system left is equipotent with Peano Arithmetic. (By *equipotent* I mean that whatever can be proved with one system can be proved with the other system and vice versa.) This can be written as

ZFC = Peano Arithmetic + Axiom of infinity.

Since ZFC can prove the consistency of Peano Arithmetic (and any other true statement that can be made in Peano Arithmetic), it follows that accepting that Peano Arithmetic is consistent is equivalent to accepting that there is a consistent way of dealing with infinite sets. Alternatively, the above equation points to the fact that if mathematicians would simply give up this belief that infinite sets exist, all of mathematics would be as consistent as basic arithmetic.

22. Or do we? A central idea in computer science is that all different types of physical computing devices can basically solve the same problems. Some devices can perform them quicker than others, but they all have the same computing ability. What is computable by one device is computable by another. More important for our discussion, what one device can never compute, another device can also never compute. This idea is called the *Church-Turing thesis*. It is a thesis as opposed to a theorem because it can never be proved. There is no way we can prove something about *every* physical device. But perhaps the Church-Turing thesis is wrong. Maybe in the distant future, scientists will develop a device that can solve problems that were previously not solvable. In that case, the unsolvable computer problems described in this book might still be solvable and the computational limitations will be relative, not absolute limitations.

Chapter 10

1. "Le silence éternel de ces espaces infinis m'effraie" (Pascal, *Pensées,* passage 206).

2. Nash 1994.

3. Quoted in Wang 1996, 9.4.3.

4. I'm reminded of a joke: analogous to "a chicken is an egg's way of making an egg," I'm stressing that "a scientist is an atom's way of knowing an atom."

5. See Yanofsky 2003 for a more comprehensive list of many different self-referential systems.

6. There are many other self-referential paradoxes that I did not touch on—for example, the *paradox of nirvana* (you can only reach nirvana if you free yourself of all desires . . . including the desire to reach nirvana) and the *paradox of tolerance* (if you want to have a tolerant society you must be intolerant of those who are intolerant). While these paradoxes are fascinating, they fall outside the scope of our discussion.

7. This is also discussed in chapter 5 of Rescher 2009. The title of the chapter is "More Facts Than Truths."

8. I am reminded of Ludwig Wittgenstein's words: "The limits of my language mean the limits of my world" ("Die Grenzen meiner Sprache bedeuten die Grenzen meiner Welt") (*Tractatus Logico-Philosophicus* 5.6).

9. I briefly touched on the comparison between computer and human abilities in sections 6.5 and 9.3.

10. There is a quote said to occur in the section on map reading in the Norwegian *Boy Scout Handbook*: "If the terrain differs from the map, believe the terrain."

11. In a sense, this is the content of Paul Feyerband's dictum that "anything goes" when it comes to the scientific method.

12. "Eppur si muove."

13. James Randi performed an interesting experiment that highlights something fascinating about human nature. Randi entered a class and asked the students to put their name and some personal information like their birthday and favorite color on a card. He then collected the cards and returned the next day. Attached to every card was a personalized description and horoscope for each student. Randi distributed the papers to the students and asked them to read their own personalized horoscope. Before discussing it with them, he asked them to judge whether their description was accurate. On a scale of "excellent," "very good," "good," or "poor," the vast majority of the students evaluated the accuracy of their personalized description as "good" or better. Randi then permitted the students to see each other's personalized horoscope and to their shock, all the horoscopes were exactly the same and filled with ambiguous pleasantries. Most students accepted the pleasant horoscope even though it did not say anything about them individually. Randi's experiment can be found online at http://www.youtube.com/watch?v=3Dp2Zqk8vHw.

14. Wood engraving from Camille Flammarion, *L'Atmosphere: Météorologie Populaire* (Paris, 1888), 163.

15. While I am criticizing this book, let me add more self-criticism. There is no doubt that I am using the words *limit* and *limitation* in many different ways. The words *reason* and *exist* are also used in diverse ways. In my defense, these words do not have clear definitions.

16. As opposed to answers to questions such as "How many teeth does a unicorn have?" or "Is the present king of France bald or not?", for which the information does not exist.

17. In section 6.4 I described a hierarchy of some computer problems that humans and computers cannot solve. Can we make a similar stratification of unknowns here?

18. The philosophical phrase is "reason is the handmaiden of will and desire." This is one of the main themes in the writings of Arthur Schopenhauer and David Hume.

Bibliography

Adams, Douglas. *The Hitchhiker's Guide to the Galaxy*. New York: Del Rey, 1995.

Al-Daif, Rashid. *Dear Mr. Kawabata*. Trans. Paul Starkey. London: Quartet Books, 2000.

Allen, Woody. *Getting Even*. London: Picador, 1993.

Aristotle. *The Basic Writings of Aristotle*. Ed. Richard McKeon. New York: Random House, 1941.

Baase, Sara. *Computer Algorithms: Introduction to Design and Analysis*. 2nd ed. Reading, MA: Addison-Wesley, 1988.

Baker, T. P., J. Gill, and R. Solovay. Relativizations of the P =? NP question. *SIAM Journal on Computing* 4, no. 4 (1975): 431–442.

Balaguer, Mark. Fictionalism in the philosophy of mathematics. In *Stanford Encyclopedia of Philosophy*. 2011. http://plato.stanford.edu/entries/fictionalism-mathematics.

Barrow, John D. *Impossibility: The Limits of Science and the Science of Limits*. Oxford: Oxford University Press, 1999.

Barrow, John D. *New Theories of Everything*. Oxford: Oxford University Press, 2007.

Barrow, John D., and Frank J. Tipler. *The Anthropic Cosmological Principle*. Oxford: Oxford University Press, 1986.

Bell, E. T. *Men of Mathematics*. New York: Simon and Schuster, 1937.

Bell, J. S. Bertlmann's socks and the nature of reality. *Journal de Physique,* colloque C2, suppl. 3, vol. 42 (1981). Reprinted in J. S. Bell, *Speakable and Unspeakable in Quantum Mechanics*. Cambridge: Cambridge University Press, 1987.

Bell, J. S. On the Einstein-Podolsky-Rosen paradox. *Physics* 1, no. 3 (1964): 195–200. Reprinted in J. S. Bell, *Speakable and Unspeakable in Quantum Mechanics*. Cambridge: Cambridge University Press, 1987.

Bell, J. S. *Speakable and Unspeakable in Quantum Mechanics*. Cambridge: Cambridge University Press, 1987.

Berlinski, David. *The Advent of the Algorithm: The 300-Year Journey from an Idea to the Computer*. San Diego: Harcourt, 2001.

Berto, Francesco. *There Is Something about Gödel: The Complete Guide to the Incompleteness Theorem*. Malden, MA: Wiley-Blackwell, 2009.

Bierce, Ambrose. *The Collected Works of Ambrose Bierce*. Reprint of the 1909 edition, Forgotten Books, 2012. http://www.forgottenbooks.org.

Bierce, Ambrose. *The Devil's Dictionary of Ambrose Bierce*. Ed. James H. Ford. Reprint of the 1906 edition, 2010. Special Edition Books. http://www.specialeditionbooks.com.

Birkhoff, Garrett, and Saunders Mac Lane. *A Survey of Modern Algebra*. Rev. ed. New York: Macmillan, 1957.

Birkhoff, Garrett, and John von Neumann. The logic of quantum mechanics. [Second Series] *Annals of Mathematics* 37 (4) (1936): 823–843.

Bohr, Neils. Can quantum-mechanical description of physical reality be considered complete? *Physical Review* (48) (1935): 696–702.

Boolos, George S., John P. Burgess, and Richard C. Jeffrey. *Computability and Logic*. 4th ed. Cambridge: Cambridge University Press, 2002.

Born, Max. *The Born-Einstein Letters: Correspondence between Albert Einstein and Max and Hedwig Born from 1916–1955*. Trans. Irene Born. London: Macmillan, 1971.

Brandenburger, Adam, and H. Jerome Keisler. An impossibility theorem on beliefs in games. *Studia Logica* 84 (2006): 211–240.

Bub, Jeffrey. *Interpreting the Quantum World*. Cambridge: Cambridge University Press, 1997.

Bunch, Bryan H. *Mathematical Fallacies and Paradoxes*. New York: Van Nostrand Reinhold, 1982.

Burtt, E. A. *The Metaphysical Foundations of Modern Science: The Scientific Thinking of Copernicus, Galileo, Newton, and Their Contemporaries*. Rev. ed. Garden City, NY: Doubleday Anchor, 1932.

Calaprice, Alice. *The New Quotable Einstein*. Princeton, NJ: Princeton University Press, 2005.

Calude, C., H. Jürgensen, and M. Zimand. Is independence an exception? *Applied Mathematics and Computation* (66) (1994): 63–76.

Carr, Bernard, ed. *Universe or Multiverse?* Cambridge: Cambridge University Press, 2007.

Casti, John L. *Paradigms Lost: Images of Man in the Mirror of Science.* New York: Morrow, 1989.

Churchill, Winston. *My Early Life: 1874–1904.* New York: Scribner, 1996.

Cohen, Paul J. Skolem and pessimism about proof in mathematics. *Philosophical Transactions of the Royal Society* 363, no. 1835 (2005).

Conway, John, and Simon Kochen. The Free Will Theorem. *Foundations of Physics* 36, no 10 (2006): 1441–1473.

Cook, Alan. *The Observational Foundations of Physics.* Cambridge: Cambridge University Press, 1994.

Cook, Stephen. The complexity of theorem proving procedures. *Proceedings of the Third Annual ACM Symposium on Theory of Computing,* 151–158. 1971.

Cook, Stephen. The P versus NP problem. 2002 http://www.claymath.org/millennium/P_vs_NP/pvsnp.pdf.

Corman, T. H., C. E. Leiserson, R. L. Rivest, and C. Stein. *Introduction to Algorithms.* 3rd ed. Cambridge, MA: MIT Press, 2002.

Cutland, Nigel. *Computability: An Introduction to Recursive Function Theory.* Cambridge: Cambridge University Press, 1980.

Dasgupta, Sanjoy, Christos Papadimitriou, and Umesh Vazirani. *Algorithms.* Boston: McGraw-Hill Science/Engineering/Math, 2006.

Dauben, Joseph W. *Georg Cantor: His Mathematics and Philosophy of the Infinite.* Cambridge, MA: Harvard University Press, 1979.

Davies, Paul. *The Goldilocks Enigma: Why Is the Universe Just Right for Life?* Boston: Houghton Mifflin, 2008.

Davies, P. C. W. *The Accidental Universe.* Cambridge: Cambridge University Press, 1982.

Davies, P. C. W. Where do the laws of physics come from? http://www.scribd.com/doc/6436157/Where-Do-the-Laws-of-Physics-Come-From. N.d.

Davis, Martin, ed. *The Undecidable: Basic Papers on Undecidable Propositions, Unsolvable Problems and Computable Functions.* Mineola, NY: Dover Publications, 2004.

Davis, Martin. The Incompleteness Theorem. *Notices of the AMS* 53, no. 4 (2006): 414–418.

Davis, Martin. What is a computation? In Lynn Arthur Steen, ed., *Mathematics Today,* 241–267. New York: Vintage Books / Random House, 1980.

Davis, Martin, and Reuben Hersh. Hilbert's 10th problem. *Scientific American*, November 1973, 84–91.

Davis, Martin D., Ron Sigal, and Elaine J. Weyuker. *Computability, Complexity, and Languages: Fundamentals of Theoretical Computer Science*. 2nd ed. Boston: Academic Press, 1994.

d'Espagnat, Bernard. The quantum theory and reality. *Scientific American*, November 1979, 159–181.

d'Espagnat, Bernard. *In Search of Reality*. New York: Springer-Verlag, 1983.

Deutsch, David. *The Fabric of Reality*. New York: Penguin, 1997.

Devlin, Keith. *The Joy of Sets: Fundamentals of Contemporary Set Theory*. 2nd ed. New York: Springer-Verlag, 1993.

Dewdney, A. K. *Beyond Reason: 8 Great Problems That Reveal the Limits of Science*. Hoboken, NJ: Wiley, 2004.

Diaconis, Persi, Susan Holmes, and Richard Montgomery. Dynamical bias in the coin toss. 2007. http://comptop.stanford.edu/u/preprints/heads.pdf.

Diacu, Florin. The solution of the n-body problem. *Mathematical Intelligencer* 18 (1996): 66–70.

Diacu, Florin, and Philip Holmes. *Celestial Encounters: The Origins of Chaos and Stability*. Princeton, NJ: Princeton University Press, 1996.

Dirac, P. A. M. The evolution of the physicist's picture of nature. *Scientific American*, May 1963, 45–53, http://blogs.scientificamerican.com/guest-blog/2010/06/25/the-evolution-of-the-physicists-picture-of-nature.

Dirac, P. A. M. Pretty mathematics. *International Journal of Theoretical Physics* 21 (1982): 603–605.

Dirac, P. A. M. *The Principles of Quantum Mechanics*. 4th ed. Oxford: Clarendon Press, 1986.

Dirac, P. A. M. The relation between mathematics and physics. *Proceedings of the Royal Society of Edinburgh* 59 (1939): 122–129.

Dyson, Freeman. *Disturbing the Universe*. New York: Harper and Row, 1979.

Eddington, Arthur. *Philosophy of Physical Science*. Ann Arbor: University of Michigan Press, 1958.

Eddington, Arthur. *Space, time and gravitation*. 1920. http://www.gutenberg.org/files/29782/29782-pdf.pdf.

Einstein, Albert. Geometry and experience. 1921. http://www.relativitycalculator.com/pdfs/einstein_geometry_and_experience_1921.pdf.

Einstein, A., B. Podolsky, and N. Rosen. Can quantum-mechanical description of physical reality be considered complete? *Physical Review* 47 (777) (1935).

Einstein, Albert. Physics and reality. *Journal of the Franklin Institute* 221 (3) (1936): 349–382.

Einstein, Albert. *Relativity: The Special and General Theory*. New York: Crown Publishers, 1961.

Eklund, Matti. Fictionalism. In *Stanford Encyclopedia of Philosophy*. 2007. http://plato.stanford.edu/entries/fictionalism/.

Enderton, Herbert B. *A Mathematical Introduction to Logic*. San Diego: Academic Press, 1972.

Eves, Howard. *An Introduction to the History of Mathematics*. 4th ed. New York: Holt, Rinehart and Winston, 1976.

Feynman, Richard Phillips. *The Feynman Lectures on Physics*. 3 vols. Reading, MA: Addison Wesley Longman, 1970.

Flammarion, Camille. *L'Atmosphere: Météorologie Populaire*. Paris: Librairie Hachette, 1888.

Fogelin, Robert. *Walking the Tightrope of Reason*. Oxford: Oxford University Press, 2003.

Friedman, Michael. *Dynamics of Reason*. Stanford, CA: CSLI Publications, 2001.

Galilei, Galileo. *Dialogue Concerning the Two Chief World Systems*. Trans. Stillman Drake. Berkeley: University of California Press, 1953.

Gamow, George. *One, Two, Three—Infinity: Facts and Speculations of Science*. New York: Dover, 1988.

Gamow, George. *Thirty Years That Shook Physics: The Story of Quantum Theory*. New York: Dover, 1966.

Gardner, Martin. *My Best Mathematical and Logic Puzzles*. New York: Dover, 1994.

Gardner, Martin. *Relativity Simply Explained*. Illustrated by Anthony Ravielli. New York: Dover, 1997.

Gardner, Martin. Review of *Science in the Looking Glass: What Do Scientists Really Know? Notices of the American Mathematical Society* 52, no. 11 (2005): 1344–1347.

Garey, Michael, and David S. Johnson. *Computers and Intractability: A Guide to the Theory of NP-Completeness*. New York: Freeman, 1979.

Gibbon, Edward. *The Autobiographies of Edward Gibbon: Printed Verbatim from Hitherto Unpublished MSS., with an Introduction by the Earl of Sheffield*. Chestnut Hill, MA: Adamant Media Corporation, 2001.

Gilder, Louisa. *The Age of Entanglement: When Quantum Physics Was Reborn*. New York: Vintage, 2009.

Gillespie, Daniel T. *A Quantum Mechanics Primer*. 1970. New York: Wiley.

Glazebrook, Trish. Zeno against mathematical physics. *Journal of the History of Ideas* 6, no. 2 (2001): 193–210.

Gleick, James. *Chaos: Making a New Science*. New York: Penguin Books, 1987.

Gödel, Kurt. What is Cantor's continuum problem? *American Mathematical Monthly* 54, no. 9 (1947): 515–525.

Godfrey-Smith, Peter. *Theory and Reality: An Introduction to the Philosophy of Science*. Chicago: University of Chicago Press, 2003.

Goodman-Strauss, Chaim. Can't decide? Undecide! *Notices of the AMS* 57, no. 3 (March 2010): 344–356.

Goodman-Strauss, Chaim. Tassellazioni. In Claudio Bartocci, ed., *La Matematica*, vol. 4, 249–285. Turin: Einaudi, 2011. An English version, "Tessellations," is available at http://mathfactor.uark.edu/downloads/tessellations.pdf.

Gorham, Geoffrey. *Philosophy of Science: A Beginner's Guide*. Oxford: Oneworld, 2009.

Greene, Brian. *The Fabric of the Cosmos*. New York: Vintage, 2004.

Greene, Brian. *The Hidden Reality: Parallel Universes and the Deep Laws of the Cosmos*. New York: Knopf, 2011.

Gribbin, John. *Schrödinger's Kittens and the Search for Reality: Solving the Quantum Mysteries*. Boston: Little, Brown, 1995.

Gribbin, John. *In Search of Schrödinger's Cat: Quantum Physics and Reality*. New York: Bantam Books, 1984.

Gribbin, John, and Martin Rees. *Cosmic Coincidences: Dark Matter, Mankind, and Anthropic Cosmology*. New York: Bantam Books, 1989.

Grim, Patrick. *The Incomplete Universe: Totality, Knowledge, and Truth*. Cambridge, MA: MIT Press, 1991.

Grünbaum, Adolf. Modern science and refutation of the paradoxes of Zeno. *Scientific Monthly* 81 (1955): 234–239. Reprinted in Wesley C. Salmon, ed. *Zeno's Paradoxes*, 164–176. Indianapolis, IN: Hackett, 1970.

Guillemin, Victor. *The Story of Quantum Mechanics*. New York: Scribner, 1968.

Hannabuss, Keith. *An Introduciton to Quantum Theory*. Oxford: Clarendon Press, 1997.

Hardy, G. H. *Ramanujan: Twelve Lectures on Subjects Suggested by His Life and Work*. Providence, RI: Chelsea, 1999.

Harel, David. *Computers Ltd.: What They Really Can't Do*. Oxford: Oxford University Press, 2003.

Hartmanis, J., and J. Hopcroft. Independence results in computer science. *SIGACT News* 8, no. 4 (1976): 13–24.

Hayes, Brian. *Group Theory in the Bedroom, and Other Mathematical Diversions*. New York: Hill and Wang, 2008.

Heisenberg, Werner. *Physics and Philosophy: The Revolution in Modern Science*. New York: Harper Perennial, 2007.

Held, Carsten. The Kochen-Specker Theorem. In Edward N. Zalta, ed. *The Stanford Encyclopedia of Philosophy*. Winter 2008 ed. http://plato.stanford.edu/archives/win2008/entries/kochen-specker.

Herbert, Nick. *Quantum Reality: Beyond the New Physics*. Garden City, NY: Anchor Press / Doubleday, 1985.

Hodges, Andrew. *Alan Turing: The Enigma*. New York: Simon and Schuster, 1992.

Hofstadter, Douglas R. *Gödel, Escher, Bach: An Eternal Golden Braid*. New York: Basic Books, 1979.

Hofstadter, Douglas R. *I Am a Strange Loop*. New York: Basic Books, 2007.

Horgan, John. *The End of Science: Facing the Limits of Knowledge in the Twilight of the Scientific Age*. New York: Broadway Books, 1996.

Huggett, Nick. Zeno's paradoxes. In *Stanford Encyclopedia of Philosophy*. 2010. http://plato.stanford.edu/entries/paradox-zeno.

Hume, David. *Dialogues Concerning Natural Religion*. Indianapolis, IN: Hackett, 1988.

Hume, David. *An Inquiry Concerning Human Understanding*. New York: Liberal Arts Press, 1955.

Hume, David. *A Treatise of Human Nature*. Oxford: Oxford University Press, 1978.

Jacobson, Nathan. *Basic Algebra I*. San Francisco: Freeman, 1985.

Jech, Thomas. *Set Theory*. San Diego: Academic Press, 1978.

Jech, Thomas. Set theory. In *Stanford Encyclopedia of Philosophy*. 2009. http://plato.stanford.edu/entries/set-theory.

John, Gribbin, and Martin Rees. *Cosmic Coincidences: Dark Matter, Mankind, and Anthropic Cosmology*. New York: Bantam New Age, 1989.

Jordan, Thomas F. *Quantum Mechanics in Simple Matrix Form*. New York: Wiley, 1986.

Kaku, Michio. *Physics of the Impossible: A Scientific Exploration into the World of Phasers, Force Fields, Teleportation, and Time Travel*. New York: Doubleday, 2008.

Kant, Immanuel. *The Critique of Pure Reason*. Trans. Norman Kemp Smith. New York: Bedford Books, 1969.

Kant, Immanuel. *Prolegomena to Any Future Metaphysics*. 1949. http://www.archive.org/stream/kantsprolegomena00kantuoft/kantsprolegomena00kantuoft_djvu.txt.

Karp, Richard M. Reducibility among combinatorial problems. In R. E. Miller and J. W. Thatcher, eds., *Complexity of Computer Computations*, 85–103. New York: Plenum, 1972.

Kilmister, C. W. *Eddington's Search for a Fundamental Theory: A Key to the Universe*. Cambridge: Cambridge University Press, 1994.

Kirby, L., and J. Paris. Accessible independence results for Peano arithmetic. *Bulletin of the London Mathematical Society* 14 (1982): 285–293.

Klein, Morris. *Mathematics and the Physical World*. New York: Dover, 1981.

Kline, Morris. *Mathematics: The Loss of Certainty*. Oxford: Oxford University Press, 1980.

Kramer, Edna E. *The Nature and Growth of Modern Mathematics*. 2 vols. New York: Fawcett, 1970.

Kuhn, Thomas S. *The Structure of Scientific Revolutions*. 2nd ed. Chicago: University of Chicago Press, 1970.

Kuhn, Thomas S. What are scientific revolutions? In L. Krüger, L. Daston, and M. Heidelberger, eds., *The Probabilistic Revolution*, 7–22. Cambridge: Cambridge University Press, 1987.

Kursunoglu, Behram N., and Eugene Paul Wigner, eds. *Paul Adrien Maurice Dirac: Reminiscences about a Great Physicist*. Cambridge: Cambridge University Press, 1990.

Laplace, Pierre Simon. *A Philosophical Essay on Probabilities*. New York: Wiley, 1902. http://archive.org/details/philosophicaless00lapliala.

Laplace, Pierre Simon. *A Philosophical Essay on Probabilities*. Translated into English from the original French 6th ed. by F. W. Truscott and F. L. Emory. New York: Dover Publications, 1951.

Lavine, Shaughan. *Understanding the Infinite*. Cambridge, MA: Harvard University Press, 1994.

Lawvere, F. William. Diagonal arguments and cartesian closed categories with author commentary. http://www.tac.mta.ca/tac/reprints/articles/15/tr15abs.html. *Lecture Notes in Mathematics* 92 (1969): 134–145.

Lederman, Leon M., and Christopher T. Hill. *Symmetry and the Beautiful Universe*. Amherst, NY: Prometheus Books, 2004.

Levin, Leonid. "A survey of Russian approaches to perebor (brute-force searches) algorithms." *Annals of the History of Computing* 6, no. 4 (1973): 384–400.

Losee, John, ed. *A Historical Introduction to the Philosophy of Science*. 4th ed. Oxford: Oxford University Press, 2001.

Lorenz, Edward. "Predictability: Does the Flap of a Butterfly's Wings in Brazil Set Off a Tornado in Texas?" Address at the Annual Meeting of the American Association for the Advancement of Science in Washington, December 29, 1972. In E. N. Lorenz, *The Essence of Chaos*, Seattle: University of Washington Press, 1993.

Maimonides, Moses. *The Guide for the Perplexed*. Trans. M. Friedländer. London: Routledge & Kegan Paul, 1904.

Makin, Stephan. Zeno of Elea. In *Routledge Encyclopedia of Philosophy*. London: Routledge, 1998.

Malin, Shimon. *Nature Loves to Hide: Quantum Physics and Reality, a Western Perspective*. Oxford: Oxford University Press, 2001.

Manin, Yuri Ivanovich, with B. Zilber. *A Course in Mathematical Logic for Mathematicians*. 2nd ed. New York: Springer, 2010.

Manson, Neil A. *God and Design: The Teleological Argument and Modern Science*. London: Routledge, 2003.

Mazur, Joseph. *Motion Paradox: The 2,500-Year-Old Puzzle behind All the Mysteries of Time and Space*. New York: Dutton, 2007.

Mendelson, Elliott. *Introduction to Mathematical Logic*. 4th ed. Boca Raton, FL: Chapman & Hall/CRC, 1997.

Mickens, Ronald E. *Mathematics and Science*. Teaneck, NJ: World Scientific, 1990.

Musser, George. *The Complete Idiot's Guide to String Theory*. New York: Penguin Books, 2008.

Nash, John. John F. Nash, Jr.—Autobiography. (Nobel Prize autobiography.) Nobelprize.org. July 29, 2011. http://nobelprize.org/nobel_prizes/economics/laureates/1994/nash-autobio.html.

Nietzsche, Friedrich. *Basic Writings of Nietzsche*. Trans. Walter Kaufmann. New York: Modern Library, 2000.

Nørretranders, Tor. *The User Illusion: Cutting Consciousness Down to Size*. New York: Viking, 1998.

Okasha, Samir. *Philosophy of Science: A Very Short Introduction*. Oxford: Oxford University Press, 2002.

Pagels, Heinz R. *The Cosmic Code: Quantum Physics and the Language of Nature*. New York: Simon and Schuster, 1982.

Pais, A. Einstein and the quantum theory. *Reviews of Modern Physics* 51 (1979): 863–914.

Pais, A. Playing with equations, the Dirac way. In Behram N. Kursunoglu and Eugene Paul Wigner, eds., *Paul Adrien Maurice Dirac: Reminiscences about a Great Physicist*, 93–116. Cambridge: Cambridge University Press, 1990.

Papadimitriou, Christos H. *Computational Complexity*. Reading, MA: Addison-Wesley, 1994.

Parikh, Rohit. Existence and feasibility in arithmetic. *Journal of Symbolic Logic* 36, no. 3 (1971).

Parikh, Rohit. Vagueness and utility: The semantics of common nouns. *Linguistics and Philosophy* 17 (1994): 521–535.

Pascal, Blaise. *Pascal's Pensées*. New York: Dutton, 1958. A French version is available at http://www.ub.uni-freiburg.de/fileadmin/ub/referate/04/pascal/pensees.pdf.

Paulos, John Allen. *Mathematics and Humor*. Chicago: University of Chicago Press, 1980.

Peat, F. David. *Einstein's Moon: Bell's Theorem and the Curious Quest for Quantum Reality*. New York: Contemporary Books, 1991.

Penrose, Roger. *The Emperor's New Mind: Concerning Computers, Minds, and the Laws of Physics*. Oxford: Oxford University Press, 1991.

Penrose, Roger. *The Road to Reality: A Complete Guide to the Laws of the Universe*. New York: Knopf, 2005.

Penrose, Roger. *Shadows of the Mind: A Search for the Missing Science of Consciousness*. Oxford: Oxford University Press, 1994.

Pickering, Andrew. *Constructing Quarks: A Sociological History of Particle Physics*. Chicago: University of Chicago Press, 1984.

Poincaré, Henri. *Science and Method*. New York: Cosimo Classics, 2010.

Popper, Karl. *Conjectures and Refutations: The Growth of Scientific Knowledge*. 2nd ed. London: Routledge, 2002.

Poundstone, William. *Labyrinths of Reason: Paradox, Puzzles, and the Frailty of Knowledge*. New York: Anchor Press / Doubleday, 1989.

Poundstone, William. *The Recursive Universe: Cosmic Complexity and the Limits of Scientific Knowledge*. Chicago: Contemporary Books, 1985.

Pour-El, Marian Boykan, and J. Ian Richards. *Computability in Analysis and Physics*. New York: Springer, 1989.

Priest, Graham. *Beyond the Limits of Thought*. 2nd ed. Oxford: Oxford University Press, 2003.

Quine, W. V. *The Ways of Paradox and Other Essays*. New York: Random House, 1966.

Rescher, Nicholas. *The Limits of Science*. Rev. ed. Pittsburgh: University of Pittsburgh Press, 1999.

Rescher, Nicholas. *Scientific Progress: A Philosophical Essay on the Economics of Research in Natural Science*. Pittsburgh: University of Pittsburgh Press, 1978.

Rescher, Nicholas. *Unknowability: An Inquiry into the Limits of Knowledge*. Lanham, MD: Lexington Books, 2009.

Rice, H. G. Classes of recursively enumerable sets and their decision problems. *Transactions of the American Mathematical Society* 74, no. 2 (March 1953): 358.

Rindler, Wolfgang. *Essential Relativity: Special, General, and Cosmological*. New York: Van Nostrand Reinhold, 1969.

Rivest, R. L., A. Shamir, and L. Adleman. A method for obtaining digital signatures and public-key cryptosystems. *Communications of the ACM* 21, no. 2 (1978): 120–126.

Ross, Kenneth A., and Charles R. B. Wright. *Discrete Mathematics*. 5th ed. Englewood Cliffs, NJ: Prentice Hall, 2003.

Rucker, Rudy. *Infinity and the Mind: The Science and Philosophy of the Infinite*. Boston: Birkhäuser, 1982.

Russell, Bertrand. *The Scientific Outlook*. London: Routledge, 2009.

Sainsbury, R. M. *Paradoxes*. 2nd ed. Cambridge: Cambridge University Press, 2007.

Sakurai, J. J. *Modern Quantum Mechanics*. Rev. ed. Reading, MA: Addison-Wesley, 1994.

Salinger, J. D. *Raise High the Roof Beam, Carpenters and Seymour: An Introduction*. New York: Back Bay Books, 2001.

Salmon, Wesley C. *Zeno's Paradoxes*. Indianapolis, IN: Hackett, 1972.

Scarani, Valerio. *Quantum Physics: A First Encounter; Interference, Entanglement and Reality*. Trans. Rachael Thew. Oxford: Oxford University Press, 2006.

Schwartz, Jacob T. *Relativity in Illustrations*. New York: Dover, 1989.

Shainberg, Lawrence. *Memories of Amnesia: A Novel*. New York: Ivy Books, 1989.

Sipser, Michael. *Introduction to the Theory of Computation*. 2nd ed. Boston: Thomson Course Technology, 2005.

Smolin, Lee. *The Life of the Cosmos*. Oxford: Oxford University Press, 1999.

Sorensen, Roy. *A Brief History of the Paradox: Philosophy and the Labyrinths of the Mind*. Oxford: Oxford University Press, 2003.

Sorensen, Roy. Epistemic paradoxes. In *Stanford Encyclopedia of Philosophy*. 2006. http://plato.stanford.edu/entries/epistemic-paradoxes.

Sorensen, Roy. *Vagueness and Contradiction*. Oxford, New York: Oxford University Press, 2001.

Stenger, Victor J. *The Comprehensible Cosmos: Where Do the Laws of Physics Come From?* Amherst, NY: Prometheus Books, 2006.

Stewart, Ian. *Galois Theory*. 3rd ed. Boca Raton, FL: Chapman & Hall/CRC, 2003.

Sudbery, Anthony. *Quantum Mechanics and the Particles of Nature: An Outline for Mathematicians*. Cambridge: Cambridge University Press, 1986.

Sudkamp, Thomas A. *Languages and Machines: An Introduction to the Theory of Computer Science*. 3rd ed. Reading, MA: Pearson / Addison-Wesley, 2006.

Tarski, Alfred. Truth and proof. *Scientific American*, June 1969, 63–77.

Tavel, Morton. *Contemporary Physics and the Limits of Knowledge*. New Brunswick, NJ: Rutgers University Press, 2002.

Torkel, Franzén. *Gödel's Theorem: An Incomplete Guide to Its Use and Abuse*. Wellesley, MA: A. K. Peters, 2005.

Truss, John. *Discrete Mathematics for Computer Scientists*. 2nd ed. Reading, MA: Addison-Wesley, 1998.

Unger, Peter. There are no ordinary things. *Synthese* 41 (1979): 117–154.

Van Heijenoort, J. *From Frege to Gödel: A Source Book in Mathematical Logic, 1879–1931*. Cambridge, MA: Harvard University Press, 1967.

Van Heijenoort, J. Gödel's Theorem. In *The Encyclopedia of Philosophy*. London: Collier Macmillan, 1967.

Vlastos, Gregory. Zeno of Elea. In *The Encyclopedia of Philosophy*, vol. 8, 369–379. New York: Macmillan / Free Press, 1972.

Von Prantl, C. *Geschichte der Logik im Abendlande*. Vol. 1. Leipzig: S. Hirzel, 1855.

Vygotsky, L. S. *Thought and Language*. Trans. Alex Kozulin. Cambridge, MA: MIT Press, 1986.

Waldrop, M. Mitchell. *Complexity: The Emerging Science at the Edge of Order and Chaos*. New York: Simon and Schuster, 1992.

Wang, Hao. *A Logical Journey: From Gödel to Philosophy*. Cambridge, MA: MIT Press, 1996.

Wapner, Leonard M. *The Pea and the Sun: A Mathematical Paradox*. Wellesley, MA: A. K. Peters, 2007.

Webb, Stephen. *If the Universe Is Teeming with Aliens . . . Where Is Everybody? Fifty Solutions to Fermi's Paradox and the Problem of Extraterrestrial Life*. New York: Springer, 2002.

Weinberg, Steven. *Dreams of a Final Theory*. New York: Vintage, 1994.

Weyl, Hermann. *Symmetry*. Princeton, NJ: Princeton University Press, 1952.

Wheeler, J. A. Law without law. In J. A. Wheeler and W. H. Zurek, eds., *Quantum Theory and Measurement*, 362–386. Princeton Series in Physics. Princeton, NJ: Princeton University Press, 1984.

Wheeler, J. A., and W. H. Zurek, eds. *Quantum Theory and Measurement*. Princeton Series in Physics. Princeton, NJ: Princeton University Press, 1984.

Whewell, William. *History of the Inductive Sciences*. 3rd ed. Vol. 1. New York: Parker, West Strand, 1858.

White, Robert L. *Basic Quantum Mechanics*. New York: McGraw-Hill, 1966.

Wick, David. *The Infamous Boundary: Seven Decades of Controversy in Quantum Physics*. Boston: Birkhäuser, 1995.

Wigner, Eugene. The unreasonable effectiveness of mathematics in the natural sciences. http://www.dartmouth.edu/~matc/MathDrama/reading/Wigner.htm.

Wittgenstein, Ludwig. *Tractatus Logico-Philosophicus*. Trans. David Pears and Brian McGuinness. London: Routledge, 1994.

Yablo, Stephen. Paradox without self-reference. 1993. http://www.mit.edu/~yablo/pwsr.pdf.

Yanofsky, Noson S. Towards a definition of an algorithm. *Journal of Logic and Computation*, 21, no. 3, (2010): 253–286. http://arxiv.org/pdf/math/0602053v3.pdf.

Yanofsky, Noson S. A universal approach to self-referential paradoxes, incompleteness and fixed points. *Bulletin of Symbolic Logic* 9, no. 3, (2003): 362–386.

Yanofsky, Noson S., and Mirco A. Mannucci. *Quantum Computing for Computer Scientists*. Cambridge: Cambridge University Press, 2008.

Index

Abel, Niels Henrik, 306, 316, 374n11
Abstract Algebra, 260–262
Adams, Douglas, 65
Adleman, Leonard, 119
Aesthetics, 38, 352
al-Daif, Rashid, 97
Aleph-null, 72–76
Algorithm, 99, 133, 136
  approximation, 126, 129–131, 134, 362n12
  binary search, 100–102
  brute-force search, 100–102, 110, 114–116, 121
  extreme pairs, 130, 131
  merge-sort, 104
  selection-sort, 102–104
al-Khowârizmi, Muhammed, 98
Allen, Woody, 12, 297, 371n41
Ambiguous statement, 51
Anaxagoras, 161
Anderson, Carl, 254
Anti-Platonism. *See* Nominalism
Antirealism. *See* Nominalism
Apollonius of Perga, 253–257, 264, 266, 269
Argument from design, 280
Aristotle, 31, 41, 48, 50, 62, 247, 248, 357n2, 358n9, 368n7
Arithmetization, 321, 329
Armstrong, Neil, 262
Artificial intelligence, 56, 157–160
Aspect, Alain, 200
Axiom of choice, 89–91, 335, 341

Baase, Sara, 133, 134
Back to the Future, 359n14
Baker, T. P., 157
Barings Bank, 33
Barr, Michael, xi, 357n7
Barrow, John D., 13, 29, 133, 295
Baxter, Jack, 51
Beauty, 242–243, 252
Bell, E. T., 337
Bell, John Stewart, 197
Bell's inequality. *See* Theorem, Bell's
Benedict XVI, Pope, 263, 370n34
Berger, Robert, 313
Berra, Yogi, 15, 204, 356n2
Bierce, Ambrose, 31, 158
Birkhoff, Garrett, 210, 337
Birth of Tragedy, The, 355n2
Bohm, David, 209, 363n8, 364n16
Bohr, Neils, 181, 186, 189, 191, 207, 214, 232
Bolai, Janos, 259
Boltzmann, Ludwig, 368n10
Bond, James, 92
Boone, William W., 318
Born, Max, 365n19
Bourbaki, Nicholas, 360n7
Brandenburger, Adam, xii, 61–63
Branes, 284

Brooklyn Dodgers, 33
"Brown Penny" (Yeats poem), 135
Bruns, Ernst Heinrich, 171
Brynner, Yul, 56
Bub, Jeffrey, 232
Burtt, E. A., 295, 368n14
Butterfly effect, 163, 166, 232, 342

Callahan, Harry, ix
Calude, C., 330
Can Quantum-Mechanical Description of Physical Reality Be Considered Complete?, 194
Cantor, Georg, 65, 72–77, 88, 94, 334, 359n4, 360n12
Cardano, Gerolamo, 260, 306
Carroll, Lewis, 161
Causality, x, 5, 161, 225, 231, 288, 365n19
Cayley, Arthur, 261
Chapin, Harry, 161
Churchill, Winston, 297
Church-Turing thesis, 372n22
Clarke, Arthur C., 135
Clauser, John, 200
Clay Institute, 127, 134, 362n6
Clinton, Bill, 31
Cohen, Paul J., 89–90, 94, 360n14
Collapse (quantum), 179, 180–185
Complementarity, 185, 186
Complete, weighted graph, 110, 111, 123
Comprehensible Cosmos: Where Do the Laws of Physics Come From?, The, 290
Conic sections, 255–257, 269
Consciousness, 193, 194, 204, 279, 285, 362n9
Conservation laws, 194, 289–290
Continental Congress, 49, 358n13
Continuum hypothesis, 88–94, 335, 341, 360n14
Conway, John Horton, 366n26
Cook, Stephen, 127, 361n8
Copenhagen interpretation, 207–208, 212
Copernicus, Nicolaus, 240, 241, 245, 247, 256, 278, 347, 367n37
Cosmological constant, 274
Cretica, 16
Crick, Francis, 169, 368n11
Cubic formula, 306, 374n10
Curie, Marie, 371n43
Cutland, Nigel, 160, 337

d'Alembert, Jean-Baptiste, 258
Dalí, Salvador, 275
Dante, 355n3
Dark energy, 347
Dark matter, 347
Darwin, Charles, 278
Dauben, Joseph W., 94
Davies, Paul, 295
Davis, Martin D., 160, 317, 337
D-Branes, 284
Dedekind, Richard, 359n4
Deduction, 53, 236, 332
Deep Blue, 159
Deity, 262, 263, 266, 280, 287, 294, 360n8, 370n37, 371n47, 373n56
Democratic Party, 33
de Sitter, William, 218
d'Espagnat, Bernard, 232, 233, 365n21
Deutsch, David, 209, 212–213
Diaconis, Persi, 363
Dialetheism, 57
Dialogues Concerning Natural Religion, 367n33
Dijkgraaf, Robbert, xii, 269
Ding an sich, 40, 357n5
Dirac, Paul A. M., 232, 242, 252, 254
Distributive law, 210, 267–268, 358n7
DNA, 118, 127, 169, 368n11
Double pendulum, 165, 174
Dreams of a Final Theory, 262
Dyson, Freeman, 294

Eddington, Arthur, 229–230, 235, 244, 246, 290–292, 353, 368n13, 371n43
Ehrenpreis, Leon, xiii, 337
Einstein, Albert, ix, 1, 11, 50, 181, 191–195, 201, 214–233, 244, 246–248, 253, 259, 269, 288, 363n8, 365n19, 367n38, 368n10, 371n43
Eldredge, Niles, 368n16
Elegance, 242, 368n10
Elements (Euclid), 257
Eliot, T. S., 356n6
Emergence, 169
Empty set, 67
End of science, 249–252
Entanglement, 194–201, 231, 264n18, 365n24
Epicycles, 241, 245
Epimenides, 16, 356n3
Epistemology, 2, 40, 137, 251, 290, 369n23
Equation
  cubic, 306
  Diophantine, 316, 317
  Dirac, 254
  linear, 305
  Maxwell's, 218
  quadratic, 306
  quartic, 306
  quintic, 306–307
Equinumerous, 68, 71–85, 88
Escher, M. C., 311
Ether, 346
Eubulides of Miletus, 54
Euclid, 257–260, 360n12, 374n5
Euler, Leonhard, 105, 108, 114
Euler cycle, 104–109
Everett, Hugh, III, 208, 283, 286
Experiment
  delayed-choice quantum erasure, 204–205, 285, 372n55
  double-slit, 176–180, 201, 365n20
  EPR, 194–197, 200, 364n17
  Kochen-Specker (*see* Theorem, Kochen-Specker)
  quantum eraser, 201–205
  Schrödinger's cat, 192–194, 208, 364n16
  Stern-Gerlach, 186–189
  Wigner's friend, 193

Falsifiability, 243–246, 250, 369n21
Feedback, 169
Ferrari, Lodovico, 306
Feyerabend, Paul K., 377n11
Feynman, Richard, 176, 370n37
Fictionalism. *See* Nominalism
Fine-tuned universe, 272–295, 372n48
Finnegans Wake, 356n6
Fixed point machine, 322–329, 341
Flammarion, Camille, 377n14
Flash, The, 45
Fogelin, Robert, xii, 353
Fractals, 168
Fraenkel, Abraham, 86
Free will, 158, 204–205, 366nn25–28
Frege, Gottlob, 85
Freud, Sigmund, 279, 371n45
Friedman, Harvey, 333, 334

Galilei, Galileo, 71, 215–216, 227, 252, 288, 347, 370n24
Galle, Johann, 254
Galois, Évariste, 304–309, 317, 337
Gamow, George, 94, 232
Gardner, Martin, 233, 264
Gauss, Karl Friedrich, 259, 304
Gelfand, Israel M., 265
Gell-Mann, Murray, 207
Gentzen, Gerhard, 332, 335
Geocentrism, 241, 247, 248, 256, 367n37
Geometry, 319
  Greek, 297–304
  noncommutative (*see* Theory, noncommutative geometry)

Geometry (cont.)
non-Euclidean, 259, 264, 269, 360n12
Gibbon, Edward, 297
Gilder, Louisa, 232
Gill, J., 157
Gleick, James, 232
Gödel, Kurt, x, 41, 50, 63, 88, 89, 94, 95, 158, 320, 339
Goethe, Johann, 364n11
Goldbach conjecture, 153, 154, 362n6
Goldilocks enigma. *See* Fine-tuned universe
Goodman-Strauss, Chaim, xii, 337, 374nn12–13
Goodstein, Reuben, 331
Gould, Stephen Jay, 368n16
Grassmann, Hermann, 261
Greene, Brian, 232, 233, 283, 295
Grossmann, Marcel, 259
Group, 307–309
Guide for the Perplexed, The, 368n7
Gurwitz, Chaya, xii

Hall, Monty, 57–60
Halt Program, 142–144
Hamilton, William R., 261
Hamiltonian cycle, 113
Hardy, Godfrey H., 26
Harel, David, 133, 134
Hartmanis, Juris, 157
Hatfield, Doug, 192
Hawking, Stephen, x
Hayes, Brian, 337
Heisenberg, Werner, 192, 232, 261
Held, Carsten, 233
Heliocentrism, 240, 241, 247, 248, 256, 367n37
Heller, Alex, xiii, 166, 370n37
Heraclitus, 33
Herschel, William, 253
Hershfeld, Shayna Leah, xiv
Hertz, Heinrich, 264
Hidden Reality: Parallel Universes and the Deep Laws of the Cosmos, The, 283
Hidden variables, 181, 196, 197, 200, 201, 209, 212, 363n8
non-local, 200
Higgs boson, 347
Hilbert, David, 69, 88, 316, 374n14
Hilbert's hotel, 69, 73, 94
Hippasus of Metapontum, 298–299
Hitchhiker's Guide to the Galaxy, The, 65, 362n4
Hofstadter, Douglas R., 135, 158, 337
Hopcroft, John E., 157
Horgan, John, 295, 355n2, 369n21
Hume, David, 62, 213, 236, 238, 243, 280, 287, 367n33, 371n43, 377n18
Humorism, 346
Hypergeometric functions, xiv

I Am a Strange Loop, 135
Ichthyology, 291
Incommensurability, 247–249
Induction. *See* Problem, induction
Infinite loop, 139–146, 157, 362n2
Infinity, 65–95, 271, 272, 331, 332, 342, 344, 375n21
countable, 72–78, 80, 151, 152, 175, 329, 345, 375n19
uncountable, 77–85, 151, 152, 175, 329, 345, 375n19
Instrumentalism, 212–213
Interference effect, 177–178, 180, 201–205
Io (moon of Jupiter), 217

Jacobi, Carl, 304
Jefferson, Thomas, 334
Jeopardy, 159
Joyce, James, 356n6
Jupiter, 217

Kac, Mark, 370n37
Kant, Immanuel, 1, 250, 287, 357n5, 361n3
Kasparov, Garry, 159
Keisler, H. Jerome, 61–63
Kepler, Johannes, 241, 255–257, 262, 266, 269, 368n11
Kirby, Laurie, 331
Kirk, Captain James Tiberius, 219
Kneiphof Island, 104, 361n3
Kochen, Simon, 188, 191, 366n26
Königsberg, 104, 361n3
Kuhn, Thomas S., 246–249, 250, 369n18

Lambert, Johann Heinrich, 258
Laplace, Pierre-Simon, 162–163, 166, 363n3
Law of excluded middle, 53
Lawvere, F. William, 29
Lawyers, 52, 276, 281
Lederman, Leon M., 295
Legendre, Adrien-Marie, 258
Length contraction, 219–223, 227
Leslie, John, 279
Let's Make a Deal, 57
Leverrier, Urbain, 245, 253–254, 263
Levin, Leonid, 127, 361n8
Libet, Benjamin, 366n28
Life of the Cosmos, The, 284
Lightyear, Buzz, 65
Limitations
  absolute, 335–336, 376n22
  of intuition, 342–343
  mental-construct, 340–343
  physical, 340, 343
  practical, 342
  relative, 335–336, 376n22
Lindermann, Ferdinand von, 303
Little Red Riding Hood, 93
Lobachevsky, Nicolai Ivanovitch, 259
Locality, 194–201, 208, 209, 213
Logarithm, 101
Logic, 320–331
  fuzzy, 56
  paraconsistent, 57, 63
  quantum, 210–211
  three-valued, 56
Lorenz, Edward, 163
Love Song of J. Alfred Prufrock, The, 356n6
Löwenheim, Leopold, 375n20
Lucian of Samosata, 363n7
Luthor, Lex, 119

Mac Lane, Saunders, 337
Magnum Force, ix
Maimonides, Moses, 368n7
Mandelbrot set, 166–168, 367n35
Manin, Yuri I., 233, 237
Manson, Neil, 287
Many-worlds hypothesis. *See* Multiverse
Marx, Groucho, 12
Marxism, 244
Mass-energy equivalence, 225–226
Materialism, 193
Mathen, Jolly, 372n51
Matiyasevic, Yuri, 317
Matryoshka doll, 274
Maxwell, James Clerk, 218
Measurement, 184, 186, 189, 191, 196, 198–200, 208
Mendelson, Elliott, 337
Mickens, Ronald E., 295
Mickey Mouse, 40, 92
Milky Way galaxy, 235, 282
Modus ponens, 53–55, 117, 327, 341
Morgenbesser, Sidney, 371n41
Morissette, Alanis, 363n2
Morphogenesis, 169, 232
Multiverse, 208–209, 212, 282–287, 294, 347, 372n53
Mutilated chessboard puzzle, 2–4, 340, 356n4

Nash, John, 339
Natural selection, 284
Nearest neighbor heuristic, 129–130
Neptune, 253–254, 263
Newton, Isaac, 1, 162, 166, 170, 171, 226, 237, 241, 244–249, 253, 263, 345
Nietzsche, Friedrich, 355n2
Noether, Emmy, 289, 290, 371n43
Nominalism, 92–94, 360n11
 classical, 39–40
 extreme, 39–40
Non-locality, 194–201, 209, 213, 364n18, 365n23
Normal science, 246–248
Nørretranders, Tor, 366n28
Novikov, Pyotr, 318
Number
 algebraic, 303–304, 345
 complex, 167–168, 181, 205, 260, 261, 264, 292, 293, 363n5, 373n4, 374n9
 constructible/Euclidean, 302–303
 Gödel, 323–327
 Hamiltonian/quaternion, 261
 imaginary, 205, 260, 347
 irrational, 70, 298–301, 337
 negative, 72–73, 91, 271–272, 347, 357n7, 360n10
 transcendental, 303–304, 345

Occam's razor, 208, 212, 240–242, 250, 264, 285, 357n4, 368n7
On the Einstein-Podolsky-Rosen Paradox, 197
Oracle, 152–157, 212–213, 252
Ornithology, 236–237, 238–240, 367n2
Outer Limits of Reason, The, 12, 377n15
Oxford English Dictionary, 5, 13
Oxymoron, 15–16

P =? **NP** question. *See* Problem, **P** =? **NP**

Parade Magazine, 57, 63
Paradigm shift, 246–249, 250
Paradox, 5–8, 11, 13, 360n12
 Achilles and the Tortoise, 45–46
 arrow, 46–47
 bald-man, 53
 Banach-Tarski, 90–94
 barber, 19, 20, 23, 340, 344, 357n4
 Berry, 26–27, 340
 Brandenburger-Keisler, 61–63
 coastline, 214–215
 crocodile's dilemma, 173
 dichotomy, 41–45
 Epimenides, 16
 Fermi, 372n49
 grandfather, 49
 Grelling's, 20–23, 340, 341
 Hempel's, 238–240
 heterological, 20–23, 340, 341
 interesting number, 26, 340
 liar, 7, 10, 15–20, 28, 60, 141, 144, 173, 340
 linguistic, 15–22, 344
 Löb, 327–328, 341
 of nirvana, 376n6
 ravens, 238–240
 reference-book, 22, 23, 340
 Richard's, 27–28, 340, 357n9
 Russell's, 22, 23, 85–86, 88, 341, 344
 self-referential, x, 7–8, 15, 60, 62, 63, 173, 343–344
 small-number, 55
 sorites, 53–56, 340
 sorites-type, 55, 340
 stadium, 47–48
 surprise-test, 60–61, 63
 time-travel, 31, 49–50, 63, 340, 344
 of tolerance, 376n6
 twins, 222
 Wang's, 55
 Yablo's, 24–26

Zeno's, 7, 31, 41–49, 62, 91, 340, 360n10
Parallel postulate, 258
Parikh, Rohit, xi, 63, 326–327, 337, 359n20, 370n38
Paris, Jeff, 331
Parmenides, 41, 48, 242
Pascal, Blaise, 65, 339, 359n1, 371n43, 371n47, 376n1
Pasteur, Louis, 245, 247, 347
Peano, Giuseppe, 320
Peano Arithmetic, 320–328, 330–335, 375n21
Penrose, Roger, 158, 160, 232
Pensées, 359n1, 376n1
Perelman, Grigori, 361n10
Philosophy of science, 235–252, 294
Phlogiston, 346
Phrenology, 346
Pickering, Andrew, 232, 295
Piggybacking, 8–9, 316, 332, 343. *See also* Reduction
Pinocchio, 93
Planck, Max, 206
Planck's energy, 373n4
Planck's length, 44, 373n4
Planck's time, 373n4
Plato, 91, 263
Platonism, 91–94, 263–265, 284
classical, 38–40
extreme, 38–40, 357n4
Plato's attic, 52, 263, 264, 357n4
Podolsky, Boris, 194
Poincaré, Henri, 161, 163, 171, 235
Poincaré conjecture, 361n10
Point of view invariance, 290
Polarization filters, 182–185, 201–205
Popper, Karl, ix, 238, 241–243, 250
Positron, 254, 272
Post, Emil, 321
Poundstone, William, 13, 29, 133
Pour-El, Marian, 337
Powerset, 67–68, 80–83
Predictability: Does the Flap of a Butterfly's Wings in Brazil Set Off a Tornado in Texas?, 163
Pressburger Arithmetic, 328
Priest, Graham, 56, 63, 353
Principle
anthropic, 278–295, 368n8, 372n48
Copernican, 279
Heisenberg's uncertainty, 185–186, 188, 189, 198
mediocrity of, 279
parsimony of (*see* Occam's razor)
participatory anthropic, 285–287, 371n46, 372n55
strong anthropic, 278, 285
weak anthropic, 278–295
Problem
decidable, 139, 152
decision, 113
doubling a cube, 302, 303, 335
Equivalent Program, 149–150
Euler Cycle, 107, 114, 128
Euler Path, 108
exponential, 116–120, 124, 127–128, 132, 157
factorial, 111, 118, 127–128, 132, 156–157
feasible, 109
Halting, 136–160, 315, 316, 343, 344, 351, 362n2
Halting Problem for Halt Oracle Programs, 154
Halting Problem for Halt' Oracle Programs, 155
Hamilton Cycle, 113–115, 123–124, 128
Hilbert's Tenth, 316–317
of identity, 32–40, 62
induction, 236–240, 250, 287
infeasible, 109
intractable (*see* infeasible)
Königsberg Bridge, 105–108, 133
measurement, 179, 208

Problem (cont.)
  Millennium, 127, 134, 361n10, 362n6
  Monty Hall, 57–60, 63, 342
  n-body, 170, 172
  NP (nondeterministic polynomial), 117–128, 133, 134, 155, 156, 361n6
  NP-Complete, 125–131, 133
  optimization, 113
  P, 109, 117, 119, 126, 133, 155, 156
  P =? **NP**, 127–128, 134, 156, 157, 361n9
  of personal identity, 34–40, 62
  polynomial, 109, 117, 119, 126, 133, 155, 156
  Printing 42 Problem, 146–147
  PSPACE, 132
  Satisfiability, 117, 127
  Set Partition, 115–116, 121, 130
  squaring the circle, 302, 303, 335
  Subset Sum, 116, 121
  superexponential, 131
  three-body, 169, 170–172, 174, 232
  tiling, 309–316
  tractable, 109
  Traveling Salesman Problem, 109–113, 119, 123–125, 129, 350
  trisecting an angle, 302–303, 335
  two-body, 170, 174
  undecidable, 139, 152, 160, 315–319
  unsolvable, 136–140, 172, 343, 345
  word problem, 318–319
  Zero Program, 147–150
Proof
  by contradiction, 6, 42, 77, 80
  by stereographic projection, 84
  diagonalization, 66, 78–83, 85, 144, 323
  necklace, 75, 76
  sunshine, 84
  zig-zag, 74, 76
Pseudoscience, 244
Psychoanalysis, 244
Ptolemy, 247, 257
Punctuated equilibrium, 368n16
Putnam, Hilary, 317
Pythagoras of Samos, 284, 298

Quantum computers, 120, 134, 336
Quantum gravity, 158, 231–232
Quantum leap, 44
Quantum mechanics, x, 5, 44–49, 120, 161, 174, 175–214, 231–232, 237, 243, 247, 249, 254–255, 260–262, 263, 277, 285, 290, 292–293, 340, 342–343, 347, 363n8, 364n16, 368nn7–8, 373n4
Quantum physics/quantum theory. *See* Quantum mechanics
Quine, Willard Van Orman, 7, 17

Rabinowitz, Avi, xii
Raise High the Roof Beams, Carpenters, 372n48
Ramanujan, Srinivasa, 26
Randi, James, 377n13
Realism, 91
Realism, naïve, 186, 209
Reductio ad absurdum, 6, 42, 77, 80
Reduction, 9, 147–150, 315, 343, 345. *See also* Piggybacking
  polynomial, 121–128 (*see also* Piggybacking)
Regular polygon, 307–308
Relative statement, 51
Rescher, Nicholas, 294, 353, 364n10
Revolutionary science, 246–247
Rice, Henry, 150
Riemann, Bernhard, 259
Riemann hypothesis, 36n6
Rivest, Ron, 119
Robinson, Julia, 317
Rømer, Ole, 217
Rosen, Nathan, 194
Rosser, John B., 375n18
RSA, 119
Rubik's cube, 309

Rucker, Rudy, 50
Ruffini, Paolo, 306, 316
Russell, Bertrand, 15, 19, 22, 85, 368n13

Saari, Donald G., 172
Saccheri, Girolamo, 258
Sainsbury, R. M., 29, 63
Salinger, J. D., 372n48
Salk, Jonas, 349
Savalas, Telly, 56
Schopenhauer, Arthur, 377n18
Schrödinger, Erwin, 192
Selective subjectivism, 290
Self-organizing, 169, 232
Self-reference, 12, 13, 18, 28, 140, 141, 150, 158, 160, 298, 320–333, 343, 345, 374n17
Semmelweis, Ignaz, 347
Sensitive dependence on initial conditions. *See* Butterfly effect
Sentence
  Gödel, 324–326, 331–333, 351
  Gödel-Rosser, 375n18
  Löb, 327
  Parikh, 326–327
  Quine's, 15, 16
  Tarski, 323–324
Shainberg, Lawrence, 160
Shakespeare, William, 355n3
Shamir, Adi, 119
Shape
  kite and dart, 312–314
  Myers, 311, 312
  rhombuses, 313
Shaw, George Bernard, 369n20
Ship of Theseus, 32–40, 52, 62, 342, 357n2
Simon, Carly, 18
Simplicity, 240–242
Simplicity of hypothesis, 241, 285
Simplicity of ontology, 241, 285
Simpsons, The, 276, 277
Simultaneity, 224–225
Sipser, Michael, 160
Six degrees of separation, 107
Smollin, Lee, 284
Solipsism, 369n23
Solipsism of the present moment, 369n23
Solovay, Robert, 157
Sorensen, Roy, 13, 29, 63
Spacetime, 223, 228–230
Space, Time and Gravitation, 291
Specker, Ernst, 188, 191
Spin (quantum), 187–191, 194–201, 208
Spock, Mister, 352
Spontaneous generation, 346
Spooky action at a distance. *See* Non-locality
Statistical mechanics, 172, 174, 232
Stellar nucleosynthesis, 274, 284
Stenger, Victor J., 290
Stewart, Potter, 359n16
Straightedge and compass, 301–304, 307, 335
Streep, Meryl, 371n43
Structure of Scientific Revolutions, The, 246
Sudbery, Anthony, 233
Sudkamp, Thomas A., 160
Sundman, Karl, 172
Superposition, 120, 176–214, 285, 347
Supersymmetry, 348
Symbolization, 320–321
Symmetry, 242, 287–293, 307, 308, 318
System
  binary star, 218
  chaotic, 164, 175, 181, 351
  deterministic, 165, 172, 174, 180–185, 206, 210, 231, 286, 363n8
  integrable, 164, 174
  predictable, 165, 166, 172, 174

System (cont.)
random, 166, 174, 180–185, 206, 209, 231, 340, 363n8, 366n25
stable, 164, 174
Szilárd, Leó, 372n49

Tartaglia, Niccolò Fontana, 306
Tegmark, Max, 209, 284, 285, 295
Teleological argument, 280
Tenenbaum, Stanley, 337
Tequila, 158, 277
Theorem
Bell's, 197–201, 209, 233, 365n21
Cook-Levin, 127
downward Löwenheim-Skolem, 375n20
Edge-of-the-wedge, xiv
Free-Will, 366n26
Gödel's first incompleteness, x, 298, 324–326, 327, 329, 330, 332, 334–335, 337, 344
Gödel's second incompleteness, x, 88, 331–334, 335, 337
Goodstein's, 330–331, 333, 337
Kochen-Specker, xiv, 186–192, 233, 365n20
Parikh's, 326–327
Pythagoras', 298
Rice's, 150, 160
Tarski's, 323, 337
Theory
chaos, 161–175, 232, 342, 351
electromagnetic, 218, 264, 363n8
Galilean relativity, 215–216
Galois, 304–309, 335
general relativity, 226–231, 237, 243, 246, 288
Grand Unified Theory (*see* Theory of Everything)
graph, 106–107, 118, 252
group, 232, 252, 262–263, 290, 317–319
loop quantum gravity, 232
noncommutative geometry, 232
quantum theory (*see* Quantum mechanics)
relativity, x, 5, 48, 49, 161, 214–233, 249, 254–255, 260, 288, 342–343, 347
set, 66–95, 328, 331, 348, 359n4, 360n12
special relativity, 217–226, 254, 271, 288
string, 232, 242–243, 250, 254–255, 269, 284, 286, 295, 347
Theory of Everything (TOE), 231, 233, 243, 250, 254–255
Zerrmelo-Fraenkel set theory, 86–94, 359n6, 362n7
Zerrmelo-Fraenkel set theory with choice, 90–91, 332–336, 375n21
Theseus, 32–40, 342
Thought and Language, 356n6
Through the Looking Glass, 10
Tierney, John, 63
Tiles, 309–316
aperiotic, 312–316
Tilings, 309–316
nonperiodic, 312–314
periodic, 312
Time dilation, 219–223, 227
Tiramisu, 371n43
Toy Story, 65
Tractatus Logico-Philosophicus, 15, 376n8
Transitivity of identity, 35, 359n17
Turing, Alan M., 139–140, 152, 155, 160, 169, 232

Unicorn, 40, 93, 377n16
Unreasonable Effectiveness of Mathematics in the Natural Sciences, The, 253
Uranus, 245, 253
Urban VIII, Pope, 347
USS Constitution, 357n1
USS Intrepid, 357n1

Vagueness, 11, 31, 40, 50–57, 63, 340, 341, 360n14

epistemic, 52, 55, 364n13
ontological, 52, 364n13
van Gogh, Vincent, 38
Verne, Jules, 262
von Neumann, John, 210, 280
vos Savant, Marilyn, 57
Vygotsky, Lev, 356n6

Wang, Hao, 376
Wang, Quidong (Don), 172
Wapner, Leonard M., 94
Washington's Ax, 33
Watson, 159
Watson, James, 169, 368n11
Webb, Stephen, 372n48
Webber, Andrew Lloyd, 371n43
Weinberg, Steven, 235, 294, 368n11, 369n19
Weyl, Hermann K. H., 242, 305, 321, 360n8
Wheeler, John Archibald, 237, 285, 355n2
Whitehead, Alfred North, 97
Wholeness Postulate, 176, 180, 200, 202, 204, 206
Wigner, Eugene Paul, 193, 253, 264, 266, 290
Wigner's unreasonable effectiveness, 253–272, 295
William of Ockham, 240, 250, 367n7
Wittgenstein, Ludwig, 175, 356n6, 376n8
Wojtowicz, Ralph, xii
Wright, Steven, 12
Wright brothers, 245

X-Files, 372n49
Xia, Zhihong (Jeff), 172

Yablo, Stephen, 24
Yanofsky, Hadassah, xiv, 227, 228, 286, 370n29, 372n54
Yanofsky, Rivka, xiv, 372n54
Yeats, William Butler, 135
Young, Thomas, 176

Zeno of Elea, 5, 41–49, 90, 359n19
Zerrmelo, Ernst, 86
Zorba the Greek, 1